中国科协学科发展研究系列报告

2016—2017
学科发展报告
综合卷

中国科学技术协会 | 主编

COMPREHENSIVE REPORT ON ADVANCES IN SCIENCES

中国科学技术出版社
· 北 京 ·

图书在版编目（CIP）数据

2016—2017学科发展报告综合卷/中国科学技术协会主编．—北京：中国科学技术出版社，2018.3

（中国科协学科发展研究系列报告）

ISBN 978-7-5046-7989-5

Ⅰ.①2… Ⅱ.①中… Ⅲ.①科学技术—技术发展—研究报告—中国— 2016–2017　Ⅳ.① N12

中国版本图书馆 CIP 数据核字（2018）第 055383 号

策划编辑	吕建华　许　慧
责任编辑	高立波　赵　佳
装帧设计	中文天地
责任校对	杨京华
责任印制	马宇晨

出　　版	中国科学技术出版社
发　　行	中国科学技术出版社发行部
地　　址	北京市海淀区中关村南大街16号
邮　　编	100081
发行电话	010-62173865
传　　真	010-62179148
网　　址	http://www.cspbooks.com.cn

开　　本	787mm×1092mm　1/16
字　　数	570千字
印　　张	23.25
版　　次	2018年3月第1版
印　　次	2018年3月第1次印刷
印　　刷	北京盛通印刷股份有限公司
书　　号	ISBN 978-7-5046-7989-5 / N・238
定　　价	98.00元

（凡购买本社图书，如有缺页、倒页、脱页者，本社发行部负责调换）

2016—2017
学科发展报告综合卷

专 家 组

组　长　李静海

副组长　项昌乐　黄　维

成　员　（按姓氏笔画排序）

　　　　王岁祥　王蕴红　史培军　包为民　任胜利
　　　　向　巧　苏小军　吴孔明　何满潮　宋　军
　　　　初景利　陈维江　范维澄　姜永茂　龚旗煌
　　　　董尔丹

编写组　（按姓氏笔画排序）

　　　　王　平　王飞跃　王玉杰　王建国　戈　峰
　　　　邝仕均　宁津生　刘小杰　刘会洲　刘细文
　　　　孙　伟　李秀珍　李晓刚　李家永　杨玉芳
　　　　杨志萍　杨维中　吴　昊　何满潮　张守攻
　　　　张志强　陆　颖　林　智　欧阳峥峥　周　涛
　　　　郑水林　赵志祥　胡振琪　侯　磊　姚建年
　　　　倪汉祥　高卫东　黄小卫　梅旭荣　彭双清
　　　　蒲嘉陵　谭一泓　谭建荣　樊春良

学术秘书　杨书宣　魏　丹　郭晓芹　严雯羽

党的十八大以来，以习近平同志为核心的党中央把科技创新摆在国家发展全局的核心位置，高度重视科技事业发展，我国科技事业取得举世瞩目的成就，科技创新水平加速迈向国际第一方阵。我国科技创新正在由跟跑为主转向更多领域并跑、领跑，成为全球瞩目的创新创业热土，新时代新征程对科技创新的战略需求前所未有。掌握学科发展态势和规律，明确学科发展的重点领域和方向，进一步优化科技资源分配，培育具有竞争新优势的战略支点和突破口，筹划学科布局，对我国创新体系建设具有重要意义。

2016年，中国科协组织了化学、昆虫学、心理学等30个全国学会，分别就其学科或领域的发展现状、国内外发展趋势、最新动态等进行了系统梳理，编写了30卷《学科发展报告（2016—2017）》，以及1卷《学科发展报告综合卷（2016—2017）》。从本次出版的学科发展报告可以看出，近两年来我国学科发展取得了长足的进步：我国在量子通信、天文学、超级计算机等领域处于并跑甚至领跑态势，生命科学、脑科学、物理学、数学、先进核能等诸多学科领域研究取得了丰硕成果，面向深海、深地、深空、深蓝领域的重大研究以"顶天立地"之态服务国家重大需求，医学、农业、计算机、电子信息、材料等诸多学科领域也取得长足的进步。

在这些喜人成绩的背后，仍然存在一些制约科技发展的问题，如学科发展前瞻性不强，学科在区域、机构、学科之间发展不平衡，学科平台建设重复、缺少统筹规划与监管，科技创新仍然面临体制机制障碍，学术和人才评价体系不够完善等。因此，迫切需要破除体制机制障碍、突出重大需求和问题导向、完善学科发展布局、加强人才队伍建设，以推动学科持续良性发展。

近年来，中国科协组织所属全国学会发挥各自优势，聚集全国高质量学术资源和优秀人才队伍，持续开展学科发展研究。从 2006 年开始，通过每两年对不同的学科（领域）分批次地开展学科发展研究，形成了具有重要学术价值和持久学术影响力的《中国科协学科发展研究系列报告》。截至 2015 年，中国科协已经先后组织 110 个全国学会，开展了 220 次学科发展研究，编辑出版系列学科发展报告 220 卷，有 600 余位中国科学院和中国工程院院士、约 2 万位专家学者参与学科发展研讨，8000 余位专家执笔撰写学科发展报告，通过对学科整体发展态势、学术影响、国际合作、人才队伍建设、成果与动态等方面最新进展的梳理和分析，以及子学科领域国内外研究进展、子学科发展趋势与展望等的综述，提出了学科发展趋势和发展策略。因涉及学科众多、内容丰富、信息权威，不仅吸引了国内外科学界的广泛关注，更得到了国家有关决策部门的高度重视，为国家规划科技创新战略布局、制定学科发展路线图提供了重要参考。

十余年来，中国科协学科发展研究及发布已形成规模和特色，逐步形成了稳定的研究、编撰和服务管理团队。2016—2017 学科发展报告凝聚了 2000 位专家的潜心研究成果。在此我衷心感谢各相关学会的大力支持！衷心感谢各学科专家的积极参与！衷心感谢编写组、出版社、秘书处等全体人员的努力与付出！同时希望中国科协及其所属全国学会进一步加强学科发展研究，建立我国学科发展研究支撑体系，为我国科技创新提供有效的决策依据与智力支持！

当今全球科技环境正处于发展、变革和调整的关键时期，科学技术事业从来没有像今天这样肩负着如此重大的社会使命，科学家也从来没有像今天这样肩负着如此重大的社会责任。我们要准确把握世界科技发展新趋势，树立创新自信，把握世界新一轮科技革命和产业变革大势，深入实施创新驱动发展战略，不断增强经济创新力和竞争力，加快建设创新型国家，为实现中华民族伟大复兴的中国梦提供强有力的科技支撑，为建成全面小康社会和创新型国家做出更大的贡献，交出一份无愧于新时代新使命、无愧于党和广大科技工作者的合格答卷！

2018 年 3 月

前言 PREFACE

距离《2014—2015中国科协学科发展研究系列报告》发布已近两年，国际科研环境又发生了新的变化，各学科领域涌现出众多新的优秀科研成果。2016年，中国科协组织中国化学会、昆虫学会、心理学会等30个全国学会，分别就化学、昆虫学、心理学、景观生态学、资源科学、感光影像学、岩石力学与岩石工程、机械工程（机械设计）、控制科学与工程、标准化、测绘科学技术、矿物加工工程、稀土科学技术、材料腐蚀、核技术应用、油气储运工程、煤矿区土地复垦与生态修复、矿物材料、建筑学、纺织科学技术、制浆造纸科学技术、光学工程、农学（基础农学）、林学科学、植物保护学、水土保持与荒漠化防治、茶学、毒理学、公共卫生与预防医学、科技政策学共30个学科的发展情况进行了系统研究，编辑出版了《2016—2017中国科协学科发展研究系列报告》。

受中国科协学会学术部委托，中国科协学会服务中心组织有关专家，在上述30个学科发展报告的基础上，编写了《2016—2017学科发展报告综合卷》（以下简称《综合卷》）。《综合卷》分为三章和附件部分：第一章以国家统计局、国家自然基金委、科技部等官方网站、WOS、ESI、OECD等数据库以及智库报告中的数据信息作为客观数据来源依据，结合各个学科领域专家的咨询建议，解读近两年来我国自然科学及工程技术领域学科发展的总体情况，从宏观层面分析我国自然科学及工程技术领域总体学科发展态势，评析学科发展存在的问题与挑战，提出了促进学科发展的启示与建议；第二章以此次《2016—2017中国科协学科发展研究系列报告》中的30个学科发展报告为基础，对30个学科近年的研究现状、国内外研究进展比较、学科发展趋势与展望等分别进行综述；第三章为30个学科发展报告主要内容的英文介绍；附件为2016年、2017年与学科发展相关的资料以及文中专业术语的注释。《综合卷》学科排序根据相关全国学会在中国科协的编号顺序排列。

为做好《综合卷》的研究工作，中国科协学会服务中心组织成立了《综合卷》专家组和编写组。专家组由中国科协学会与学术工作专门委员会委员及有关专家组成，编写组由中国科学院文献情报中心、中国科学院成都文献情报中心的科技智库专家以及30个学会选派的学科专家组成。其中：第一章和附件由中国科学院文献情报中心、中国科学院成都文献情报中心负责组织开展研究工作并完成编写任务；第二章和第三章由相关学科对应的30个学会负责组织开展研究工作并完成编写任务。

《综合卷》第一章内容是对我国自然科学及工程技术领域学科发展的宏观概述，由于各个数据源的学科划分标准、体系、细粒度不尽相同，不同指标的数据难以放在相同的学科分类标准下进行比较，相关指标数据均采用原始数据的学科分类予以呈现。另外，由于各个学科成果众多，而篇幅有限，所以第一章中的"学科发展成果与动态"一节仅列举了近两年来各学科少量的成果，无法将各学科重大成果一一列举，敬请见谅。第二章和第三章内容主要是在本次30个学科发展报告的基础上综合而成，仅概括相关学科领域的重要进展和总体情况，不能完整地反映我国所有学科全貌。《综合卷》涉及的学科面广，任务量大，数据统计口径众多，编写时间仓促，难免有疏漏之处，敬请广大读者谅解并指正。

本书编写组

2018年1月

目 录

序 …………………………………………………………………… 韩启德
前言 ………………………………………………………………… 本书编写组

第一章　学科发展综述

第一节　学科发展概况 ……………………………………………………… 3
第二节　学科发展成果与动态 ……………………………………………… 31
第三节　学科发展问题与挑战 ……………………………………………… 41
第四节　学科发展启示与建议 ……………………………………………… 49
参考文献 …………………………………………………………………… 53

第二章　相关学科进展与趋势

第一节　化学 ………………………………………………………………… 59
第二节　昆虫学 ……………………………………………………………… 65
第三节　心理学 ……………………………………………………………… 71
第四节　景观生态学 ………………………………………………………… 75
第五节　资源科学 …………………………………………………………… 82
第六节　感光影像学 ………………………………………………………… 86
第七节　岩石力学与岩石工程 ……………………………………………… 89
第八节　机械工程（机械设计） …………………………………………… 94
第九节　控制科学与工程 …………………………………………………… 98
第十节　标准化 ……………………………………………………………… 104
第十一节　测绘科学技术 …………………………………………………… 108
第十二节　矿物加工工程 …………………………………………………… 117

第十三节	稀土科学技术	123
第十四节	材料腐蚀	128
第十五节	核技术应用	134
第十六节	油气储运工程	138
第十七节	煤矿区土地复垦与生态修复	143
第十八节	矿物材料	149
第十九节	建筑学	156
第二十节	纺织科学技术	160
第二十一节	制浆造纸科学技术	166
第二十二节	光学工程	172
第二十三节	农学（基础农学）	177
第二十四节	林学科学	182
第二十五节	植物保护学	187
第二十六节	水土保持与荒漠化防治	193
第二十七节	茶学	197
第二十八节	毒理学	204
第二十九节	公共卫生与预防医学	210
第三十节	科技政策学	214

第三章 相关学科进展与趋势简介（英文）

1. Chemistry ······ 223
2. Entomology ······ 226
3. Psychology ······ 230
4. Landscape Ecology ······ 233
5. Resources Science ······ 238
6. Photoimaging ······ 242
7. Rock Mechanics and Rock Engineering ······ 244
8. Mechanical Engineering ······ 250

9. Automation Science ··· 256
10. Standardization Discipline ·· 263
11. Surveying, Mapping and Geoinformatic Science ·· 267
12. Mineral Processing Engineering ·· 270
13. Rare Earth Science ·· 275
14. Corrosion Electrochemical ·· 280
15. Nuclear Technology ··· 282
16. Oil and Gas Storage and Transportation ·· 286
17. Land Reclamation and Ecological Restoration in Coal Mining Areas ············ 289
18. Mineral Materials ··· 292
19. Architecture ··· 297
20. Textile Manufacturing ··· 299
21. Pulp and Paper Science and Technology ·· 302
22. Optical Engineering ·· 305
23. Basic Agronomy ··· 307
24. Forestry ··· 311
25. Plant Protection Science ··· 314
26. Soil and Water Conservation & Desertification Combating ························· 318
27. Tea ·· 321
28. Toxicology ··· 325
29. Public Health and Preventive Medicine ·· 330
30. The Science of Science and Technology Policies ······································ 334

附件

附件1：118个学科优先发展领域 ··· 341
附件2：48个重大专项 ··· 342
附件3：2016—2017年香山科学会议学术讨论会一览表 ··································· 343
附件4：2016—2017年未来科学大奖获奖者 ·· 345

附件5：2016年度"中国科学十大进展"……………………………………………345
附件6：2016—2017年度"中国十大科技进展新闻"……………………………345
附件7：2016—2017年度国家科学技术进步奖获奖项目目录…………………346
附件8：2016—2017年度国家自然科学奖项目名单……………………………353
注释……………………………………………………………………………………356

第一章

学科发展综述

第一章

绪论及文献综述

第一节 学科发展概况

学科发展是一个动态持续的过程，随着国家对科技创新的高度重视，学科发展总体上取得了长足的进步。本部分从学科建设投入、学科研究成果、学科平台建设、学科队伍发展、国际合作交流等能够有力反映学科发展概况的数据出发并进行定量分析，力求客观反映我国学科的整体发展现状。

一、学科建设投入稳步增长

1. 研发经费持续增长

随着《国家创新驱动发展战略纲要》[1]《国家中长期科学和技术发展规划纲要（2006—2020年）》[2]和《"十三五"国家科技创新规划》[3]等一系列国家战略、规划的推动，政府引导和政策环境的不断优化，我国科技经费投入力度加大，研究与试验发展（R&D）经费投入、国家财政科技支出均实现较快增长，研究与试验发展（R&D）经费投入强度稳步提高。2016年，全国共投入R&D经费15676.7亿元，比2015年增加1506.9亿元，增长10.6%，增速较2015年提高1.7个百分点，如图1所示。R&D经费投入强度为2.11%，比2015年提高0.05个百分点。虽然与以色列（4.25%）、韩国（4.23%）、日本（3.49%）等以及经济合作与发展组织（OECD）成员国平均水平（2.40%）相比还有很大差距，但已经超过欧盟15国的平均水平（2.08%）[4]，并且与发达国家的差距正在逐年缩小。按照《国家中长期科学和技术发展规划纲要（2006—2020年）》[2]，2020年我国研发投入强度要达到2.5%，而按照近年来我国研发投入强度的平均增幅来看，要实现这一目标，未来的经费投入还需要更大幅度地增长。

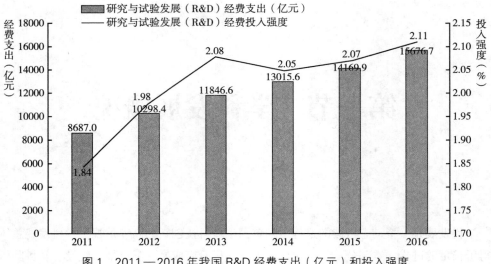

图 1　2011—2016 年我国 R&D 经费支出（亿元）和投入强度

数据来源：《2016 年中国统计年鉴》[5]

按活动类型看，2016 年全国基础研究经费 822.9 亿元，比 2015 年增长 14.9%；应用研究经费 1610.5 亿元，比 2015 年增长 5.4%；试验发展经费 13243.4 亿元，比 2015 年增长 11.1%。基础研究、应用研究和试验发展经费所占比重分别为 5.2%、10.3% 和 84.5%，如图 2 所示。

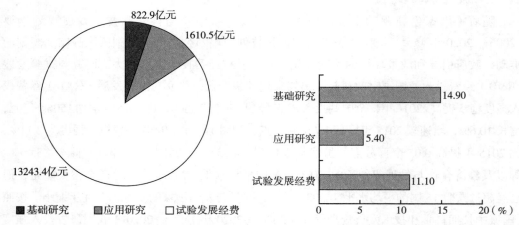

图 2　2016 年不同活动类型 R&D 经费占比（左）及增长百分比（右）

数据来源：《2016 年全国科技经费投入统计公报》[6]

从活动主体看，2016 年各类企业经费支出 12144 亿元，比 2015 年增长 11.6%；政府属研究机构经费支出 2260.2 亿元，比 2015 年增长 5.8%；高等学校经费支出 1072.2 亿元，比 2015 年增长 7.4%。企业、政府属研究机构、高等学校经费支出所占比重分别为 77.5%、14.4% 和 6.8%，如图 3 所示。

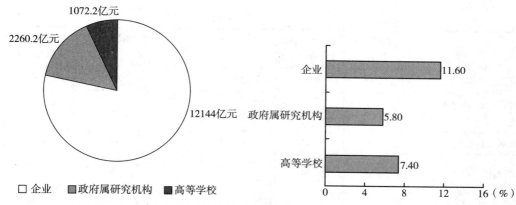

图3 2016年不同活动主体R&D经费占比（左）及增长百分比（右）
数据来源：《2016年全国科技经费投入统计公报》[6]

从产业部门看，高技术产业研发投入不断提高。2016年高技术制造业R&D经费2915.7亿元，投入强度（研究经费与主营业务收入之比）为1.9%，高于2015年1.88%的水平。高技术制造业研发投入的不断提高，对我国经济实现提质增效转型升级起到了重要的支撑和带动作用。在规模以上工业企业中，R&D经费投入超过500亿元的行业大类有7个，这7个行业的经费占全部规模以上工业企业R&D经费的比重为60.2%；R&D经费投入在100亿元以上且投入强度超过规模以上工业企业平均水平的行业大类有9个，分行业情况详见表1所示。

表1 2016年分行业规模以上工业企业R&D经费情况

行业	R&D经费（亿元）	R&D经费投入强度（%）	行业	R&D经费（亿元）	R&D经费投入强度（%）
合　计	10944.7	0.94	化学原料和化学制品制造业	840.7	0.96
采矿业	267.8	0.56	医药制造业	488.5	1.73
煤炭开采和洗选业	132.1	0.59	化学纤维制造业	83.8	1.08
石油和天然气开采业	63.9	0.99	橡胶和塑料制品业	278.8	0.86
黑色金属矿采选业	10.4	0.17	非金属矿物制品业	323.1	0.52
有色金属矿采选业	27.1	0.44	黑色金属冶炼和压延加工业	537.7	0.87
非金属矿采选业	11.1	0.21	有色金属冶炼和压延加工业	406.8	0.76
开采辅助活动	23.1	1.49	金属制品业	326.3	0.81
制造业	10580.3	1.01	通用设备制造业	665.7	1.38
农副食品加工业	249.7	0.36	专用设备制造业	577.1	1.54

续表

行 业	R&D经费（亿元）	R&D经费投入强度（%）	行 业	R&D经费（亿元）	R&D经费投入强度（%）
食品制造业	152.8	0.64	汽车制造业	1048.7	1.29
酒、饮料和精制茶制造业	100.6	0.54	铁路、船舶、航空航天和其他运输设备制造业	459.6	2.38
烟草制品业	21.4	0.25	电气机械和器材制造业	1102.4	1.50
纺织业	219.9	0.54	计算机、通信和其他电子设备制造业	1811.0	1.82
纺织服装、服饰业	107.0	0.45	仪器仪表制造业	185.7	1.96
皮革、毛皮、羽毛及其制品和制鞋业	59.0	0.39	其他制造业	28.1	1.02
木材加工和木、竹、藤、棕、草制品业	52.9	0.36	废弃资源综合利用业	11.0	0.27
家具制造业	42.9	0.49	金属制品、机械和设备修理业	17.8	1.47
造纸和纸制品业	122.8	0.84	电力、热力、燃气及水生产和供应业	96.6	0.15
印刷和记录媒介复制业	46.8	0.58	电力、热力生产和供应业	81.6	0.15
文教、工美、体育和娱乐用品制造业	91.9	0.54	燃气生产和供应业	7.7	0.13
石油加工、炼焦和核燃料加工业	119.6	0.35	水的生产和供应业	7.4	0.35

注：数据来源：《2016年全国科技经费投入统计公报》[6]。

2. 基金资助稳中有变

2014年，我国开始部署国家科技计划管理改革，逐步将原有的国家重点基础研究发展计划（"973"计划）、国家高技术研究发展计划（"863"计划）等100多个科技计划整合成国家自然科学基金、国家科技重大专项、国家重点研发计划、技术创新引导专项（基金）、基地和人才专项五大类。其中国家自然科学基金、国家科技重大专项、国家重点研发计划是高校和科研单位获批国家级科研项目的重要来源。截至2017年8月，科技部和国家自然科学基金等相继公布或公示了2017年国家重点研发计划、国家自然科学基金和国家科技重大专项等部分项目的立项情况，三类项目总经费达到452.0亿元[7]。

（1）国家自然科学基金

国家自然科学基金坚持支持基础研究，逐渐形成和发展了由探索、人才、工具、融合四大系列组成的资助格局。2011—2016年间，国家自然科学基金投入由140.44亿元增长至246.6亿元，资助投入年均增长率超过25%（如图4所示），虚线为模拟增长曲线。2016年国家自然科学基金面上项目资助16934项，直接费用101.75亿元；重点项目资助

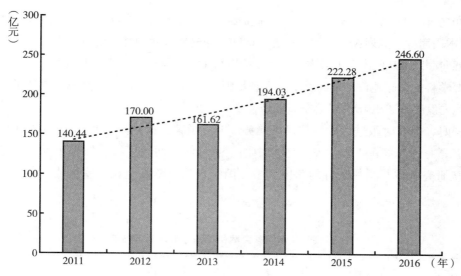

图 4　2011—2016 年国家自然科学基金中央财政拨款

数据来源：2011—2016 年《国家自然科学基金委年度报告》[8]

612 项，直接费用 17.15 亿元。

2015 年和 2016 年两年国家自然科学基金各个学部批准项目资助（注：统计口径为面上项目、青年科学基金项目、地区科学基金项目、重点项目、杰青项目、海外及港澳台合作研究基金以及优秀青年基金项目）基本保持一致，最多的三个学部分别是生命科学部、工程与材料学部和医学科学部，图 5 为 2016 年国家自然科学基金各个学部批准项目资助资金分布情况。

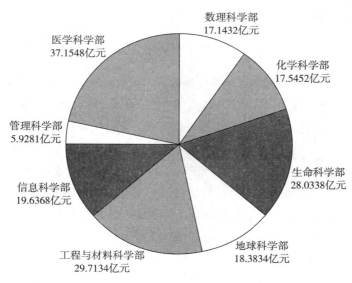

图 5　2016 国家自然科学基金各个学部批准项目资助分布情况

数据来源：2016 年《国家自然科学基金委年度报告》[8]

2016年6月14日,国家自然科学基金委发布《国家自然科学基金"十三五"发展规划》(本段简称《规划》)[9],"十三五"期间,通过支持我国优势学科和交叉学科的重要前沿方向,以及从国家重大需求中凝练可望取得重大原始创新的研究方向,进一步提升我国主要学科的国际地位,提高科学技术满足国家重大需求的能力。《规划》遴选了118个学科优先发展领域和16个综合交叉领域,详细情况请参见附件1。

同时,为及时跟进世界科技发展趋势,响应国家科技战略,2018年国家自然科学基金新增"人工智能"与"交叉学科中的信息科学"两大领域,如表2所示。这也意味着从研究领域与学科划分的角度来看,国家自然科学基金在这些领域上将加大投入力度[10]。

表2 2018年国家自然科学基金申请新增领域

新增领域	管理部门	研究内容
人工智能 (代码F06)	信息科学部	强调围绕人工智能领域的核心科学问题与关键技术,进行原创性、基础性、前瞻性和交叉性研究;鼓励在人工智能基础、机器学习、机器感知与模式识别、自然语言处理、知识表示与处理、智能系统与应用、认知与神经科学启发的人工智能等方向的理论与方法研究。支持人工智能领域的科研人员与其他自然科学、人文社会科学等领域的研究人员密切合作。还特别鼓励和支持科研人员研究解决国际公认难度大、有重大影响、探索性强的基础性问题
交叉学科中的信息科学 (代码F07)	信息科学部	信息科学与其他学科交叉领域的基础理论、基本方法和关键技术。主要领域包括:教育信息科学与技术、信息与数学交叉问题,如电子通信与数学交叉、计算机与数学交叉、自动化与数学交叉等

注:数据来源:国家自然科学基金委员会[10]。

(2)国家重点研发计划

国家重点研发计划由原来的国家重点基础研究发展计划("973"计划)、国家高技术研究发展计划("863"计划)、国家科技支撑计划、国际科技合作与交流专项、产业技术研究与开发基金和公益性行业科研专项等整合而成,是针对事关国计民生的重大社会公益性研究,以及事关产业核心竞争力、整体自主创新能力和国家安全的战略性、基础性、前瞻性重大科学问题、重大共性关键技术和产品,为国民经济和社会发展主要领域提供持续性的支撑和引领。截至2018年1月,科技部已公布2017国家重点研发计划重点专项48项[7],这些专项也代表着国家重点发展的学科领域。每个专项下设多个项目,入选项目约1200余个,总经费超过215亿元。48个重大专项如附件2所示。其中资助金额超过7亿元的重点专项如表3所示。

表3　国家重点研发计划部分重点专项学科领域及经费资助情况

重点专项名称	经费（万元）
重点基础材料技术提升与产业化	120770
公共安全风险防控与应急技术装备	99122
干细胞及转化研究	94021
化学肥料和农药减施增效综合技术研发	92902
大气污染成因与控制技术研究	91647
纳米科技	88397
量子调控	87747
战略性先进电子材料	79927
数字诊疗装备研发	78557
深地资源勘查开采	75896
蛋白质机器与生命过程调控	73531
新能源汽车	71646

注：数据来源：国家科技管理信息系统公共服务平台[7]。

3. "双一流"视角下的学科建设

2015年10月，国务院印发《统筹推进世界一流大学和一流学科建设总体方案》[11]，提出"推动一批高水平大学和学科进入世界一流行列或前列"。一流学科是一流大学的生成基础，建设一批接近或达到世界先进水平的学科，是"双一流"建设阶段性目标的基本要求。"双一流"建设是国家中长期教育规划的重要组成部分，也是中国走向国际化的重要一步，必将对中国未来几十年内的高等教育、学科发展布局、人才培养、科研体制等产生一系列深远的影响。

2017年9月，教育部正式公布世界一流大学和世界一流学科（简称"双一流"）建设高校及建设学科名单，全国共137所高校的465个学科入选，其中"双一流"建设专家委员会认定学科421个，高校自主确定学科44个，这465个学科分布于108个学科领域中。从入选学科数统计来看，超过六成的学科均为理学、工学学科。材料科学与工程、化学、生物学、计算机科学与技术、数学、生态学、机械工程，入选数量都在10个以上。而地理学、动力工程及工程热物理、航空宇航科学与技术、公共管理等入选数量则相对较少。"双一流"建设学科门类分布如表4所示。

表4 2017中国高校"双一流"建设学科门类分布

名次	学科门类	双一流建设学科数	入选高校数	双一流建设高校最多的学科
1	工学	188	75	材料科学与工程、计算机科学与技术、机械工程、化学工程与技术、环境科学与工程
2	理学	104	50	化学、生物学、数学、生态学、物理学、统计学（并列）
3	医学	42	21	基础医学、临床医学、药学、中药学、口腔医学
4	法学	24	15	法学、马克思主义理论、政治学、民族学、社会学
5	农学	24	14	作物学、畜牧学、植物保护、草学、林学
6	管理学	21	13	管理科学与工程、工商管理、农林经济管理、图书情报与档案管理、公共管理
7	文学	19	13	外国语言文学、中国语言文学、现代语言学、新闻传播学、语言学
8	艺术学	13	12	戏剧与影视学、音乐与舞蹈学、美术学、设计学、艺术学理论
9	经济学	10	8	应用经济学、理论经济学、经济学和计量经济学
10	历史学	9	7	中国史、世界史、考古学
11	教育学	6	5	教育学、体育学、心理学
12	哲学	5	5	哲学

注：数据来源：《中国大学及学科专业评价报告2017—2018》[12]。

在"双一流"建设高校及建设学科名单出台后，各大高校也纷纷提出了面向未来的学科发展计划，加强传统优势学科的引领与辐射作用、发挥自身特色优势学科、建设科学群、加强学科交叉融合、分层次有序推进一流学科建设与发展。清华大学将构建"4个学科领域+20个学科群+8个独立学科"的学科建设体系，将现有11个学科门类整合为工程科学与技术、自然科学、人文社会科学与艺术和生命科学与医学4个学科领域。同时将集成学科领域内相近的学科，形成包括建筑学科群，土木水利学科群，核科学技术与安全学科群等在内的20个相互支撑、协同发展的学科群，并着力建设包括电气工程、力学、动力工程与工程热物理等在内的8个自身具有很强竞争力，且学科知识体系相对独立的学科[13]。北京大学启动"30+6+2"学科建设项目布局，即面向2020年，重点建设30个国内领先、国际一流的优势学科；面向2030年，部署理学、信息与工程、人文、社会科学、经济与管理、医学等6个综合交叉学科群；面向更长远的未来，布局和建设以临床医学+X、区域与国别研究为代表的前沿和交叉学科领域，带动学科结构优化与调整，培育新的学科增长点[14]。浙江大学实施学科分类分层发展，实施"高峰学科建设""一流骨干基础学科建设""优势特色学科发展"计划。推进人文学科板块、社会科学学科板块、理学学科板块、工学学科板块、信息学科板块、农业生命环境学科板块、医药学科板块等7大学科板块建设。建设会聚型学科领域，发挥学科综合优势，面向2030优化战略布局，通过对突破性事项、颠覆性技术、革命性产业的前瞻以及国家战略、全球重大科技和智力问题的带动，促进学科板块、学科领域的联动发展和会聚造峰[15]。中国人民大学在方案中呈现了"珠峰"—"高峰"—"高原"的学科生态规划。其中，学科"珠峰"目标为进入世界一流，包括马克思主义理论、理论经济学、应用经济学、法学等学科在内。学科"高峰"目标为冲击世界一流，如哲学、农林经济管理等学科。学科"高原"建设，目标为引领相关学科发展[16]。各大高校除了调整学科建设方案外，也都加快出台配套的相应的教学科研、人才培养、人才引进、科研教学管理、科研评价、交流合作、财务资产管理等一系列体系、政策与制度。

然而，"双一流"建设名单的出台也引起了很大的争议。首先是对遴选规则的质疑，一些高校的传统优势学科未被选入，例如复旦大学的王牌学科新闻传播学，南京大学的社会学，西北工业大学、哈尔滨工业大学、南京航空航天大学的航空航天学等均未入选。其次是对"双一流"建设导致"新一轮不公平竞争"的担忧，认为"双一流"建设更多起到"助强"的作用，并不利于中西部地区、地方院校、弱势学科的发展[17]。另外，有学者认为自从"双一流"概念被抛出后，在人才引进与学科发展等方面开始呈现出某种无秩序、不和谐现象。在人才引进上"砸钱抢人"愈演愈烈，欠发达地区人才流失严重；一些高校在学科发展、建设和规划方面，为迎合"双一流"而"自断经脉"，以调整为名"砍掉"了本校一些具有较长办学传统，而且就业未必很差的学科和专业[18]。

二、学科研究成果进步显著

随着科学研究和学科建设投入的不断增加，我国重大科技成果产出丰硕，许多关键科学技术指标进入世界前列，并改变着世界的研发格局。国家统计局最新发布的《中国创新指数（CII）》[19]测算结果指出，2016年中国创新指数为181.2，比2015年增长5.7%，呈现稳步提升态势。康奈尔大学、英士国际商学院和世界知识产权组织共同发布的《2017年全球创新指数》[20]显示，中国在创新质量上再进一步，位列第16位，这也是中国连续第5年成为中等收入国家（区域）创新排行的领头羊，和高收入经济体的差距进一步缩小。2017年3月，汤森路透发布的第二届"全球最具创新力政府研究机构25强"[21]中，中国科学院荣登榜单，位居第11位，比2016年上升5位。这些数据表明，我国在落实创新驱动发展战略和"大众创业、万众创新"方面取得了显著成效，我国创新能力和创新质量稳步提升，创新体系不断完善，创新型国家建设持续推进。

1. 论文产出进入高速通道

中国近年来科研论文产出进入高速通道，总体论文数量持续增长，多个学科高水平论文加速涌现，但整体论文质量有待提升。

根据多个数据库显示，我国近两年的发文量稳步增长，SCI（Science Citation Index

Expanded)、CPCI-S(Conference Proceedings Citation Index-Science)中国论文所占份额已排在世界第二位、EI(The Engineering Index)中国论文所占份额则排名世界第一位。Scopus 中国论文所占份额世界排名由 2015 年的第二位跃升为第一位。两年间，SCI、EI 数据库中，中国第一作者论文数量占总发文量的比例基本保持不变，而 CPCI-S 数据库中该比例大幅度上升，如表 5 所示。值得注意的是，根据 2016 年数据显示，我国 SCI 论文虽然排名世界第二位，但排在第一位的美国，其论文数量为 50.23 万篇，占世界份额的 26.5%，是我国的 1.5 倍。2015 年，中国作者发文数量在 SCI、CSCD(Chinese Science Citation Database)数据库中处于持平的情况，而 2016 年，SCI 中国作者发文数已经远远高于 CSCD，这表明中国作者更多地在国外期刊上发表论文。

表 5 各大数据库中中国科技论文收录情况

数据库	论文数量（万篇）		世界份额		世界排名	第一作者论文数量（万篇）		第一作者数量论文比例	
年份	2015	2016	2015	2016	2015/2016	2015	2016	2015	2016
SCI	29.68	32.42	16.30%	17.10%	2/2	26.55	29.06	89.45%	89.64%
EI	21.73	22.65	32.00%	33.20%	1/1	20.74	21.62	95.44%	95.45%
CPCI-S	7.12	8.63	15.24%	15.30%	2/2	3.82	6.31	53.65%	73.12%
Scopus	43.9	44.04	15.83%	18.60%	2/1	—	—	—	—
国内-CSCD	26.82	26.08	—	—	—	—	—	—	—
国内-CSTPCD	56.95	49.42	—	—	—	—	—	—	—

注：数据来源：2017《中国科技论文统计结果》[22]、Nature[23] 及 CSCD 数据库。

为更全面了解我国论文的学科分布以及影响力，我们分别选取 SCI、ESI、Nature Index 三个数据源的论文数据进行分析呈现。这三个数据源各有其特点，其中 SCI 数据库论文最多，学科领域分类最细，涵盖了理、工、农、医等 252 个学科领域[24]；ESI 数据库则是收录全球高被引论文，并将其分成了生物学与生物化学、化学、计算机科学、工程学等 22 个学科；Nature Index[25] 则由 Nature 集团定期发布，Nature 集团遴选了化学、物理学、生命科学、地球与环境科学 4 个基础学科相关的 68 种一流期刊（如 Nature、Science、Cell 等），并对这些期刊发表的论文进行统计分析，形成 Nature Index。

（1）SCI 论文学科分布及影响力

从 2007 年至 2017 年（截至 2017 年 10 月），中国科技人员共发表国际论文 205.82 万篇，继续排在世界第二位，数量比 2016 年统计时增加了 18.1%；论文共被引用 1935.00 万次，增加了 29.9%，排在世界第二位，比 2016 年上升了 2 位。中国国际科技论文被引用次数增长的速度显著超过其他国家。中国平均每篇论文被引用 9.40 次，比 2016 年度统计时的 8.55 次提高了 9.9%。与世界整体篇均被引用次数 11.80 次/篇相比，中国虽然还有一定的差距，但提升速度相对较快，中国各十年段科技论文被引用次数世界排位如表 6 所示。

表 6 中国各十年段科技论文被引用次数世界排位变化

时间	1998—2008 年	1999—2009 年	2000—2010 年	2001—2011 年	2002—2012 年	2003—2013 年	2004—2014 年	2005—2015 年	2006—2016 年	2007—2017 年
世界排位	10	9	8	7	6	5	4	4	4	2

注：数据来源：SCI 数据库[26]，检索时间为 2017 年 9 月。

近两年来，我国 SCI 论文分布最多的学科除了药学代替数学进入前 10 名外，发文数最多的前 10 位基本保持不变，其中仅有个别学科发文量排名发生了微调，如图 6 所示。

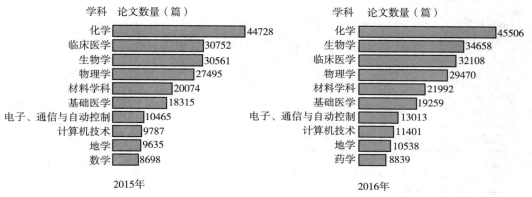

图 6 2015 年（左）和 2016 年（右）SCI 论文数最多的十个学科
数据来源：2017《中国科技论文统计结果》[22]

从世界份额占有率看，我国材料学、化学、物理学位列三甲。计算机科学、地学、数学、微生物学等学科，虽然发文量相对不高，但是其占世界份额百分比却较高，相反临床医学虽然发文量很高，但是其占世界份额百分比较低，仅为 7.88%。影响力方面，得益于各学科发文量的提升，各学科总引用量均在稳步上升，但是从篇均被引及其相对影响（我国篇均被引与其世界篇均被引平均值之比）来看，我国各个学科论文的影响力仍较薄弱，除农业科学与世界平均影响力水平相当外，其他学科均低于世界平均影响力水平，具体如表 7 所示。

表7 2007—2017年我国各学科产出SCI论文与世界平均水平比较

学科	论文数量（篇）	占世界份额（%）	被引用次数	占世界份额（%）	世界排位	位次变化趋势	篇均被引用次数	相对影响
农业科学	45005	11.40	378519	11.39	2	—	8.41	1.00
生物与生物化学	93579	13.33	936249	8.24	4	↑1	10	0.62
化学	392086	24.03	4902936	21.53	2	—	12.5	0.90
临床医学	203954	7.88	1641481	5.07	10	—	8.05	0.64
计算机科学	63906	19.01	347922	15.74	2	—	5.44	0.83
经济贸易	11888	4.66	66599	3.34	9	↑1	5.6	0.72
工程技术	229606	19.93	1497844	18.71	2	—	6.52	0.94
环境与生态学	60129	14.00	569489	10.60	2	↑1	9.47	0.76
地学	69797	16.53	672127	13.47	3	↑1	9.63	0.81
免疫学	18260	7.46	201196	4.39	11	—	11.02	0.59
材料科学	221817	29.56	2354095	27.52	1	↑1	10.61	0.93
数学	77379	19.06	316571	18.53	2	—	4.09	0.97
微生物学	22357	11.45	191928	6.58	5	—	8.58	0.57
分子生物学与遗传学	62476	14.15	734204	6.96	6	↑1	11.75	0.49
综合类	2590	13.26	31237	11.54	3	—	12.06	0.87
神经科学与行为学	36310	7.37	358487	4.12	10	—	9.87	0.56
药学与毒物学	51966	13.97	461777	9.97	2	—	8.89	0.71
物理学	220506	20.25	1877360	15.72	2	—	8.51	0.78
植物学与动物学	66691	9.53	543058	8.63	4	—	8.14	0.91
精神病学与心理学	8671	2.28	64550	1.41	15	↑1	7.44	0.62
空间科学	12258	8.58	144872	5.79	13	—	11.82	0.67
社会科学	18855	2.22	121886	2.20	11	↑1	6.46	0.99

注：统计时间截至2017年9月。"↑1"的含义是：与上年度统计相比，位次上升了1位；"—"表示位次未变。相对影响：我国篇均被引用次数与该学科世界平均值的比值。
数据来源：2017《中国科技论文统计结果》[22]。

（2）ESI 论文学科分布及影响力

从 ESI 各学科领域论文数量排名看，在 22 个学科领域中，中国有 16 个学科领域位于世界前 5 位，其中化学、材料科学和工程科学 3 个学科领域更是位于世界第一位；物理学、临床医学等 13 个学科领域紧随美国之后，位列世界第二位。从发表论文总体被引用数量排名看，中国有 18 个学科领域论文被引数量进入世界前 10 位；其中材料科学论文被引数量首次位列世界第一位；化学、工程科学、物理学等 8 个学科领域论文被引数量位居世界第二位；生物及生物化学、地球科学和多学科等 3 个学科领域论文被引数量位居世界第三位，如表 8 所示。

表 8　我国在 ESI 22 个学科领域中论文发文量和引文量世界排名

学科领域	发文量	引文量	学科领域	发文量	引文量
化学	1	2	农业科学	2	2
材料科学	1	1	计算机科学	2	2
工程科学	1	2	数学	2	2
物理学	2	2	微生物学	2	5
临床医学	2	10	多学科	2	3
生物及生物化学领域	2	3	神经科学及行为学	4	10
分子生物与遗传学	2	6	免疫学	4	11
地球科学	2	3	空间科学	7	13
环境生态科学	2	2	社会科学	10	11
植物学与动物学	2	4	神经病学与心理学	10	15
药理学与毒理学	2	2	经济与商业学	11	10

注：数据来源：ESI 数据库[27]，检索时间为 2017 年 9 月。

（3）Nature Index 论文学科分布及影响力

根据 Nature Index 对各国化学、物理学、生命科学、地球与环境科学四大基础学科高水平论文的分析，反映了我国在这四个学科领域上的高水平论文正在加速涌现。2017 年，中国在这 4 个学科上高水平论文 WFC 值（加权分数式计量，通常以此值进行比较）位居世界第二，仅次于美国。其中，中国化学高水平论文占比最多，物理学其次，生命科学、地球与环境科学高水平论文占比相对较少。WFC 值排名前三位的国家高水平论文的学科分布如图 7 所示。值得一提的是，如果以发文机构进行比较，中国科学院最近一年来（2016 年 10 月 1 日—2017 年 9 月 30 日）在一流期刊上发文的 WFC 值已经远超哈佛大学、马普学会、法国国家科学研究中心（CNRS）、斯坦福大学、麻省理工学院等，位居世界第一。

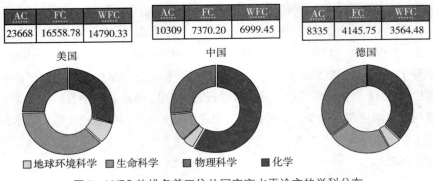

图 7　WFC 值排名前三位的国家高水平论文的学科分布

数据来源：2017 Nature Index[25]，检索时间为 2017 年 9 月

2. 专利成果持续激增

随着知识经济的兴起和经济全球化不断加深，知识产权已经成为各国参与国际竞争的重要资源、对外投资的重要资本和国家发展战略的重要组成部分。2006 年世界知识产权组织（WIPO）首次发布《世界专利报告》，根据全球专利统计数据介绍发明专利申请、授权、分布情况以及专利制度发展趋势。2009 年，WIPO 在《世界专利报告》的基础上，将其扩展为《世界知识产权指标》报告，并逐年发布。截至 2017 年 12 月，WIPO 共发布了 9 份《世界知识产权指标》报告。

中国持续成为全球申请量增长的巨大驱动力。中国的专利发展虽然起步较晚，但是从 2000 年后，其申请量呈指数级增长趋势。根据 WIPO 最新发布的《世界知识产权指数 2017》[28] 报告，2016 年，世界各地提交的专利申请量首次突破 300 万件，比 2015 年增长 8.3%，是连续第七年保持增长势头。在新增的近 24 万件专利申请中，中国占 23 万余件，占总增量的 98%。中国受理的专利申请量为 130 余万件，授权量上升 8.9%，申请量超过了美国（60.6 万）、日本（31.8 万）、韩国（28.9 万）和欧洲专利局（15.9 万）的总和，持续位列世界第一。五大专利申请国专利申请年度变化趋势如图 8 所示。

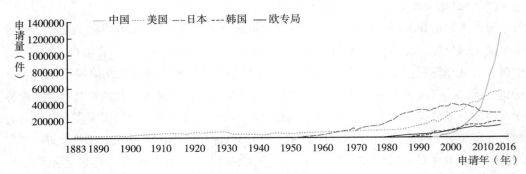

图 8　五大专利申请国专利申请年度变化趋势

数据来源：《世界知识产权指数 2017》[28]

海外申请数量不足,专利申请覆盖地理范围较小。近年来,我国在海外申请专利数量有所增加,但总量仍然不足。2016 年,在向国外申请专利方面,美国(21.6 万件)继续领先,中国的申请量(5.2 万)仅为美国的四分之一。从海外申请量占总申请量的比例来看,中国海外申请占比仅为 4%,远远低于美国和日本。

专利家族是一件专利在不同国家的申请集合,也是其后续衍生专利的申请集合,反映了专利权人对某市场区域的布局和专利的重视程度,是专利重要性和质量的体现。在 2012—2014 年间,以色列居民申请的专利家族占比最高(92.4%);其后依次是瑞士(89.3%)、荷兰(85.4%);中国申请的专利家族占比仅为 2.9%,在前 20 名国家中所占比例最低,详情如图 9 所示。

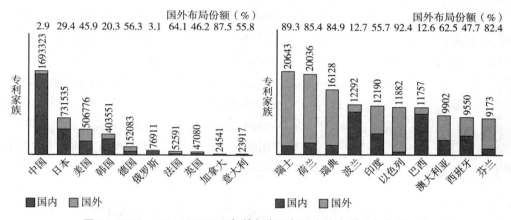

图 9 2012—2014 TOP20 专利申请国家国内外专利家族布局占比
数据来源:《世界知识产权指数 2017》[28]

从专利申请的领域技术分布来看,2015 年,计算机技术是全球公布的专利申请中最常用的技术,大约有 18.7 万件已公布的申请。其次是电器机械、设备、能源(17.6 万件),测量技术(12.4 万件),数字通信(12.3 万件)和医疗技术(11 万件)。这 5 个领域占全球所有已发表申请的 28.6%。变化趋势如图 10 所示。食品化学(+10.9%),数字通信(+8.7%),材料冶金(+8.1%)和基础材料化学(+7.7%)是 2005—2015 年间增长较快

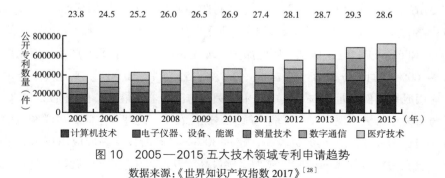

图 10 2005—2015 五大技术领域专利申请趋势
数据来源:《世界知识产权指数 2017》[28]

的技术领域。相比之下，光学（-0.9%），视听技术（-1.5%）和电信（-1.8%）的公开专利申请略有下降。中国、日本、韩国在电器设备上专利布局最多，美国则是在计算机技术上布局最多。

3. 高技术领域科技成果应用转化突出

成果转移转化的主力军仍然是企业。根据国家科技成果网[29]最新数据显示，2016年度登记的51728项应用技术科技成果中，产业化应用的成果数达到29805项，其中由企业完成的科技成果占57.56%；小批量或小范围应用成果数14304项，其中由企业完成的占34.51%；试用成果数4569项，其中由企业完成的占23.55%；应用后停用成果数70项，未应用成果数2980项，独立科研机构和大专院校在这两项中所占的比例总计达到53.64%。相较而言，独立科研机构和大专院校的成果应用转化能力偏弱。

科学研究和技术服务业等行业未应用成果占比较高。从2016年各行业产业化应用成果占比例来看，租赁和商务服务业、金融业、制造业成果的产业化应用比例较高，分别为85.00%、77.36%和76.01%；采矿业，批发和零售业的这一比例在60%～70%之间；电力、热力、燃气及水的生产和供应业，居民服务、修理和其他服务业，信息传输、软件和信息技术服务业，建筑业，交通运输、仓储和邮政业，农、林、牧、渔业，住宿和餐饮业等行业的产业化应用成果比例居于50%～60%之间；而其他行业的产业化应用成果比例在50%以下。在各行业应用技术成果未应用统计中，国际组织、学科研究和技术服务业、卫生和社会工作这三大行业所占比例相对较高，具体如表9所示。

表9 各行业产业化应用技术成果未应用三大行业及其主要原因

行业	未应用的成果占比	主要原因
国际组织	16.67%	技术问题和市场问题
科学研究和技术服务业	12.62%	管理问题和技术问题
卫生和社会工作	8.04%	技术问题和资金问题

注：数据来源：国家科技成果网[29]。

高新技术领域先进制造产业化应用水平高，生物医药与医疗器械应用水平有待加强。在2016年高新技术领域成果中，先进制造领域的产业化应用成果比例最高，达到78.62%；新材料，新能源与节能领域的产业化应用成果比例在60%～70%之间，分别为65.83%和61.56%；电子信息、环境保护、现代农业、航空航天的产业化应用成果比例在50%～60%之间，分别为：57.74%、53.67%、51.75%、50.00%。其他高新技术领域的产业化应用成果比例不足50%，其中，生物医药与医疗器械最低，比例为32.99%。此外，生物医药与医疗器械、现代交通、航空航天、现代农业、地球、空间和海洋领域的小批量使用或小范围应用的成果比例也较高，超过30%。在高新技术领域应用技术成果未应用统计中，核应用技术领域的未应用成果比例最高，其次是生物医药与医疗器械、新材料，具体如表 10 所示。从成果未应用的主要原因来看，核应用技术主要是管理问题（90% 以上）；电子信息、先进制造、现代农业、地球、空间与海洋领域成果未应用或停用的最主要原因是资金问题，航空航天、新材料、生物医药与医疗器械领域的成果未应用或停用的最主要原因则是技术问题。

表 10　高技术领域应用技术成果未应用三大行业及其主要原因

行业	未应用的成果占比	主要原因
核技术	25.45%	管理问题
生物医药与医疗器械	11.27%	技术问题
新材料	8.20%	技术问题

注：数据来源：国家科技成果网[29]。

三、学科平台建设对标国际一流

1. 引领型、突破型、平台型国家科学中心启动建设

2016 年 12 月印发的《国家重大科技基础设施建设"十三五"规划》[30]（本段以下简称《规划》）明确提出，要建设"服务国家战略需求、设施水平先进、多学科交叉融合、高端人才和机构汇聚、科研环境自由开放、运行机制灵活有效"的综合性国家科学中心。《规划》明确，要将"综合性国家科学中心"打造成为"全球创新网络的重要节点、国家创新体系的基础平台以及带动国家和区域创新发展的辐射中心"。综合性国家科学中心将以重大科技基础设施群为依托，集聚全球高端创新资源，实现基础科学突破和引领未来技术发展，为其他各类创新主体提供支撑、开放与协同服务，力争成为具有广泛国际影响的引领型、突破型、平台型的科技创新基地。截至 2017 年 10 月，上海张江综合性国家科学中心、安徽合肥综合性国家科学中心、北京怀柔综合性国家科学中心陆续获批，并积极投入建设。

（1）上海张江综合性国家科学中心

2016 年 2 月，国家发改委、科技部批复同意建设上海张江综合性国家科学中心[31]。张江综合性国家科学中心建设主要包括 4 个方面任务：一是依托张江地区已形成的国家重大科技基础设施，积极争取超强超短激光、活细胞成像平台、海底长期观测网、高效低碳燃气轮机试验装置等一批科学设施落户上海，打造高度集聚的重大科技基础设施集群；二是聚焦生命、材料、环境、能源、物质等基础科学领域，发起设立多学科交叉前沿研究计划；三是建设世界一流科研型大学和学科，汇聚培育全球顶尖科研机构和一流研究团队，大力吸引海内外顶尖实验室、研究所、高校、跨国公司来沪设立全球领先的科学实验室和研发中心；四是探索实施科研组织新体制，建立符合科学规律、自由开放的科学研究制度环境，探索改革国家重大科技基础设施运行保障制度。在建设不到两年的时间内，张江综合性国家科学中心的优势与重要作用已经初显，例如：小卫星中心的高新技术成果不断涌现；超强超短激光实验装置（SULF）研究工作取得重大突破并达到国际同类研究领先水平；上海光源支持用户在结构生物学、催化、凝聚态物理、材料科学、地球与环境科学、流行病毒、地质考古等相关领域的研究中取得了重要突破，并在 Science、Nature、Cell 等国际顶级刊物上发表多篇论文；生命科学领域大科学装置吸引了国内外近两百家单位、一万多人次优秀科学家，进行了两千多项重大前沿创新课题研究；此外，海底长期科学观测系统、转化医学等大科学设施正在加快推进，硬 X 射线自由电子激光装置、高效低碳燃气轮机实验装置、国家生物医学大数据等项目也在争取落地。

（2）合肥综合性国家科学中心

2017 年 1 月，国家批复设立合肥综合性国家科学中心，9 月发布的《合肥综合性国家科学中心实施方案（2017—2020 年）》[32]提出，中心将聚焦信息、能源、健康、环境 4 大领域，以国家实验室为基石，依托世界一流重大科技基础设施集群，布局一批交叉前沿创新平台和产业创新转化平台，建设若干双一流大学和学科，打造多类型、多层次的创新体系，形成中国特色、世界一流的综合性国家科学中心及产业创新中心。《方案》提出建设"2+8+N+3"多类型、多层次的创新体系，其中，"2"是指争创量子信息科学国家实验室，并积极争取新的国家实验室。"8"是指争取新建聚变堆主机关键系统综合研究设施、合肥先进光源（HALS）及先进光源集群规划建设等 5 个大科学装置，提升拓展现有的全超导托卡马克等 3 个大科学装置性能。"N"是指依托大科学装置集群，建设合肥微尺度物质科学国家科学中心、人工智能、离子医学中心等一批交叉前沿研究平台和产业创新转化平台，推动大科学装置集群和前沿研究的深度融合，提升我国在该细分领域的源头创新能力和科技综合实力。"3"是指建设中国科学技术大学、合肥工业大学、安徽大学 3 个"双一流"大学和相关学科。比如，合肥工业大学将着眼"工程管理与智能制造"及"电气工程"等优势方向，布局建设一流学科；安徽大学将重点建设"物质科学与信息技术"学科群。建设一年来，合肥综合性国家科学中心也取得了一些重要成果，例如：中科院合肥研究院承担的 ITER 计划首个超导磁体系统部件——馈线（FEEDER）采购包 PF4 过渡馈线成功研制并顺利交付；锥形束 CT（CBCT）图像引导定位系统顺利研制成功；科大讯飞在人工智能的最高阶段认知智能方面也取得了国际领先水平等。

（3）北京怀柔综合性国家科学中心

2017 年 5 月，国家发展改革委、科技部联合批复了《北京怀柔综合性国家科学中心建设方案》[33]，同意建设北京怀柔综合性国家科学中心。该中心成为继上海张江、安徽合肥后批复的第三个综合性国家科学中心。怀柔综合性国家科学中心将建立与国际接轨的管理运行新机制，推动央地科技资源融合创新发展，重点建设高能同步辐射光源、极端条件实验装置、地球系统数值模拟装置等大科学装置群，创新运行机制，搭建大型科技服务平台，支持多学科、多领域、多主体、交叉型、前沿性研究，代表世界先进水平的基础科学研究和重大技术研发的大型开放式研究基地。在学科布局上，北京怀柔综合性国家科学中心将重点布局物质科学、空间科学、地球科学等重大科学领域。面向 2030 年，物质科学领域，力争在新型高温超导体和金属氢的发展、非常规超导机理、量子计算物理机制、新型量子材料的复杂相变、物性的超快调控和晶格振动实时成像等重大科学问题上取得新的突破；在清洁能源、特种功能、高温合金和高强轻质复合材料等关键新材料技术上取得突破；在推动新能源汽车、智能电网、航空航天、先进制造等产业上取得突破。空间科学领域，将有力支撑我国的载人航天、探月工程、北斗导航等太空计划，力争在宇宙起源演化与物质构成等前沿科学课题，以及天体表面生存和探索等方面取得重大科学发现，显著提升我国空间灾害应对能力，以及空间科学实验基础能力。地球系科学领域，将构建地球巨系统从地核到地球空间环境的不同尺度的物理、化学、生命过程及其相互作用的科学和技术研发体系，力争在深地资源探测、地球环境演变、气候变化与预估、生态系统变化、灾害预测与评估等领域取得重大突破，为国家资源可持续利用和生态文明建设提供科技支撑。

2. 国家级科研平台逐步优化整合

国家实验室主要围绕国家使命，从事基础性和战略性科研任务，通过多学科交叉协助，解决事关国家安全和经济社会发展全局的重大科技问题。根据国家重大战略需求，在新兴前沿交叉领域和具有我国特色和优势的领域，依托国家科研院所和研究型大学，建设若干队伍强、水平高、学科综合交叉的国家实验室。

2016 年 12 月发布的《高等学校"十三五"科学和技术发展规划》[34] 明确指出，将在高校培育、建设若干高水平国家实验室以支撑"双一流"建设。这种"强强联合"的战略布局，将进一步推动学科的发展。2017 年 8 月，科技部、财政部、国家发改委三部委联合印发《国家科技创新基地优化整合方案》[35]。方案明确表示，将根据整合重构后各类国家科技创新基地功能定位和建设运行标准，对现有试点国家实验室、国家重点实验室等国家级基地和平台进行考核评估，通过撤、并、转等方式，进行优化整合。实验室以其命名为主要学科研究方向，但是更注重学科间的交叉融合，调整后的国家实验室与国家研究中心如表 11 所示。

表 11 国家实验室与国家研究中心建设情况

序号	国家实验室/研究中心名称	年份	依托单位
1	国家同步辐射实验室	1984	中国科学技术大学
2	正负电子对撞机国家实验室	1984	中国科学院高能物理研究所
3	北京串列加速器核物理国家实验室	1988	中国原子能科学研究院
4	兰州重离子加速器国家实验室	1991	中国科学院近代物理研究所
5	沈阳材料科学国家研究中心	2000	中国科学院金属研究所
6	北京凝聚态物理国家研究中心	2003	中国科学院物理研究所
7	合肥微尺度物质科学国家研究中心	2003	中国科学技术大学
8	北京信息科学与技术国家研究中心	2003	清华大学

续表

序号	国家实验室/研究中心名称	年份	依托单位
9	北京分子科学国家研究中心	2003	北京大学、中国科学院化学研究所
10	武汉光电国家研究中心	2003	华中科技大学、中国科学院武汉物理与数学研究所、中国船舶重工集团公司第七一七研究所
11	青岛海洋科学与技术国家实验室	2006	中国海洋大学、中科院海洋研究所等
12	磁约束核聚变国家实验室（筹）	2006	中科院合肥物质科学研究院、核工业西南物理研究院
13	洁净能源国家实验室（筹）	2006	中国科学院大连化学物理研究所
14	船舶与海洋工程国家实验室（筹）	2006	上海交通大学
15	微结构国家实验室（筹）	2006	南京大学
16	重大疾病研究国家实验室（筹）	2006	中国医学科学院
17	蛋白质科学国家实验室（筹）	2006	中国科学院生物物理研究所
18	航空科学与技术国家实验室（筹）	2006	北京航空航天大学
19	现代轨道交通国家实验室（筹）	2006	西南交通大学
20	现代农业国家实验室（筹）	2006	中国农业大学

注：数据来源：教育部科技发展中心[36]。

3. 学科期刊国际影响力迅速扩大

2012年起，中国科协启动"科技期刊国际影响力提升计划"[37]，首期三年时间，投入近亿元打造我国高水平英文科技期刊。该计划力争通过一个时期的努力，引导一批重要学科领域的英文科技期刊提升其学术质量和国际影响力，并在学科国际排名中进入前列。该计划支持创办一批代表我国前沿学科、优势学科，或能填补国内学科英文科技期刊空白的高水平英文科技期刊，初步形成具有我国自主知识产权的国际一流科技期刊群，增强科技期刊服务创新能力。2013年公布的"提升计划"第一期（2013—2015年）获资助期刊中共有105种正式出版的英文期刊分别获得A类（15种，每种期刊每年获200万元资助）、B类（40种，每种期刊每年获100万元资助）、C类（50种，每种期刊每年获50万元资助）项目支持；另有30种拟创办英文期刊获"提升计划"的D类支持项目（每年评选出10种，在其取得刊号后再拨付一次性资助经费50万元，自2016年始D类计划每年增加至20种）。"提升计划"是国内迄今为止对英文科技期刊资助力度最大、目标国际化程度最高、影响力最深远的专项支持项目。"提升计划"的第一期（2013—2015年）业已实施完毕，其实施效果备受国内科技期刊界和相关管理部门的关注。

与SCI收录的全部期刊相比较，2013—2016年间我国的SCI收录期刊的数量及学术影响力指标进步更为快速：SCI收录期刊的总数增幅为4.51%，总被引频次和影响因子的增幅分别为20.50%和11.50%，同期中国大陆（含香港地区）SCI收录期刊的数量增加了16.67%，总被引频次和影响因子增幅分别为42.63%和64.88%，如表12所示。

表 12 2013—2016 年 SCI 收录期刊及中国大陆 SCI 期刊的主要计量指标变化

年份	SCI 收录期刊（平均值）			中国大陆 SCI 期刊（平均值）		
	刊数/种	总被引	影响因子	刊数/种	总被引	影响因子
2013	8474	5089	2.174	162	1187	1.193
2014	8618	5358	2.206	173	1330	1.448
2015	8802	5565	2.238	185	1416	1.599
2016	8856	6132	2.424	179	1693	1.967

注：数据来源：中国科技期刊发展蓝皮书[38]。

2016 年，中国科协、财政部、教育部、国家新闻出版广电总局、中国科学院、中国工程院六部委决定继续共同实施新一期"提升计划"。数据显示，2016 年我国英文版科技期刊数量的增加幅度依然维持在相对高位，2016 年共创办了 19 种英文版科技期刊。根据 2017 年 JCR（Journal Citation Reports）数据库显示，我国入围 SCI 的期刊约 180 种，影响因子增长迅速，不少期刊影响因子均创新高，具体如表 13 所示。

表 13 我国部分外文期刊影响因子发展情况

期刊	2007 年	2008 年	2009 年	2010 年	2011 年	2012 年	2013 年	2014 年	2015 年	2016 年
Cell Research	4.217	4.535	8.151	9.417	8.190	10.526	11.981	12.413	14.812	15.606
Light-Science & Applications	—	—	—	—	—	8.476	14.603	13.600	14.098	
Fungal Diversity	3.593	2.279	3.803	5.074	4.769	5.319	6.938	6.211	6.991	13.465
Bone Research	—	—	—	—	—	—	—	1.310	3.549	9.326
National Science Review	—	—	—	—	—	—	—	—	8.000	8.843
Molecular Plant	—	—	2.784	4.296	5.546	6.125	6.605	6.557	7.142	8.827

注：数据来源：JCR 数据库[39]，检索时间 2017 年 9 月。

中国科学院上海生命科学研究院生物化学与细胞生物学研究所与中国细胞生物学学会共同主办的 *Cell Research* 影响因子再攀新高。过去几年中，*Cell Research* 影响因子相继赶超国际知名学术期刊如 *EMBO J*、*PNAS*、*Developmental Cell* 和 *PLoS Biology*，2016、2017 年继续发力，连续两年影响因子超过 *Nature Structural & Molecular Biology* 和 *Molecular Cell*，这奠定了同国际顶级期刊 *Nature* 及 *Cell* 子刊有效竞争的基础。标志着 *Cell Research* 这一我国自主知识产权的学术期刊国际品牌"精品"，正向着更加深远坚实的目标迈进。

中国科学院长春光学精密机械与物理研究所与自然出版集团（Nature Publishing Group，

NPG）合作出版的开放获取全英文光学学术期刊 *Light: Science & Applications*，仅创刊第四年，其影响因子便达到 14.098，在 JCR 收录的 92 种光学类期刊中排名第三位，在 8000 余种 SCI 期刊中排名第 103 位。期刊发表了包括：2014 年诺贝尔物理学奖得主 Shuji Nakamura 的文章、入选 OSA2013 年度三十大重大突破的文章以及若干 ESI 高水平论文。

值得注意的是，*Fungal Diversity*、*Bone Research* 两本期刊 2017 年影响因子几乎增长一倍，《中国科学》杂志社旗下的期刊 *Science Bulletin*，更名后首次获得的影响因子就达到 4 分。在中国 SCI 影响因子前 50 位期刊中，除 5 个期刊影响因子较去年相比下降外，其余均呈现增长态势。

由此可以看出，近年来我国本土创办的 SCI 期刊正不断加快国际化步伐，尤其是在生命医学领域，发展势头十分迅猛，得到了越来越多国际科研学者的关注和认可，同时也逐步推动着我国更多学术期刊办刊水平的进一步提升。

四、学科队伍持续壮大

在全球化和知识经济时代，科技人才特别是高层次科技人才已成为衡量和决定一个国家综合实力和国际地位的重要因素之一。

1. 研发人员规模不断增加

R&D 人员的规模和素质决定国家创新活动的质量，也是实现我国科技发展规划目标的前提条件。2006—2015 年我国 R&D 人员折合全时当量数据，接近线性增长趋势，增幅约为 13%；2015 年为 375.88 万人，比 2006 年的 150.2 万人增加了 1.5 倍多，具体如图 11 所示。

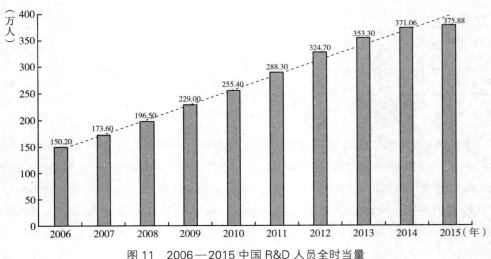

图 11　2006—2015 中国 R&D 人员全时当量

数据来源：《中国科技统计年鉴 2016》[40]

尽管我国 R&D 研究人员的规模在不断增长，但由于我国人口基数大，相比其他国家，我国 R&D 研究人员占总人口数比例较低。图 12 是根据 OECD 对主要国家 R&D 研究人员数量与人口基数的占比比较。可以看出，我国还需要持续扩大 R&D 研究人员规模。

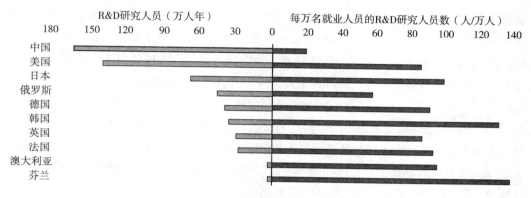

图 12　主要国家 R&D 研究人员全时当量与每万名就业人员的 R&D 研究人员数对比

数据来源：OECD, Main Science and Technology Indicators，August 2017[4]

2. 学科生力军不断成长

结合第四轮全国学科评估数据[41]，2016 年，全国有研究生培养单位 793 个、学科 11328 个；在学研究生人数 198.11 万，授予博士、硕士学位人数 56.39 万，分别比 2012 年增长了 15% 和 16%。第四轮学科评估首次对用人单位开展大规模满意度调查，结果显示，用人单位对研究生的整体满意度高达 98%。

《中国科技统计年鉴 2016》[40] 最新数据显示，2015 年我国共有研发机构 3650 个，从业人员 78.3 万人；共有高等学校 2650 个，R&D 人员共计 83.9 万人；有 R&D 企业数 73570 个，R&D 人员 364.6 万人。研发机构与高等学校 R&D 人员占比与往年相比均呈现增长的趋势，其中硕士、博士毕业生已占一半以上，这有利于创新能力的提高和高水平科研成果的产出，是我国科研创新的生力军，如图 13 所示。

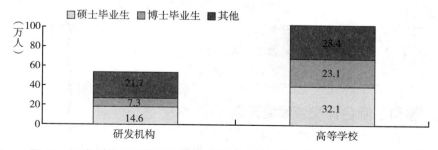

图 13　研发机构与高等学校 R&D 研究人员中硕士、博士毕业生分布情况

数据来源：《中国科技统计年鉴 2016》[40]

3. 海外高层次人才助力学科发展

随着中国国际影响力的不断扩大、人才引进计划政策的实施，高学历、高层次人才回流占比明显提升，中国正在经历新中国成立以来最大规模的海外人才回归潮。饶毅与施一公作为"千人计划"第一批引进的人才，也是这一波回国潮里国际声望很高的科学家。而潘建伟则带领其团队在量子通信方面引领中国走到世界前沿，获得重大突破，并被著名学术期刊 Nature 选为 2017 年度十大人物。这些国际范围内的高层次创新人才，引领我国在包括深海深测、载人航天工程、高温超导、高空望远镜等重大项目和高科技领域在内的多个学科领域跻身世界前列，并取得重大突破。有研究者对我国以"长江学者""百人计划""千人计划（青年千人）"为代表的海外高层次引进人才的学科领域分布进行了统计分析[42]，三大人才计划引进最多的学科较为接近，前三甲均为生物学、化学、物理学。尤其是生物学领域引进人才的数量远远高于其他学科领域的，而工程技术、基础医学、药学等领域人才引进数量偏少，具体如图 14 所示。这也是我国近两年在化学、物理以及生命科学领域取得较为显著的突破与成果的重要原因。

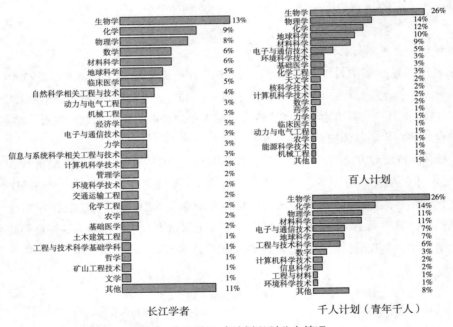

图 14　高层次人才计划学科分布情况

五、学科国际合作交流频繁深入

新一轮科技革命正在重塑世界经济结构和竞争格局，在开放创新、全球协同的趋势下，中国的科研与创新在经济高速发展和日益开放中也逐步走向国际化。中央和地方政府各有关部门，通过制订与强化国际科技合作政策机制，加大国际合作经费投入力度，调动科技界和社会力量等措施，积极推动中国的国际科技合作。

1. 国际合作规模与影响力双双提升

当前创新要素开放性、流动性显著增强，我国国际科研合作"走出去"道路不断拓展。国际合作论文中受中国国内经费资助的论文比例，从"十一五"的31.6%上升到"十二五"的65.2%。2015年，中国国际合作发文量约为7.1万篇，占全球国际合作发文量的18.6%，同比2006年提升4.4倍。

从"十一五"到"十二五"的10年间，国际科研合作不仅合作规模创出新高，还发生了质的飞跃。学科引文影响力显示，各学科国际合作论文的引文影响力均显著高于学科论文整体引文影响力，如图15所示。另一方面，根据2017年最新Nature Index[25]显示，其追踪的中国论文中，涉及国际合著者的论文比例也在逐年上升，2016年其追踪的中国论文中，国际合作论文已超过一半。这些均表明国际科研合作对提升我国学科全球科研影响力有积极的促进作用。

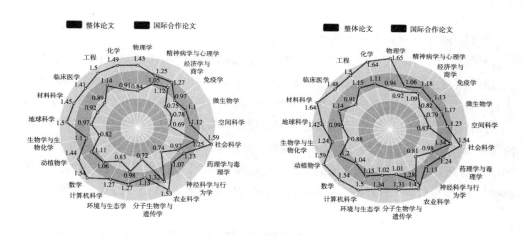

图15 "十一五"（左）和"十二五"（右）时期中国科研论文学科引文影响力
数据来源：《中国国际科研合作现状报告》[43]

美国仍然在国际科研合作中发挥着绝对的主导地位作用，中国的科研合作中心度从2006—2010年全球第十位，提升到2010—2015年的第七位，如表14所示。"国际科研合作中心度"是用来评测一个国家国际合作在全球科研合作网络中的地位和重要性的指标。随着与科研伙伴的合作不断深化，中国同科技发达地区及"一带一路"沿线国家的合作关系得到全面巩固。科研人员的研究在国际社会中的关注度显著提升，中国逐步成为国际科研合作网络中各个国家重要的合作伙伴。

表 14 国家国际科研合作中心度变化

国家	2006—2010 年中心度	2011—2015 年中心度	绝对增幅（%）	相对增幅（%）
美国	11.9	12.1	0.2	1.7
英国	6.3	7.0	0.7	11.1
德国	6.1	6.7	0.6	9.8
法国	4.3	4.9	0.6	14.0
意大利	3.1	4.2	1.1	35.5
西班牙	2.7	3.9	1.2	44.4
中国	2.0	3.4	1.4	70.0
荷兰	2.3	3.0	0.7	30.4
澳大利亚	2.0	2.9	0.9	45.0
加拿大	2.6	2.9	0.3	11.5

注：数据来源：《中国国际科研合作现状报告》[43]。

2. 国际合作学科分布较为集中

2006—2015 年间，中国的化学、物理学、工程科学、临床医学和材料科学 5 个学科领域既是论文发文量最大也是国际科研合作规模最大的学科领域，其国际合作论文总数在所有学科国际合作论文总数中占比超过 50%；化学学科的国际合作论文从 1.5 万篇增长至 3.1 万篇，增长了 1 倍多；国际合作论文发文量在万篇规模以上的学科中，地球科学是学科内国际合作论文占比（38.3%）最高的学科，显示其学科国际科研合作比较活跃；2010—2015 年间，工程科学和化学领域的合作规模超越物理学，提升至前两位，在国际科研合作规模前 10 位学科中，计算机科学国际科研合作规模提升幅度最大，如表 15 所示。

表 15 2006—2015 中国国际合作论文学科分布、比重及 2006—2010 年和 2010—2015 年间变化

学科领域	学科论文（篇）	学科国际合作论文				学科国际合作论文比重（%）	学科内国际合作论文比重（%）
		合计（篇）	2006—2010（篇）	2011—2015（篇）	增幅（%）		
化学	327210	46291	14946	31345	109.7	12.3	14.1
物理学	199222	44745	16245	28500	75.4	11.9	22.5
工程科学	179069	44485	12424	32061	158.1	11.8	24.8
临床医学	155536	35611	10274	25337	146.6	9.4	22.9
材料科学	175166	30444	8817	21627	145.3	8.1	17.4

续表

学科领域	学科论文（篇）	学科国际合作论文				学科国际合作论文比重（%）	学科内国际合作论文比重（%）
		合计（篇）	2006—2010（篇）	2011—2015（篇）	增幅（%）		
地球科学	54898	21038	6487	14551	124.3	5.6	38.3
生物及生物化学领域	73004	18477	5707	12770	123.8	4.9	25.3
计算机科学	50302	16560	4069	12491	207.0	4.4	32.9
植物学与动物学	53719	16542	5504	11038	100.5	4.4	30.8
分子生物与遗传学	44793	14633	3889	10744	176.3	3.9	32.7
数学	68038	14607	5340	9267	73.5	3.9	21.5
环境生态科学	44484	14567	4054	10513	159.3	3.9	32.7
农业科学	35270	10525	3379	7146	111.5	2.8	29.8
神经科学及行为学	28440	9368	2794	6574	135.3	2.5	32.9
药理学与毒理学	41052	8206	2671	5535	107.2	2.2	20.0
社会科学	14350	7056	1932	5124	165.2	1.8	49.2
微生物学	17863	5312	1692	3620	113.9	1.4	29.7
经济与商业学	8947	5086	1187	3899	228.5	1.3	56.8
空间科学	10483	5069	1863	3206	72.1	1.3	48.4
免疫学	14532	4753	1475	2378	122.2	1.3	32.7
神经病学与心理学	6210	3516	830	2686	223.6	0.9	56.6
多学科	1979	642	89	553	521.3	0.2	32.4

注：数据来源：《中国国际科研合作现状报告》[43]。

学科国际科研合作相对活跃度（某学科在某国国际科研合作中的相对规模）指标显示，材料学、工程学、计算机科学是中国国际科研合作相对活跃度较高的3个学科，化学、地球科学、数学、物理学和农业等6个学科的合作相对活跃度也超过了中国的平均水平。与全球主要国家学科国际合作格局相比，十年间，中国发文量最大的5个学科中，材料学、工程学、化学、物理学国际科研合作相对活跃度超过主要国家水平或与其持平，而临床医学合作相对活跃度与主要国家水平差距较大。值得注意的是，在主要国家国际科研合作最为活跃的空间科学上，中国国际合作相对活跃度却很低。美国、西班牙、加拿大的学科国际合作活跃度相对均衡，而中国的学科国际合作活跃度在个别学科上较为集中，不够均衡，如图16所示。

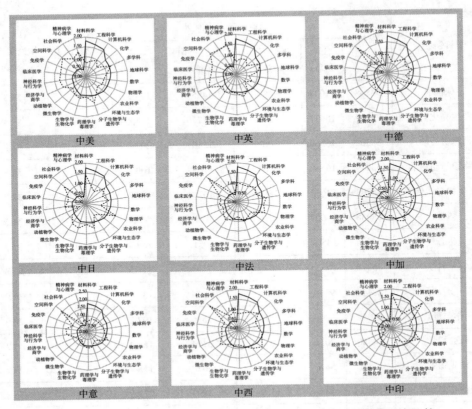

图 16　中国及其他主要国家在各学科国际科研合作的相对活跃度分布比较

注：实线代表中国各学科国际科研合作的相对活跃度，虚线代表主要国家学科国际科研合作的相对活跃度。

数据来源：《中国国际科研合作现状报告》[43]

3. 国际交流学习选择更加多元

近年来，我国留学回国与出国留学人数"逆差"逐渐缩小，出国留学与来华留学人数同步增长，成为世界最大的留学输出国和亚洲重要留学目的国。2016 年度我国出国留学人员总数为 54.45 万人，较 2015 年增长 3.9%。留学回国人数为 43.25 万人，较 2012 年增长 58%。逾八成留学人员学成后选择回国发展[44]。2016 年，在华留学生生源国家（地区）总数为 205 个，创历史新高，共有 44.3 万名各类外国留学人员来华学习，比 2015 增长 11.35%。

国家公派留学始终紧密围绕国家教育、人才、科技三个规划纲要，《全国留学工作会议精神》《关于进一步做好新时期教育对外开放工作的若干意见》[45]《共建"一带一路"教育行动计划》[46]等相关文件精神，特别是在人才培养已形成"面上铺开"格局之后，强化"高端引领"，在若干关键领域形成"人才高地"，重点服务教育、国防、外交、卫生、体育、地震、海洋、航空航天等部门、行业，培养急需紧缺人才。十八大以来，国家公派留学培养了一大批具有国际视野和竞争能力的紧缺人才和战略后备人才，据统计，国家公派留学生总数为 107005 人，其中派出访问学者 44814 人，占派出总人数的 41.88%；博士生、硕士生和本科生 62191 人，占派出总人数的 58.12%。从学科分布看，国家公派留学人员主要选择了国家发展建设急需的理、工、农、医等学科。其中攻读工科的占 36.54%、理科占 15.47%、医科占 6.68%、农科占 3.17%。

来华学习的专业也打破了以汉语学习为主的格局，学科分布更加合理。相比 2012 年，教育、理科、工科和农学学生数量显著增加，增幅均超过 100%；占比增长最快的学科为工科，比 2012 年增长了 5.2 个百分点。

第二节 学科发展成果与动态

一、学科前沿研究领跑态势初显

建设科技强国，必须审时度势，面向世界科技前沿，开展前瞻性研究。只有抓住科技革命的机遇，才能使经济实力、科技实力、国防实力迅速增强，综合国力快速提升。经过多年努力，科技发展已进入由提高数量到提升质量的跃升期，在天文学、生命科学、量子信息科技等一些前沿学科领域已经逐渐从跟跑变为领跑的态势，取得了重大原创突破。

天文学是孕育重大原创发现的前沿科学，也是推动科技进步和创新的战略制高点。在 FAST（500 米口径球面射电望远镜）之父南仁东的带领下，其团队建成具有我国自主知识产权、世界最大单口径、最灵敏的射电望远镜，被誉为"中国天眼"。利用天然的喀斯特洼坑作为台址，洼坑内铺设数千块单元组成冠状主动反射面，采用轻型索拖动机构和并联机器人实现接收机高精度定位，这是"中国天眼"的三大自主创新。与德国波恩 100 米望远镜相比，"天眼"的灵敏度提高了约 10 倍；与美国阿雷西博 350 米望远镜相比，"天眼"的综合性能也提高了约 10 倍。"天眼"能够接收到 137 亿光年以外的电磁信号，观测范围可达宇宙边缘。FAST 将在未来 20~30 年保持世界一流设备的地位。

在量子通信领域，潘建伟团队成功研制出全球首颗量子卫星，量子科学实验卫星"墨子号"圆满实现千公里级的星地双向量子纠缠分发、星地量子密钥分发、地星量子隐形传态等 3 大科学目标，具有里程碑意义。这不仅是我国量子保密通信领域"撒手锏"技术研发的重大突破，实现了从跟随创新到引领创新、从集成创新到原始创新的跨越，同时也是世界量子通信技术的重要创新，它有望使人类科技发展史上"最安全的通信手段"具备覆盖全球的能力。Science 杂志几位审稿人称赞该成果是"兼具潜在实际现实应用和基础科学研究重要性的重大技术突破"，并断言"毫无疑问将在学术界和广大的社会公众中产生非常巨大的影响"。被誉为"量子之父"的潘建伟也因此荣获 Nature 杂志 2017 年全球十大科学人物、2017 年未来科学大奖"物质科学奖"等多个奖项。

在生命科学领域，清华大学生命科学学院施一公研究团队在解析真核信使 RNA 剪接体这一关键复合物的结构，揭示其活性部位及分子层面机理的研究中做出重大贡献。其团队通过单颗粒冷冻电子显微技术（冷冻电镜）解析了酵母剪接体近原子分辨率的三维结构，阐述了剪接体对前体信使 RNA 执行剪接的基本工作机理。至此，遗传信息传递过程中所涉及的 3 个重要分子 RNA 聚合酶、剪接体和核糖体的结构和功能——得到了解析。他和团队的杰出科学成就，使我国结构生物学研究跻身国际领先行列。施一公也因此获得了 2016 年何梁何利最高奖"科学与技术成就奖"、2017 年未来科学大奖"生命科学奖"等多个奖项。值得一提的是，RNA 聚合酶和核糖体的结构解析曾分别获得 2006 年和 2009 年的诺贝尔化学奖。国内也有同行专家称施一公团队的此项研究是生命科学领域的重要突破，有望获得诺贝尔奖。

在信息技术领域，中国科学院陈天石、陈云霁兄弟研发的寒武纪深度神经网络处理器，是世界上第一款模拟人类神经元和突触进行深度学习的处理器。"寒武纪"目前正走在产业化的道路上，未来将主打人脸识别、声音识别等人工智能方向，必将引起高性能服务器、高效能终端芯片、机器人芯片等多领域的深刻变革。由国家并行计算机工程技术研究中心研发、放置在无锡中心的"神威·太湖之光"超级计算机以峰值性能为 12.5 亿亿次/秒，持续性能为 9.3 亿亿次/秒的超快运算速度成功登顶全球超级计算机 500 强榜单，并首次斩获戈登·贝尔奖和吉尼斯世界纪录。"神威·太湖之光"是我国首次使用自主研发的处理器芯片制成的世界一流超级计算机系统，巩固了我国在超级计算机领域的最高地位。

二、学科基础研究高水平成果不断

基础研究是提高国家原始性创新能力、积累智力资本的重要途径，是跻身世界科技强国的必要条件，是建设创新型国家的根本动力和源泉。当前，我国创新驱动发展已具备发力加速的基础。近两年，随着《国家创新驱动发展战略纲要》[1] 实施，我国基础研究成果进步显著。在脑科学、物理学、中微子振荡、先进核能、干细胞与基因编辑等多项基础研究领域取得了丰硕的科研成果，在国际上产生了重大的影响，同时也为基础研究的成果转化应用提供了新的思路，进而成为推动经济社会发展的原始动力。

在脑科学领域，由中科院自动化所脑网络组研究中心蒋田仔领衔的研究团队，联合国内外其他团队，通过6年努力，成功绘制出全新人类脑图谱——脑网络组图谱，第一次建立了宏观尺度上的活体全脑连接图谱。目前，脑网络组图谱已经引起国内外同行的高度关注，国际神经信息学协调委员会（International Neuroinformatics Coordinating Facility，INCF）已在第一时间在线发布了人类脑网络组图谱，欧盟人脑计划（Human Brain Project，HBP）即将在其神经信息平台（Neuroinformatics Platform，NIP）公开发布该图谱。

在物理学领域，薛其坤团队在利用分子束外延技术发现量子反常霍尔效应和单层铁硒超导等新奇量子效应方面做出开拓性工作。这被著名物理学家杨振宁称之为"诺贝尔奖级"的科研成果，薛其坤也获得了2016年未来科学大奖"物质科学奖"。中国科学院院士、北京大学教授王恩哥和北京大学教授江颖带领的科研团队通过在超高真空低温扫描隧道显微镜（STM）下记录到的水分子氢核隧穿现象，首次揭示了水的核量子效应。研究成果被*Science*（2016年）杂志收录，并且受到了审稿人的盛赞。2017年，国家大科学装置——全超导托卡马克核聚变实验装置东方超环（EAST）实现了稳定的101.2秒稳态长脉冲高约束等离子体运行，创造了新的世界纪录。东方超环是世界上第一个实现稳态高约束模式运行持续时间达到百秒量级的托卡马克核聚变实验装置，对国际热核聚变试验堆（ITER）计划具有重大科学意义。这一重要突破标志着我国磁约束聚变研究在稳态运行的物理和工程方面将继续引领国际前沿。此外，中国科学院物理研究所的科研团队在拓扑物态研究领域取得重大进展，首次发现了突破传统分类的新型费米子——三重简并费米子，为固体材料中电子拓扑态研究开辟了新的方向；中科院上海光机所利用超强超短激光获得反物质；大亚湾中微子实验测得迄今最精确的反应堆中微子能谱；中国暗物质粒子探测卫星"悟空"在太空中测量到电子宇宙射线的一处异常波动等成果，都反映出了我国的科技实力。

在数学领域，许晨阳团队在双有理代数几何上做出了极其深刻的贡献，解决了一系列代数几何学中很多不同领域的重要几何问题，得到国际同行的高度评价，同时为代数几何学在中国的发展做出了重大的贡献，并获得了2017年度未来科学大奖"数学与计算机科学奖"。上海交通大学廖世俊团队原创性地提出了求解强非线性方程解析近似解的一般性方法 - 同伦分析方法，并历经20余年不断完善，逐步形成了一个较为完整的理论体系。率先提出"广义同伦"概念，引入"收敛控制参数"，提出了一个确保级数解收敛的有效途径，使同伦分析方法适用于强非线性方程；并应用同伦分析方法成功求解了力学中的许多非线性问题，不仅能更好地求解某些经典力学问题，而且获得了一些全新的、被其他方法遗漏的解。该方法已得到国际学术界的普遍认同和广泛应用。廖世俊也多次入选全球"高被引科学家"。

在化学领域，中国科学技术大学的谢毅、孙永福课题组设计了一种杂化模型体系用来研究金属表面氧化物对其自身金属电催化性能的影响，成果显示钴在位于特定的排列方法和氧化价态时，具有更高的催化二氧化碳的活性，即超薄二维结构和金属氧化物的存在提高了催化还原二氧化碳的能力。研究成果发表在*Nature*（2016年）杂志上，对推动电催化还原二氧化碳机理研究具有重要的意义，并荣登2016中国科学十大进展榜首。此外，我国在苄位碳氢键的不对称氰化反应、甘露霉素的高效化学全合成、乙烯、乙炔的高效分离等研究方向也取得了重要进展。

在遗传学领域，对于 CRISPR/Cas9 基因编辑技术的研究呈现火热的态势，我国学者在该领域也屡屡实现跨越性突破。中国科学院遗传与发育生物学研究所高彩霞研究员利用 CRISPR/Cas9 技术，在六倍体面包小麦中成功实现了同时编辑三个同源等位基因，并由此赋予了小麦对白粉菌的遗传抵抗力。之后，她又利用 CRISPR/Cas9 特异识别病毒和外源 DNA 的特性，将该 CRISPR 切割系统引入植物，在植物中建立了高效的 DNA 病毒防御体系。这一系列成果被多家国际顶级期刊报道，高彩霞也入选 2016 年 *Nature* 评选的"中国科学之星"。中国科学院动物研究所周琪、段恩奎研究组与中国科学院上海营养科学研究所翟琦巍研究组合作，第一次从精子 RNA 角度为研究获得性性状的跨代遗传现象开拓了全新的视角，提出精子 tsRNAs 是一类新的父本表观遗传因子，可介导获得性代谢疾病的跨代遗传。其成果发表在 *Science*（2016）杂志上，并引起国际各大媒体的关注。中国科学院上海生物化学与细胞生物学研究所徐国良团队与美国威斯康星大学孙欣、北京大学汤富酬等团队合作，发现了 TET 三个成员之间功能上相互协作，介导的 DNA 去甲基化与 DNMT 介导的 DNA 甲基化相互拮抗，通过调控 Lefty-Nodal 信号通路控制胚胎原肠运动。该工作第一次系统地揭示了胚胎发育过程中关键信号通路的表观遗传调控机理，为发育生物学的基本原理提供了崭新的认识。

不仅如此，我国在多个基础学科领域都取得了突破性进展，非线性最优化计算方法得到国际公认；磁约束核聚变研究，使我国长脉冲高参数等离子体物理实验跃居国际领先水平；中生代鸟类研究填补了多项国际空白，彰显国际领先地位；超分子体系构筑、调控与功能转化，对国际分子组装领域做出了重要贡献；病原菌感染和宿主天然免疫的分子机制取得了多项原创性成果；国际上最完善的中国及世界地方猪种资源基因组 DNA 库、国际上唯一的家猪嵌合研究群体为家猪育种提供了平台……所有这些，有力地提升了我国科技的国际影响力。

三、"顶天立地"服务国家重大需求

两年来，我国科研方向明确了"上天入地下海"的战略布局，以深海、深地、深空、深蓝领域为主攻方向"决战深部"，加强国家重要战略资源和能源的开发利用。通过不懈努力取得了多项研究成果，以"顶天立地"之态服务国家重大需求。

在航空航天领域，近两年也取得了巨大成就。2016 年 8 月发射了第一颗量子卫星，旨在测试将安全量子通信扩展到太空的方法。2016 年 11 月发射了"长征五号"运载火箭，"长征五号"代表了我国运载火箭科技创新的最高水平，填补了大推力无毒无污染液体火箭发动机的空白，实现了异型发动机起飞技术的重大突破。它也是实现未来探月工程三期、载人空间站、首次火星探测任务等国家重大科技专项和重大工程的重要基础和前提保障。2016 年 10—11 月，执行"天宫二号"和"神舟十一号"载人飞行任务的两名中国航天员在太空驻留一个月，打破了中国航天员执行太空任务的最长纪录。"神舟十一号"所获成果进一步提高了我国在航天领域的国际地位，为人类探索宇宙空间做出了巨大的贡献。此外，中国大型客机项目 C919 在上海浦东机场实现顺利首飞。C919 大型客机是我国按照国际民航规章自行研制、具有自主知识产权的大型喷气式民用飞机，是我国民用飞机研发的一个重要里程碑。空空导弹、飞航导弹、先进武器装备等成果为能打仗、打胜仗提供了技术支撑。精密区域数字高程基准确定的多项理论、长寿命航天肼分解催化剂、航天器自主导航技术、空间润滑材料技术体系等对国家重大战略工程做出了重要贡献。

在海洋科学领域，2016 年 6 月—8 月，中国科学院"探索一号"科考船在马里亚纳海沟海域开展了我国海洋科技发展史上第一次综合性万米深渊科考活动。我国自主研发的万米级自主遥控潜水器（ARV）"海斗号"成功下潜到 10767 米，不仅创造了我国水下机器人的最大下潜深度纪录，也在深海水样环境、地震信号、生物特性等方面取得了一系列宝贵的原始数据，是我国海洋科技的又一里程碑，标志着我国深潜科考进入万米时代，使我国成为继日、美两国之后第三个拥有研制万米级无人潜水器能力的国家。我国首次万米深渊科考的成功宣示了我国深海科技创新能力正在实现从"跟踪"为主向"并行""领先"为主的转变，为全面实现国家"十三五"重点研发计划部署的万米载人/无人深潜的战略目标迈出了第一步。2017 年，由中国科学院沈阳自动化所研制的"海翼"号水下滑翔机，是我国具有完全自主知识产权的新型水下观测平台，突破之前由美国科学家创下的水下滑翔机最大下潜深度 6000 米的世界纪录。

在地球科学领域，2015 年 12 月 29 日，中国石化宣布，国内第一个大型页岩气田涪陵气田建成 50 亿立方产能。这意味着每年可为我国减排二氧化碳 600 万吨，相当于植树近 5500 万棵、近 400 万辆经济型轿车停开一年，并可减排二氧化硫 15 万吨、氮氧化物近 5 万吨。这个在我国能源开发史上具有里程碑意义的大气田，正是在中国石化勘探分公司总经理郭旭升创建的"二元富集"理论指导下发现的。2000 年以来，郭旭升和团队持续攻关，发现了涪陵、元坝、普光等 6 个大中型气田，带来了四川盆地新一轮天然气发现和储量增长的高峰。郭旭升也因此获得了 2016 年度何梁何利基金科学与技术进步奖的地球科学奖。中国海洋石油总公司副总工程师姜伟发明了循环通道和承载结构双约束条件的海上钻井隔水管下深设计方法和控制技术，使深水钻井导管入泥深度的控制精度比现行国际标准精度提高 20%，使我国成为全球在深水领域掌握该核心技术的少数国家之一。他发明的海上化学驱油高效配置技术装置，与国内外陆上油田相比，重量减轻 80%，面积减少 86%，溶解效率提高 67%。

在能源领域，2016 年，中科院大连化物所包信和院士和潘秀莲研究员团队的"开创煤制烯烃新捷径"成果利用纳米催化基本原理，创造性地将控制反应活性和产物选择性的两类催化活性中心有效分离，实现了煤基合成气一步法高效生产烯烃这一突破性技术，从原理上创造了一条低耗水、低耗能的煤基合成气转化制烯烃的新途径，成功地回答了李克强总理一直关心的"能不能不用水或少用水进行煤化工"的问题。Science 杂志以"令人惊奇的选择性"为题刊发了专家评论和展望，称赞该研究在原理上的突破将带来工业上的巨大竞争力，被产业界同行誉为"煤转化领域里程碑式的重大突破"。中科院大连化学物理研究所也将与国内重要化工企业和国外著名化学公司达成初步协议，着手在催化剂制备和工艺过程开发等方面共同合作，实现工业示范和产业化，努力将这一原创性成果转变为真正的生产力。值得一提的是团队在研发该成果的 9 年期间，除了申报了多件中国发明专利和国际 PCT 专利以外，没有公开发表过一篇相关研究的文章。另外，由国家电网公司等单位完成的特高压 ±800 千伏直流输电工程，取得了一系列重大技术突破和创新。首次揭示了特高电压下的污秽外绝缘沿面放电机理，首创多因素多目标控制的外绝缘配置方法，提出了双 12 脉动阀组串联主接线型式，实现了 800 千伏特高电压、复杂环境下的绝缘可靠配置；提出了极导线任意布置下同走廊多回线路离子流场计算方法，首创交、直流电磁环境叠加模拟试验方法，实现了工程技术方案最优与生态环境友好的和谐统一；首次成功研制世界上通流能力最大的换流阀，电压最高、容量最大的换流变压器等特高压直流成套设备；构建了完备的特高压直流试验技术体系，建立了国际领先的试验研究平台。特高压 ±800 千伏直流输电技术是世界上电压等级最高、输送容量最大、送电距离最远、技术水平最先进的输电技术，是解决我国能源资源与电力负荷逆向分布问题、实施国家"西电东送"战略和电力跨区域大范围输送的核心技术。该技术也获得了 2017 年科技部国家科学技术进步奖特等奖。2017 年，由国土资源部中国地质调查局组织实施的南海神狐海域天然气水合物试采工程全面完成了海上作业，这标志着我国首次海域天然气水合物试采圆满结束。试采取得了持续产气时间最长、产气总量最大、气流稳定、环境安全等多项重大突破性成果，创造了产气时长和总量的世界纪录。获取科学试验数据 647 万组，为后续的科学研究积累了大量的翔实可靠的数据资料。

四、学科应用推动国民经济建设

科技创新要服务于国民经济主战场，就需要推进科研成果转化，实现科技创新与国家需要、人民要求、市场需求的对接，从而成为推动经济社会发展的现实动力。对此，我国在生物、医学、农业、计算机、电子信息、材料等诸多学科领域都取得了具有现实意义的创举。

在生物学领域，中国科学院生物物理所的研究团队在光合作用研究中获得重要突破，在国际上率先解析了高等植物菠菜光合作用超级复合物的高分辨率三维结构。该项研究发表在 Nature（2016）杂志上。基于结构的光合作用机理研究具有重要的理论意义，同时也将为解决能源、粮食、环境等问题提供具有启示性的方案。同时，研究团队率先破解了光合作用超分子结构之谜，获得了其与外周捕光天线之间相互装配原理和能量传递过程相关的重要结构信息，为实现光能向清洁能源氢气的转换提供具有启示性的方案。此外，我国昆虫学和植物保护学研究始终面向国家需求，解决我国重大病虫害成灾与控制的关键问题，水稻条纹叶枯病和黑条矮缩病灾变规律与绿色防控技术和作物多样性控制病虫害关键技术获得了国家科技进步奖二等奖，创制的全球首个抗病毒蛋白质植物免疫诱导剂"阿泰灵"，已在全国 28 个省、市、区推广应用，获得了中国产学研合作创新成果一等奖，为保障国家粮食安全和生态安全发挥了重要作用；矮抗 58 等系列高产小麦品种已经成为我国小麦育种的骨干亲本，累计增产 119.1 亿千克；重要病原感染与免疫机制研究领域获得重大突破，有效解决家猪链球菌病的防治难题；完成西藏生物多样性考察，为西藏濒危野生动物保护做出重要贡献。

在医学领域，清华大学生命科学学院戴俊彪团队，利用化学物质合成了 4 条人工设计的酿酒酵母染色体，标志着人类向"再造生命"又迈进一大步。未来可以加快在基因组重排、环形染色体进化领域的研究进度，为人类环形染色体疾病、癌症和衰老等提供研究与治疗模型。香港大学的卢煜明团队基于孕妇外周血中存在胎儿 DNA 的发现，在无创产前胎儿基因检查方面做出了开拓性贡献，利用第二代基因测序来定量测量胎儿 DNA 的方法可用于唐氏综合征检测。这个革命性的方法为全球 90 多个国家的孕妇提供了无创产前诊断，仅在中国，每年就有超过一百万孕妇接受这项测试。卢煜明也因此项研究获得了 2016 年未来科学大奖"生命科学奖"。浙江大学医学院附属第一医院等单位在以防控人感染 H7N9 禽流感为代表的新发传染病防治体系中取得重大创新和技术突破。创立了以深度测序和高通量数据分析技术为核心的我国新发突发传染病病原早期快速识别技术体系；创立了分子分型和溯源为特色的新发突发传染病预测预警技术体系和防控模式；创立了从蛋白结构到哺乳动物模型精确解析新发突发传染病感染和发病机制研究新体系；创建了引领世界的新发突发传染病危重症患者救治的"中国技术"；创建了我国流感疫苗快速研发新技术体系；创建了我国新发突发传染病诊断试剂高效快速研发平台。中国科学院上海神经科学研究所仇子龙、孙强研究组合作，发现 MECP2 转基因猴表现出类似于人类自闭症的刻板动作与社交障碍等行为。他们首次在灵长类中成功通过精巢异体移植的方法加快猴类繁殖周期，历时三年半得到了携带人类 MECP2 基因的第二代转基因猴，且发现其在社交行为方面表现出了与亲代相同的自闭症样表型。这是世界上首个自闭症的非人灵长类模型，为深入研究自闭症的病理与探索可能的治疗干预方法做出了重要贡献。此外，我国在分子遗传、脏器移植、干细胞治疗、肿瘤靶向诊疗方面涌现出了一大批优秀的科研成果，为奠定现代生物医学基础研究，保障大众健康做出了卓著贡献。

在农业领域，科技部、中国科学院联合河北省、山东省、辽宁省和天津市共同实施的国家科技支撑项目——"渤海粮仓科技示范工程"面向国家粮食安全重大需求，在环渤海低平原区建立粮食增产增效示范区，研发、集成、示范推广抗逆作物品种，以及盐碱地改良利用与快速培肥、微咸水安全灌溉与雨水高效利用、中低产区粮食增产和棉田增粮等技术，建立企业与新型农业合作组织参与的示范机制，构建适度规模经营的现代农业生产技术体系与示范样板，规模化示范推广粮食增产增效技术，大幅度提升环渤海中低产田粮食增产能力。中国科学院遗传与发育生物学研究所、中国科学院上海生命科学研究院、中国水稻研究所组成的联合团队围绕"水稻理想株型与品质形成的分子机理"这一核心科学问题，鉴定、创制和利用水稻资源，创建了直接利用自然品种材料进行复杂性状遗传解析的新方法；揭示了影响水稻产量的理想株型的分子基础，发现了理想株型形成的关键基因；阐明了稻米食用品质精细调控网络，用于指导优质稻米品种培育；示范了高产优质为基础的设计育种，攻克了水稻高产优质协同改良的科学难题。该项目将极大推动作物传统育种向高效、精准、定向的分子设计育种转变，荣获了2017年度国家自然科学奖一等奖。此外，中国农业大学等关于玉米重要营养品质优良基因发掘与分子育种应用的研究；中国林业科学研究院林产化学工业研究所等关于农林生物质定向转化制备液体燃料多联产的研究；中国农业科学院蔬菜花卉研究所对于早熟优质多抗马铃薯新品种选育与应用的研究；中国科学院动物研究所对于飞蝗两型转变的分子调控机制的研究；清华大学等对于植物油菜素内酯等受体激酶的结构及功能研究；扬州大学等关于促进稻麦同化物向籽粒转运和籽粒灌浆的调控途径与生理机制的研究；华南农业大学等研发的水稻精量穴直播技术与机具；中国农业大学等关于生鲜肉品质无损高通量实时光学检测关键技术及应用的研发；中国农业科学院蜜蜂研究所等关于优质蜂产品安全生产加工及质量控制技术的研发，等等，都为我国农业发展起到了很好的推动作用。

在计算机科学领域，2017年，由中国科学技术大学、中国科学院－阿里巴巴量子计算实验室、浙江大学、中科院物理所等协同完成参与研发的世界第一台超越早期经典计算机的光量子计算机在中国诞生，这标志着我国的量子计算机研究领域已迈入世界一流水平之列。该原型机的取样速度比国际同行类似的实验加快至少24000倍，通过和经典算法比较，也比人类历史上第一台电子管计算机和第一台晶体管计算机运行速度快10～100倍。清华大学计算机系郑纬民团队在国内率先开展存储系统关键技术研究，提出一种与结构无关的快速容灾恢复理论及实现方法，大幅降低容灾成本并实现灾后即时恢复；同时还提出一种基于社区/群组概念的云存储系统及数据共享机制，提高了云存储的易用性。这些存储系统的部署和应用取得了可观的社会经济效益，推进了网络存储技术在国内的普及、发展和应用，也迫使国外存储产品的价格下降。

在电子信息领域，中国科学院沈阳自动化研究所于海斌团队作为我国工业控制网络技术研究的主要开拓者之一，攻克了制约我国工业自动化系统发展的系列关键技术，主持开发的现场总线芯片，改变了我国仪表行业长期无"芯"的局面。团队还参与制定了IEC 61508工业以太网国际标准，为我国在工业自动化技术标准制定方面取得话语权发挥了重要作用。其团队开发的工业通信芯片、智能仪表模块、网络化控制系统，不仅支持了多家仪表企业通过了国际认证，还为我国仪表企业的技术进步、产品出口做出了积极贡献，为企业带来了显著的经济效益。工业和信息化部电信研究院在第四代移动通信系统（TD-LTE）方面攻克了我国通信产业在芯片、仪表等薄弱落后环节，使我国移动通信行业跻身国际先进行列；构建公共试验验证平台，推进产业链整体研发和产业化进程；克服规模组网应用中的挑战，构建了全球领先的TD-LTE精品网络，推动TD-LTE在全球规模应用,首次实现我国主导的移动技术标准走向世界。我国主导的 TD-LTE 在与美国主导的 WiMAX 的全球 4G 产业竞争中胜出,成为全球两大主流 4G 标准之一,使得无线移动通信成为我国少数具有国际竞争力的高科技领域之一。

在材料科学领域,北京大学郭雪峰团队原创性地发展了以石墨烯为电极、通过共价键连接的稳定单分子器件的关键制备方法,解决了单分子器件制备难、稳定性差的难题。通过功能导向的分子工程学成功地克服了二芳烯分子与石墨烯电极间强耦合作用的核心挑战性问题,从而突破性地构建了一类全可逆的光诱导和电场诱导的双模式单分子光电子器件。这项研究工作使得在中国诞生了世界首例真实稳定可控的单分子电子开关器件。在未来高度集成的信息处理器、分子计算机和精准分子诊断技术等方面具有巨大的应用前景。此外,清华大学石墨材料科研团队开发的微膨改性鳞片石墨负极材料、微晶石墨负极材料、低温负压解理石墨烯及石墨烯基导电剂应用技术成果实现了在可快速充电和宽使用温度范围锂离子电池中的应用,推动了我国天然石墨资源深加工技术和锂离子电池材料技术的发展;中国科学院兰州化学物理研究所凹土科研团队开发了对辊处理–制浆提纯–高压均质–乙醇交换一体化工艺,实现了棒晶束的高效解离和纳米棒晶的分散,从应用基础、关键技术到高值产品开发,形成了具有自主知识产权的技术创新链;中国矿业大学(北京)非金属矿物材料团队攻克了中低品位硅藻土物理选矿提纯产业化关键技术以及用硅藻精土为原料的医药化工高浓度含盐废水和油田采油污水处理的污水处理剂以及纳米 TiO_2/硅藻土复合光催化材料产业化制备与应用关键技术,并建成了年产 12000 吨硅藻精土示范生产线、6000 吨/年污水处理剂示范生产线和 1000 吨纳米 TiO_2/硅藻土复合光催化材料示范生产线。香港科技大学唐本忠团队在实验数据和理论模拟的基础上,提出了分子内运动受限(RIM)的聚集诱导发光工作机制和结晶诱导发光的聚集诱导发光(AIE)衍生概念,并发展了结晶诱导的纯有机高效室温磷光体系。在大量 AIE 材料体系制备的基础上,实现了其在高效光电转化器件、高灵敏传感器和智能响应体系以及高分辨特异性成像和精准医疗等领域的高技术应用,并展示了其显著优于传统有机发光材料的工作特性。此项工作在光材料研究领域取得了重大原创突破,并获得了 2017 年度国家自然科学奖一等奖。

在核科学技术领域,平面硅工艺探测器技术的重大突破,派生出一系列先进的硅半导体探测器,这些探测器技术在现代高分辨率成像、防爆安检、空间探测、科学研究、武器装备、国民经济等领域发挥了重要作用。钢铁研究总院等单位围绕先进压水堆核岛关键设备材料技术开展攻关,自主研制实验室研究装备,实现大型先进压水堆核岛关键设备材料的自主化和大批量生产,成功填补了国内外核岛主设备材料技术空白。其创新技术处于国际领先水平,彻底实现了我国百万千瓦压水堆核岛主设备材料技术自主化,显著提升了国家高端装备制造业核心能力,为我国成为世界核电技术和产业中心奠定了坚实基础。

在机械工程学领域,中国科学院长春光学精密机械与物理研究所突破了精密机械加工、精密光学加工、精密检测、高精度微位移控制、高精度环境恒温和高精度防微振等一系列关键技术,研制出了刻划面积(400 毫米 ×500 毫米)属世界之最、技术水平达国际领先的大型高精度衍射光栅刻划机，证明了我国的精密机械加工技术达到了国际前沿水平。盛瑞传动股份有限公司等单位研发了世界首款前置前驱 8 挡自动变速器（8AT），突破和掌握了 AT 自动变速器正向研发和产业化核心技术，打破了国外专利封锁，培育了自主的产业链和高水平的研发团队，新增产值 41 亿元，利润 3 亿元。

在标准化学科领域，浙江大学詹爱岚等借助科学的社会学研究理论 – 行动者网络理论分析我国通信领域如何战略性地推行本土标准，提出标准的技术路线要在性能与兼容性、开放与封闭间找到平衡点；湖南大学侯俊军等发现行业标准化水平等要素与产业国际竞争力存在显著的正向关系。浙江工业大学曾耀艳等认为我国（标准后发国）的标准创新应该在战略上推进标准技术的选择、创新与再创新，在战术上实施自主标准政策，努力打造国际型标准商用联盟。

在油气储运工程领域，我国化学驱油田采出液处理技术取得重大突破，大庆油田建成了世界上规模最大、年处理原油 1300 万吨的聚合物驱采出液处理系统，以及年处理原油 400 万吨的三元复合驱采出液处理系统；创新了低渗透和非常规天然气集输处理工艺，实现了经济、有效开发；形成了高酸性气田地面工程系列技术，以及高压、高产、高酸性"三高"气田地面工程安全设计系列技术，保障了"三高"气田的安全开发；创新了地面工程标准化建设技术体系，形成了地面建设新模式。同时，我国还建成亚洲最大的全尺寸钢管爆破试验场，填补我国在高压、高钢级天然气管道全尺寸断裂行为和管道爆炸环境影响研究领域的空白；复杂地质条件下管道施工技术进一步配套完善，满足了中缅油气管道等各种复杂环境下管道施工的需要；油气管道大型装备国产化取得重大进展，30 兆瓦级燃驱机组、20 兆瓦级高速直连电驱机组、48 英寸 Class900 全焊接球阀、2500 千瓦级 10 兆帕输油泵机组等一批大型装备研制成功并开始应用。

在建筑学领域，中西方学者在城市设计理论和方法研究上齐头并进，中国学者从跟跑到并跑再到部分领跑。西方学者继续在城市形态分析理论、城市设计方法论等相关领域展开探索。中国城市设计出现了一些体系性的新发展。理论探索主要反映在城市设计对可持续发展和低碳社会的关注、数字技术发展对城市设计形体构思和技术方法的推动，以及当代艺术思潮流变对城市设计的影响等方面。与西方相比，中国城市设计实践呈现出后发的活跃性、普遍性和探索性。

在制浆造纸工业领域，纸基功能材料领域的技术创新和产品开发取得突破性进展，一些技术含量高的产品填补了国内空白，如芳纶纸、空气换热器纸、热固性汽车滤纸等。以纳米纤维素为原材料，通过化学、生物、物理、复合等方式可以制备附加值更高、性能更加优异、生物可降解性与生物相容性极佳的纳米纤维素基功能材料，是纳米纤维素未来发展的一大热点。我国目前拥有世界上领先的制浆造纸技术和装备，相关企业开始国际化布局，"走出去"稳步推进。"中国技术""中国制造"越来越受到国外造纸企业的青睐。

在资源科学领域，土地、水、生物、能源矿产、气候等五大传统资源研究不断深化，为解决国家社会经济发展中的资源问题发挥了重要作用。攻克了一批重点行业废水深度处理关键技术，构建了我国水环境治理基础技术体系和监测预警网络；中国地质科学院等单位借鉴国内外矿产预测经验，创新性地提出了矿床模型综合地质信息矿产预测方法体系，建立了全国矿产资源潜力评价预测数据库。海洋资源研究异军突起，海域资源调查和海洋资源勘探技术实现跨越发展，海水淡化工程设计、海洋生物资源应用技术开发取得显著进步。中国科学院地理资源所发展了自然资源资产负债表编制的理论与方法，从分类到综合、从实物到价值、从存量到流量，突破了自然资源资产负债表原型设计与表式结构、负债核算与价值化技术，并率先编制完成湖州市/安吉县和承德市自然资源资产负债表。

在纺织学科领域，我国常规纤维的差别化、多功能化水平显著提升，生物基纤维原料的产业化、绿色化生产技术取得突破，碳纤维、间位芳纶、超高分子量聚乙烯纤维、聚苯硫醚纤维等高性能纤维稳步发展；国产纺织装备高速化、连续化和自动化水平进一步提升，纺织产品开发力度加大，产品品种增加、品质提升；产业用纺织品制备技术取得显著进步，产业化步伐加快，应用范围扩大，增长较为迅速；印染加工领域积极开展绿色生产技术创新，少水及无水印染加工、短流程工艺、生态化学品应用等清洁生产技术得到推广应用；依托工业物联网技术，纺织智能制造正在给行业带来深刻的变化，自主研发的互联网针织 MES 系统于 2017 年 4 月通过中国纺织工业联合会组织的科技成果鉴定。

在煤矿区土地复垦与生态修复领域，我国学者结合露天采矿工艺和我国的实际情况，研究了露天煤矿剥离－采矿－回填－复垦的一体化工艺过程，提出了横跨采场倒堆内排工艺的露天煤矿采复一体化的基本原理及定量化表达。针对井工煤矿创新性地提出了采复一体化的"边开采边复垦"的概念和技术原理，实现采矿与复垦同步进行的一种复垦思想，是对以往沉陷稳定后再复垦技术的重要革新。

第三节 学科发展问题与挑战

2016 年，国务院印发《"十三五"国家科技创新规划》[3]，这个首次以"科技创新"命名的规划，确立了迈进创新型国家行列、为建成世界科技强国奠定坚实基础的总目标。我国科研投入逐年增加，基本保持与 GDP 增长一致，全社会研发投入已经超过 1.5 万亿元，科技人力资源和研发人员总量居世界第一，科技成果产出总量稳居世界前列，综合性国家科学中心、双一流学科建设对标世界一流水平，科技整体水平呈现"三跑"并存的格局。然而令人欢欣鼓舞的进步背后，依然存在基础学科投入严重不足、学科研究环境尚待优化、学科产出质量不佳、服务国民经济建设能力不够、高端人才十分紧缺、关键核心技术受制于人、深层次体制机制改革亟待推进等问题。

一、学科研究环境问题与挑战

随着科研投入的不断加大，我国对于基础研究的战略需求也越来越明显。科技部部长万钢认为，基础研究领域的差距直接影响到未来产业技术水平的提升，如果一些中间领域的基础问题不能得到有效解决，一些共性核心关键技术就不能突破，我国的产业将面临长期锁定中低端的风险。当然，我国已经在量子通信、生命科学、物理、超级计算机等领域处于并跑甚至领跑的地位。与此同时，我国相关学科领域的国际影响力大幅提升，在国际高影响力期刊上的发文也呈现出"井喷"态势。然而，相比我国整体科技投入体量、其他创新型国家基础学科投入情况、学科整体产出质量等情况，发现我国在基础研究投入与产出方面尚有较大差距。此外，科研投入渠道比较单一、学科发展不平衡、学科国际合作不均衡、科研管理体制机制不灵活等也是学科研究发展中日益突出的问题。

研发投入强度较低，基础研究经费严重不足。虽然科研投入多年来都快速增长，但是研发经费投入强度只达到2.11%，要完成2020年达到2.5%的目标还存在着较大的差距。与某些国家比较，差距则更为显著，比如以色列（4.25%）、韩国（4.23%）、日本（3.49%）等。因此，总体来看，我国科研经费的整体投入还显不足。从经费投入结构看，我国基础研究经费虽然实现两年连增，但仍处于较低水平，与发达国家15%～25%的占比水平相比有很大差距，如图17所示。虽然基础经费研究不足是一个"老问题"，但是其对于科技创新及学科发展意义深远，因此还需给予足够的重视。

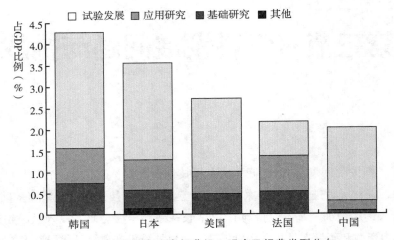

图17 主要国家研发经费投入强度及经费类型分布

数据来源：*China by the numbers*：*Nature News & Comment*，2016[47]

地方、企业投入占比低，基础研究投入渠道单一。我国基础科学投入主要来自于中国科学院、国家自然科学基金、科技部等机构的财政拨款，而地方、企业投入比例低。长期以来国家层面的基金稳定性有余，灵活性不足。如新兴的人工智能领域，国外企业在2015年便开始部署并争夺控制权，而我国在2018年初才将人工智能领域新增到基金委资助范围内。2017年最新数据显示，近5年来全世界发表的高影响力论文中中国占18.1%，其中受科学基金资助的论文占62.4%。而在对ESI数据库核心论文的持续分析中发现，中国政府资助学科领域的核心论文产出率也不均衡，如"973"计划，与国家自然科学基金资助的研究领域有明显的重叠[48]。以中央财政为主的科研投入结构，有时候仅能起到锦上添花的作用，无法更好地支持对基础前沿的探索。

学科建设投入"见物不见人"。学科投入重视实验设备和条件的改善，忽视智力的价值与投入，在人才的引进和培养上缺少制度性的财政保障。经过近十年的高强度投入，我国现在大部分重点学科建设在实验室的装备上早已达到世界一流水平，钱与设备已不是制约重点学科发展的主要原因，而人才特别是学科的领军人物成为学科高水平成果产出的关键因素。然而，在我国现有的学科建设制度中缺少用于人力的经费以及相应的制度安排。当前，重点建设的学科设备经费"花不完"，年底"突击花钱"已成普遍现象。现行制度下，人与资源不相匹配，因而出现了学术能力不足与资源相对过剩的矛盾[49]。

量化考核过紧，违背科学研究发展规律。科学研究，尤其是基础研究往往存在投入大、周期长、风险高的特点，有些研究需要持续多年的积累，比如FAST项目，从开始到建成持续长达20余年；美国LIGO项目投入100亿持续30年，其中20多年都没有任何成果。而在当前的科研体制下，从项目申请到结题，从科学研究到财务支出，往往需要通过层层申请、审批与考核。在考核力度越来越大、细致程度越来越高的科研管理体制下，基础研究缺乏宽松的科研环境、信任的支持体系、自主高效的管理机制，这往往导致科学家为申请项目、应对项目任务、财务制度、绩效职称等多重考核疲于奔命，或者直接导致一些学者不得不放弃基础研究，转向做"短平快"的研究。

二、学科发展布局问题与挑战

无论是教育部学科评估、国家双一流建设，还是国家综合性科学中心建设等都体现了国家对于学科发展与科技创新的重视。我国学科的国际合作与交流越来越频繁，学科交流的层级也越来越高，国际合作论文影响力显著大于我国整体论文影响力。但是纵观目前学科发展布局，还存在学科间发展不平衡、学科国际合作发展不均衡、前瞻性学科发展布局滞后、交叉学科融合还需进一步推进等问题。

学科发展存在区域、机构、学科之间不平衡现象。部分学科布局和研究规模存在失衡现象，重点学科往往集中在少数学科门类中。历次学科评估数据显示重点学科中理工科占比超过50%，双一流学科建设中，同样显示超过六成的学科均为理工科，这种学科高度密集布局虽然有利于部分学科短期内冲击世界一流,但从长远来看,学科整体发展也可能产生马太效应,强势学科越来越强,而弱势学科发展空间逐渐变小。从地域看,北京、江苏、上海的高校在一流学科建设布局高度集中,一流学科超过半数(56.25%),中西部地区布局严重缺乏。而机构之间学科布局依然存在上述问题。另一方面,同一个一级学科或二级学科被同时建在多所学校当中,这往往导致知识创新资源在学科内重复配置,这种分散建设也可能导致学科内资源竞争加剧而协同不足。

国际合作学科分布不均衡,合作广度和深度有待提高。我国国际合作学科分布不均衡现象非常突出,除材料学、工程学、化学、物理学国际科研合作相对活跃度超过或与主要国家水平持平外,其他学科尤其是空间科学的相对活跃度较低。同时,我国许多实质性国际合作,主要研究对象和研究区域大部分都分布在国内,在全球影响力还是有限。国际合作交流的深度同样不足,国际合作基本上是个人行为,缺乏组织导向,难以形成国家层面的特色和优势,也难以补强自身短板。大部分国际交流仍然处于简单的学术交流,而深层次的学术研究合作、学术管理合作、学术项目合作、人才培养合作等开展得相对较少。

学科发展前瞻性布局尚需加强。随着科学研究的细化,颠覆性技术不断涌现,需要进一步加强对学科的前瞻性布局,特别是对于新兴、前沿的学科需要快速反应和布局。"兵马未动,粮草先行",学科布局要走在科学研究之前,实现布局引领科技发展。当前在一些新兴重大基础前沿,如新一代人工智能、量子通信和量子计算机、脑科学与类脑研究等领域上,支持不够及时,布局偏晚。特别是对于颠覆性的科技领域,美国通过50多年时间已经形成颠覆性技术的常态化研究机制、英国通过打好"军产学研环境"牌积极预测支持颠覆性技术、日本则在 2013 年设立"颠覆性技术创新计划",而我国在该方面的布局稍显滞后。

学科交叉研究环境不够成熟,理论基础薄弱。学科之间的交叉融合成为目前学科发展的一个主要趋势,大量原创性科研成果的产生、新的学科增长点的培育以及优势学科方向的凝练,都依赖于多学科之间的相互交叉与融合。最近 25 年,交叉性合作研究获得诺贝尔奖项的比例已接近一半(49.07%)。然而,目前国家在基金资助时,对于交叉学科的支持、对于交叉学科的判断和标准都显不足。不同学科都想从交叉学科上谋求新的增长点,但促进交叉学科发展的环境机制还处于发展阶段,有待健全,比如成立专门的学院、专门实验室、会议交流和讨论等发展还在探索阶段。另一方面,在行政主导的学科专业设置体制机制下,学科与专业成了资源分配的依据,各个学科专业为了自身利益最大化,都在不断扩张,学科与专业划分越来越细致,数量庞大。但各学科又自成一派,彼此之间相对封闭,学科群建设特别是学科优化与交叉融合非常困难。此外,部分交叉学科实践走在了理论前面,如在计算生物学等数据环境下催生的新学科,其研究的基础性和方法学均需要加强[50]。

学科平台建设重复叠加较多,且缺乏有效监管。经过多年的学科建设,我国已经形成了国家重点实验室、国家重点工程中心等一系列的学科建设平台。随着新时期的发展，我国又提出了建设上海、合肥、北京三个综合性国家科学中心的战略规划，并对国家实验室做出了调整，进一步强化国家学科平台的建设，优化资源配置。但是，我国学科平台建设仍然存在着一些问题。一方面，之前我国建设的国家重点实验室等学科平台之间的重复性比较多，同时各个地区、省部级单位也建设有各自的学科平台，这些学科平台之间容易形成非良性竞争，同时也造成了资源的分散，无法形成高影响力的科研成果。而且这些学科平台主要集中于经济富裕的省份和城市，依托于科研实力本身就很强的研究所和高校，这会在一定程度上产生"木桶效益"，导致难以在短期内提高我国科学研究的整体水平。另一方面，我国学科平台缺乏开放性。目前虽然有一些仪器共享的政策，但在实际操作方面还存在着一定困难，难以从仪器开放测试层面的共享上升到围绕某一学科问题形成解决方案的共享与协同。

三、学科支撑能力问题与挑战

作为《"十三五"国家科技创新规划》（本段简称《规划》）[3]全程参与编制者，科技部创新发展司司长许倞强指出，《规划》最显著的特征，关注点不仅仅是科学技术研究本身的改革和发展，更加关乎国民经济主战场，关乎面向科技前沿，关乎面向重大需求。虽然在基础研究重大前沿上我国已取得一些突破，面向国家重大需求在深海、深地、深空、深蓝领域取得了傲人成绩，在生物、医学、农业、计算机、电子信息、材料等诸多学科领域取得的成果成为推动经济社会发展的现实动力。但是相比我国战略与经济发展需求，学科研究成果支撑国家重大战略需求以及经济建设的力度还远远不够，论文导向的科研评价体系、盲目追求国际影响力、缺乏核心关键技术等问题，仍然对我国整体科技竞争力的进一步提高形成严重掣肘。

论文导向的科研评价体系需要反思。虽然在第四轮学科评估、"双一流"学科建设遴选上，已经考虑了除论文、引用数以外的很多其他指标来进行综合评价。但从当前整体的科研评价环境来看，"数文章、看点数"仍然是各大高校以及科研院所的主流方式。"原有的评价体系使高校、科研院所对科技成果转化既无主动投入的义务，又无自觉投入的动力，科研机构的研究成果主要用于论文发表，在完成职称评定的'使命'后，或束之高阁，或埋入抽屉。"全国政协经济委员会副主任、北京市政协副主席王永庆指出，"论文是发表了，但只是对自己有用。"在北京大学纳米科学与技术研究中心主任、中科院院士刘忠范看来，发表论文是奔着毕业、拿奖、头衔而来，而不是凭着自己的兴趣爱好去做研究、出成果，这样的论文出再多也没有用，这也是当下科研学术界心浮气躁的原因之一。北京大学工学院院长张东晓指出，近年来国内以发表 CNS（Cell、Nature、Science 杂志的简称）论文为目的的研究风气"有些过头了"，在高水平的学术期刊发表研究成果本是一件好事，但以此为目的去做科研，甚至去迎合期刊的"口味"，这与研究的本质有所背离。同时，这种导向导致的"创新焦虑"难以形成自由、开拓、敢于探索、静心钻研、宽容失败的创新研究氛围，不利于科研创新，尤其不利于具有引领性超越性的研究。

重西方论文，轻本土论文，缺乏科学文化自信。近年来，中国自然科学各个领域论文发表数量、总被引数量和高被引论文数量屡创新高。但是，大部分高被引论文都是在国外科技期刊上发表，鲜有在国内科技期刊上刊载。科研工作者不仅需要在基础科研领域"显山露水"，也需要在经济和民生需求中体现价值。清华大学副校长施一公指出，当前大部分优秀成果选择发表在西方杂志上，而这些杂志订阅费用昂贵，获取成本高，加之大部分工程师很难看懂英文文章，导致了我们的工程师无法学习国内最新成果。我们的大学和研究所的科研工作实际是在为西方免费劳动，而且有时还付费在西方发表文章，等于倒贴为西方服务。中国科学院院士、《科学通报》主编高福日前也撰写发表了一篇题为《写在青蒿素论文发表40周年》的文章以纪念屠呦呦所在的中国中医研究院等几家单位以"青蒿素结构研究协作组"名义在《科学通报》上首次发表有关青蒿素化学结构及相对构型的论文《一种新型的倍半萜内酯——青蒿素》。文章指出，当前已经到了用我们自己的科技期刊引导我国研究方向的时候，我们期待更多中国科学家的原创性科研成果，能够发表在中国人自己的学术期刊上。中科院院士方精云明确指出，"大家不分领域地只想着发表SCI论文，被国外的评价指标牵着走，看不到国家的发展需求并研究解决问题，这是非常可怕的，会把中国的科技发展引入歧途。"导致这种现象的主要原因有两点：一是与国外知名科技期刊相比，国内大部分科技期刊无论从质量方面还是影响力方面确实还存在很大差距；二是由于目前国内科研评价体系导向存在一些偏颇，使得大家对SCI收录期刊过度膜拜。但无论哪种原因，背后隐藏的一个重要病灶还是我们的科研人员缺乏科学文化自信。

专利申请大国不等于专利强国，专利数量激增应引起警惕。中国发明专利申请受理量自2011年起连续6年位居世界第一，2016年中国的受理量更是超过了美国、日本、韩国和欧洲专利局四方受理量的总和。但反观我国专利在海外的布局，却远远低于美国、德国、日本、韩国等国家。我国虽然已成为专利申请大国，但仍未进入专利强国阵营。我国专利质量有待进一步提升，专利布局需要进一步优化，专利运用水平还需提高，专利对创新驱动发展的支撑保障和引导作用尚未充分发挥。同济大学朱雪忠教授[51]指出，目前的专利数量已经脱离了我国创新能力的真实情况，必须高度警惕。专利本是市场竞争的利器，因此申请专利的目的主要是服务于市场需要。然而，片面追求专利数量，盲目鼓励申请"荣誉"性专利，异化了取得专利的动机，将专利沦为完成学业、晋升职称、争取资助、获取奖励、享受优惠政策等的工具，使专利与市场的关系渐行渐远。这种低价值专利严重浪费社会资源，且容易让决策层、科学界、大众过高估计我国创新能力，形成错误判断。由于"非市场"动机的影响，当前很多专利本来就不是为了转化实施而申请，因此，"专利转化率低"是必然的结果。在没有深入探讨影响专利转化的供给侧因素情况下,"专利转化率低"这个伪问题误导了政府部门,盲目出台各种政策,并提供大量财政资金设立专利转化运营基金等。在专利转化率低真正原因不明的情况下,这些举措不仅难以取得预期成果,甚至还容易形成恶性循环,浪费大量的政策资源和财政资金。

政策不够细化,缺乏核心技术,成果转移转化艰难。近年来我国高度重视成果转移转化,希望各学科领域的研究能够发挥好服务国民经济与国家重大战略需求的作用。国务院、科技部、财政部、教育部、中科院等在2016年密集出台了很多相关政策,如《实施〈中华人民共和国促进科技成果转化法〉若干规定》[52]《促进科技成果转移转化行动方案》[53]《促进高等学校科技成果转移转化行动计划》[54]《中国科学院关于新时期加快促进科技成果转移转化指导意见》[55]等。但是仍然存在成果转移转化困难问题。原因之一是上文提到的科研评价体系导向,使得科研院所缺乏成果转移转化的动力。同时,成果转移转化相关政策缺少实施细则,不具有可操作性,落地困难。上海市教科院副院长胡卫指出,"由于职务发明权属关系等原因,体制内科研成果在技术转移的获利分配方面极易'擦枪走火',甚至'犯错',如可能涉及贪污或侵占。因此除论文发表外,'多做多错,不做不错',就成了体制内科研人员对于成果转移心照不宣的普遍态度,这在事实上造成了我国成果转移的巨大障碍"。此外,成果转化难究竟是政策支持不足还是成果本身就达不到转化条件,也是需要深入思考的问题。清华大学副校长施一公就直接指出,成果转移转化难是因为基础研究能力不够,导致没有太多可以转化的成果。只有大力加强学科基础研究,加强原始创新能力,才能产生出能够转化成生产力的成果。

四、学科队伍建设问题与挑战

人才计划名目繁多,阶梯效应明显。从院士、"千人计划""万人计划""杰出青年""长江学者",再到省市的各类计划,如黄河学者、泰山学者、孔雀计划等,全国各级各类的创新人才计划多达百余项。这容易导致青年科研学者将其注意力从学术问题转移到人才头衔上。在现行的人才与科研评价体系下,各种人才头衔、奖项一定程度上反映了科研的"原始资本",在科研人才发展过程中,"阶梯"效应已经显现,既如果在科研生涯起始阶段,没有获得人才或相关奖项,将不利于后期重量级人才计划的申请甚至学术生涯的发展。这使得许多青年学者无法踏实地在其领域深耕细作,而逐步向"资源型学者"发展。由于一部分人才计划的作用是锦上添花,因此"资源型学者"就更加容易获取更多、更高的人才头衔、奖项等学术资源,集多个人才头衔于一身的学者并不少见,而那些"输在起跑线"上的青年学者,往往由于一开始错过某个"阶梯"导致无法继续攀登到更高的"阶梯"。

人才评价导向单一,人才评价体系尚未健全。第一评价内容片面,定性因素少于定量因素,论文导向、专利导向、课题级别导向的评价滋生了学术界的功利主义和浮躁之风,形成了很多学术泡沫。尤其是对于应用性强的学科，这种评价导向对于真正核心人才的发展具有很大的负面作用。第二，在评价方法上还有待完善，虽然使用了同行评议、科学计量、经济分析、综合评价等多重技术和评价手段，可是实际效果并不理想。第三，对科研人才的评价很少有外部科学共同体参与，缺乏反映外部意见的条件和取得社会认可的途径。第四，评价组织化程度不高，机制不健全。评审小组临时设置，没有形成对评价本身的问责，评价的公正性受到怀疑，影响到社会对科研人员的认可。

学科发展加剧人才无序流动，"携帽子流动"现象突出。在"双一流"建设的背景下，人才较之前更是受到了空前的重视，各高校、地方政府把吸引人才作为首要任务。客观上讲，人才争夺战有利于促进建立一个人才自由流动的环境，而这对于人才发展不无益处。但是当优秀的国内学者日趋集中于少数"双一流大学"的时候，意味着非"双一流大学"的人才流失。一旦人才流失，非"双一流大学"的发展要么缓慢，要么停滞不前，而"双一流大学"由于更多优秀人才的加盟，将会获得更加快速的发展，同时由于地理原因也势必会拉大东西部之间的差距，形成马太效应。另外"帽子人才"是政府行政管理的产物。国家推出各类"人才计划"项目，其初衷是政府管理部门以人才计划项目形式对学者加以鼓励和扶持，以促进其更好地成长和服务于社会经济的发展，其客观上也发挥了很大的作用。但"帽子"的固化和滥用，使原本针对人才管理和培养的行政权力泛化，产生或多或少的不良连锁效应，争抢"帽子人才"背后的实质是学术发展受制于行政活动。

缺乏对海外高层次人才分布情况的准确把握，海外高层次人才引进同质化严重。在海外人才引进时，大部分机构在人才引进上缺乏前期规划和系统调研，从而不了解目标学科海外高层次人才整体分布的情况。不少机构通常采取网上发布信息、媒体发布广告以及通过海外校友会、机构成员出国交流等渠道来获取人才信息，这些引才方式较为被动且信息来源不稳定，还会出现引进人才与学科发展目标契合度不足等问题。另一方面，我国海外高层次人才引进同质化明显，以我国"长江学者""百人计划""千人计划（青年千人）"为例，其引进最多的学科较为接近，前三甲均为生物学、化学和物理学，尤其是生物学领域的引进人才数量远远高于其他学科领域的人才数量，而工程技术领域、基础医学、药学等领域人才引进偏少。在电信行业，高级人才占全行业专业技术人员比例仅有0.14%；在海洋领域，我国在世界海洋专家数据库中登记的专家不足百人，不到全球总量的1%，仅有美国的1/20；在电子信息产业中，技师、高级技师占技术工人比例为3.2%，而发达国家一般在20%～40%之间，这些领域大多数都属于工程技术领域[42]。目前，我国已经开始注意到该领域人才短缺的问题，并加大了人才引进力度，如2017年青年千人工程领域人才引进有大幅度提升，位列第一位。工程科学领域、信息科学领域引进人数已经超过传统的生命科学、化学与物理。

第四节 学科发展启示与建议

一、学科研究环境发展建议

高度重视基础科学研究，大力加强基础研究投入。我国当前科技快速发展，特别是在大科学装置、量子通信等方面取得了重大突破，但是在基础理论研究方面的突破还比较少，在出思想方面还比较困难，需要大力加强对基础研究的支持。国家应高度重视战略前瞻，加强顶层设计，关注颠覆性技术，在发挥导向作用的基础上发挥科学家的自主创新能力。

调动地方、企业积极性，形成多元投入新格局。通过与地方、企业协同，引入社会资金，在稳步增加国家投入的同时，使得基础研究投入总量得到快速增加。在这一点上，浙江省、浙江大学、阿里巴巴三方先行大胆尝试，组建开放协同的混合所有制新型研发机构"之江实验室"，旨在聚焦网络信息技术前沿，开展重大前沿基础研究和关键技术攻关。随后，阿里巴巴建立探索人类科技未来的实验室"达摩院"，并斥资1000亿人民币，用于涵盖基础科学和颠覆式技术创新的研究[56]。

进一步加强我国大科学装置的建设。作为重大科技基础设施，大科学装置在提高我国自主创新能力方面占据重要地位，是产出大量前沿科技成果、集聚海内外人才、加强学科交叉融合、参与国际合作与竞争的必要保障。在加强大科学装置建设时应从国民经济发展、国家安全战略需求和世界科技前沿需求出发，考虑我国有优势符合国情的项目进行建设。充分考虑学科、地域等因素进行合理布局，形成一定的聚集效应。对于评审通过并实施建设的多期工程，应开通"绿色通道"，简化申请、审批程序。完善大科学装置建设过程中与建设完成后人员投入的经费保障机制。定期对大科学工程的科学目标和应用成果的实践进行评估。

保持科技投入要素稳定性、均衡性，加强科研经费使用的灵活性。科技投入要素需要相对稳定，在科技设备、科技人才、科技政策等方面需要一定时期的稳定。比如科技政策需要保持顶层与操作性政策的一致性，保持人才投入的稳定性，防止人才投入的大幅度波动。对基础研究的考核要摒弃"短平快"的做法，不能急功近利。建立有利于基础研究积累的长效的科学评价机制，对学者的量化考核可适度松绑。评价机制应在督促科研成果产出与保障学者研究自主权之间寻求平衡。如在美国，学者在成长过程中也面临着严格的量化考核，但评上终身教授之后，考核压力就会大大降低，可以花5年甚至10年时间做一项研究。鼓励科学家自主选题，自由探索，宽容失败，科学家无论从事的是"热门"还是"冷门"的研究，都可以使其无忧无虑地潜心研究。在科技硬件设施方面也需要结合实际，减少设施建设的盲目性。在科研经费的管理方面，避免出现"今天拨款，明天交差"现象。同时，需要进一步解放思想，利用当前信息技术与大数据管理等方式，解放科研人员日常管理事务。经费管理模式需要进一步优化，增加总额控制，增强经费使用灵活性，允许科研人员在合理范围进行经费的调整。

二、学科发展布局发展建议

加强学科发展顶层设计与前瞻性布局。在集中力量建设一流优势学科的情况下，应考虑通过制定其他相关政策、投入充足的学科资源要素、加强产业、学科协同合作等来促进学科在区域、机构及学科之间的平衡发展。在重点学科遴选上，应从单一的"竞争选优"到与"择需布局"相结合转变。虽然"竞争选优"是政府投入支持的主要依据，也是体现公开公平的必要程序，在早期的学科建设中发挥了重要的作用。然而在我国学科经过近20年的投入和建设后，学科建设的终极目标应更多考虑为国家重大战略需求和社会经济发展提供有效的创新创业的人才和智力支持，因此需要与"择需布局"结合起来。面对科学技术的日新月异，经济产业发展的转型升级，世界局势的复杂多变，世界一流学科建设更需要用全球视野来谋划，因此需要加强顶层设计和布局。面向新兴前沿领域以及颠覆性技术，应该充分发挥科技智库的作用，通过对各个学科最新研究发现、技术、市场等多种信息的及时跟踪，建立长期稳定的学科前沿信息监测体系，及时洞察全球科学技术研究前沿与发展动态，提前为学科发展布局、基金项目设置、学科资源要素分配等学科发展顶层设计提供信息支持。

加强国际化、综合性学科研究平台建设，聚合发展要素。建设高水平的科研平台将会大力推动学科交叉融合，促进新兴学科发展，进一步提高科学研究水平，为学术队伍建设、高水平人才培养以及国际合作创造有利条件。未来在学科发展布局方面需要进一步聚集政策、管理、人才、资金等各方面的要素，打造国际化、综合性学科研究平台，支撑具有全球影响力的重大产出。近两年，北京怀柔、上海张江、安徽合肥三个综合性国家科学中心已经陆续建成，并已经开始展现其平台的优势。在学科研究平台建设上，未来还需进一步梳理国家层面、地方层面、企业等各个方面的学科平台建设情况，整合国家与地方两个层面的学科规划，有针对性地引导各类学科平台有序发展，节约资源。另外，应以平台为牵引展开合作，加强协同与共享，从仪器测试层面的共享上升到围绕重大学科问题的方法、技术共享，从而进行更加高效的协同创新。

三、学科支撑能力发展建议

建立面向不同研究类型的多元科研评价体系。对于基础研究，短期内很难有用于转化并产生经济价值的成果，同时短期内也很难分清研究成果与贡献的高低。对于基础科研评价制度应更加宽容，容许失败，容许短期内没有成果。同时要提高基础研究者的待遇，降低为了生存和金钱利益去做科研的功利性。加长考核周期，考核周期内不考核，到期进行科研成果与财务的严格审计。对此，北京大学等已经率先开始了预聘制（Tenure-Track）的尝试。应用研究的目的在服务国民经济，应降低其论文考核的要求，增强其对技术创新与保护的要求。避免因缺少论文而无法获取经费从而终止研究的现象。基础研究、应用研究、交叉研究以及不同学科之间都存在自身的研究特点与成果产出特点，需要根据不同的研究类型及学科特点有针对性地进行设置。除论文、专利、获奖等定量指标外，纳入同行评议、科研在该领域的推动引领作用、服务业绩、社会贡献等定性指标，形成科学的、多元的、学科特色的科研评价体系。当科研人员为科学而努力，而不是为文章点数而努力的时候，中国科学发展才会持续取得胜利。

树立自信，加强本土学科研究传播利用，填补研究与产业的鸿沟。习总书记指出："广大科技工作者要把论文写在祖国的大地上，把科技成果应用在实现现代化的伟大事业中。"我们在为SCI论文量达到全球第二而激动的同时，也应意识到我国在农业、能源、环境、健康、高端制造等事关国计民生的重大学科领域上还存在很多现实问题。李亚栋院士曾指出我国虽然科技论文已居世界第二位，但科技成果在促进企业新技术发展方面收效甚微，这与我国绝大多数科研创新成果以英文发表在国外期刊上不无关系。在第十三期中国科协科技期刊主编（社长）沙龙上，欧阳钟灿院士呼吁："改变科研评估规则，真的是时候了！"；高福院士则强调只有管理部门对高水平中文刊物认可了，才能促进优秀原创性成果向国内中文刊物投稿。要推动我国本土学科研究成果的传播和利用，填补我国研究与产业的鸿沟，首先是需要广大研究者认识到其科研之初心，意识到当前研究与产业之间的巨大鸿沟，转变思想，树立科研自信。其次，需要决策、管理部门认识到本土期刊发展传播的重要性，加大投入，培育本土期刊发展的土壤，制定相应的政策，促进优秀原创性成果在国内中文刊物投稿。最后，加强对出版社和期刊的监管，大力提升本土期刊的质量与影响力，促进本土期刊良性发展。

让专利回归市场竞争属性，合理定位专利在科技创新评价中的作用。虽然国家知识产权局已经意识到专利质量问题并在2017年年底提出"实施专利质量提升工程"，从严格提升专利申请质量、专利代理质量、专利审查质量以及坚持市场导向、价值驱动方面来全面提升专利质量，但是如果追求数量的指导思想、评价体系、激励机制等没有根本改变，在专利数量持续暴增的情况下，仍无法保证投入足够资源要素来有效提升专利质量。因此，应该让专利回归市场竞争属性，压缩或取消专利申请费用的财政资助，淡化对专利数量指标的考核，更要淡化专利数量作为衡量我国科技创新水平决策依据的作用。另外，需要深入思考专利转化率低的真正原因，避免盲目出台政策和投入资金，建议加强对相关政策实施效果的评估，根据评估结果及时调整优化政策内容。

细化成果转移转化政策，加强技术创新原动力，形成科技生产新动能。建议加强技术创新支撑，对技术创新进行资金的支持，加强社会资金的进入，通过社会资金的进入促进技术创新的发展，并更好地与市场进行对接，形成良性循环。国家与地方在政策制定上应该适当鼓励一些高风险、高收益的产学研项目，允许失败探索，重视长期产出和评估。切实改变片面将论文、项目及经费、专利数量等与科研人员评价和晋升直接挂钩的做法。从服务国家发展战略出发，引导科研院所和高校的应用科研瞄准国内外市场需要，主动走向市场，科研项目从选题立项到成果转化必须紧密围绕产业发展，尤其是与新兴产业的发展紧密结合。除少数科研院所、研究型高等学校从事重大基础研究外，大多数高等学校、科研院所应把科技成果转移作为硬性指标，并将科技成果市场贴近度、转化率、产学研合作的项目数量、项目带动产业发展的产值等纳入科研评价体系。最后，不同学科、不同机构根据自身特点，在宏观政策的指导下，进一步细化落实成果转移转化政策，有针对性地制定具有可操作性的政策与实施细则，进一步加强对成果转化的管理和操作层面的指导，真正实现对科技成果转化的支撑。

四、学科队伍建设发展建议

加强科技管理顶层设计，科学统筹各类人才计划，按需设立人才计划，加强人才后期考核。一是加强科研院所、高校原有科研岗位制度本身对于人才引进和管理的功能，加大其人才引进的力度和精准性，降低科研院所在人才引进工作中对于各类人才计划的依赖性，减少人才计划种类及名目，让更多的学者能够潜心深耕科研，而非疲于积累学术资本。二是淡化人才阶梯效应与马太效应，让人才项目更多地发挥雪中送炭的功能。如一些面上项目应该适当向没有任何头衔、任何光环的青年学者、普通教授、副教授、讲师等倾斜。人才遴选也应根据其科研的创新水平与成果价值来进行衡量，而非已有的头衔与成就。三是真正按需引进人才。在一些重点领域，产业领军人才、高层次技术专家和高技能人才匮乏，已经严重制约了我国相关领域的进一步发展，建议开展全国高层次学科领域人才的需求调查，特别关注一些与经济发展密切相关的学科领域如电子信息产业、装备制造业等的高层次人才需求状况，为国家下一步有针对性的人才引进提供基础数据支撑。四是实行跟踪评估和动态淘汰制度，对于不达标的人才予以"摘帽"，从"一次性奖励"转向"分期培养"模式。

建立多元价值导向，多层面、多类别的人才评价体系。一是建立更加关注质量、学科影响力、社会贡献度、应用价值与经济价值等的多元价值导向，而非简单地看学历、资历、论文、专利、成果的数量、课题的级别等。二是建立不同层面、学科、行业的人才评价标准和指标体系，实行科学的分类评价。如对于基础研究的科研人员，减少定量考核，增加国际同行匿名评议比重；对于应用研究型人才，应增加社会效益、潜在经济效益、自主知识产权及工作业绩等指标的权重；从事科技成果转化与产业化的人才，则应重点考察其推动科学技术成果转化和产业化的能力，以及取得的经济和社会效益；对于从事条件保障与实验技术的人才，则应重点考察其为研究与发展活动提供服务的能力和水平、工作质量、服务的满意度等。三是加强评价组织化程度，完善评价体制机制建设。针对参与评价的各个主体，进行职能、职责的界定，了解风险、责任，下放评价权力，提升责任意识并且建立法律约束机制。充分发挥专业组织、第三方评价机构的作用，对人才进行独立、专业评价，还应实行"谁评价、谁负责"机制，保障评价的专业性以及公平公正。

完善人才管理与薪酬激励制度，改变以巨额物质奖励为主导的人才争夺局面。管理部门、科研院所、研究者本身应正确对待人才有序流动，人才流动的初衷是各个科研机构优化学科布局，发展建设特色学科，而不是盲目的数字上的人才争夺。管理部门不应再片面依赖高薪酬高待遇争夺人才，简单以"学术头衔""人才头衔"确定薪酬待遇、配置学术资源，而应积极地从人才的可持续发展以及为其提供良好的科研空间的角度出发，完善人才管理与薪酬激励制度，避免人才无序流动，保持学科队伍的稳定性。另外还需要培育一个良性的、公平的竞争环境，一个健全的、合理的资源配置，一个公正的、健康的考核制度。

整合海外高层次人才信息渠道，建立海外学科人才数据库。整合高校、科研院所现有海外高层次人才信息渠道，依托科技智库等第三方机构，通过人才身份、学历、科研水平、社会贡献、获奖荣誉等系统的数据分析与调研，客观全面反映各学科、子学科全球人才的分布情况，并为目标人才建立信息管理档案，实时维护更新，形成海外学科人才信息数据库。机构结合自身学科发展规划，随时调取海外目标学科人才信息作为重要参考依据，从而使引才工作有的放矢，充分发挥机构自主性，将海外人才引进由被动引进转变为主动出击的局面。

参考文献

［1］科技部. 中共中央国务院印发《国家创新驱动发展战略纲要》［EB/OL］. http://www.most.gov.cn/yw/201605/t20160520_125675.htm［2017-12-06］.
［2］科技部. 国家中长期科学和技术发展规划纲要（2006—2020年）［EB/OL］. http://www.most.gov.cn/mostinfo/xinxifenlei/gjkjgh/200811/t20081129_65774.htm［2017-10-15］.
［3］科技部. 国务院关于印发"十三五"国家科技创新规划的通知［EB/OL］. http://www.most.gov.cn/mostinfo/xinxifenlei/gjkjgh/201608/t20160810_127174.htm［2017-12-08］.
［4］OECD.OECD［EB/OL］. Stat http://stats.oecd.org/［2017-09-26］.
［5］国家统计局. 2016年中国统计年鉴［EB/OL］. http://www.stats.gov.cn/tjsj/ndsj/2016/indexch.htm［2017-10-17］.

［6］国家统计局. 2016年全国科技经费投入统计公报［EB/OL］. http://www.stats.gov.cn/tjsj/zxfb/201710/t20171009_1540386.html［2017-10-17］.

［7］国家科技管理信息系统公共服务平台. 计划专项公示［EB/OL］. http://service.most.gov.cn/jhzxgs/［2018-01-25］.

［8］国家自然科学基金委员会. 年度报告［EB/OL］. http://www.nsfc.gov.cn/publish/portal0/tab535/［2017-10-19］.

［9］国家自然科学基金委员会. 国家自然科学基金"十三五"发展规划［EB/OL］. http://www.nsfc.gov.cn/publish/portal0/tab405/info50064.htm［2017-11-02］.

［10］国家自然科学基金委员会. 代码查询 >> 信息科学部［EB/OL］. http://www.nsfc.gov.cn/publish/portal0/tab556/［2018-01-06］.

［11］国务院. 国务院关于印发统筹推进世界一流大学和一流学科建设总体方案的通知［EB/OL］. http://www.gov.cn/zhengce/content/2015-11/05/content_10269.htm［2017-11-03］.

［12］中国科学评价研究中心. 中国大学及学科专业评价报告2017—2018［EB/OL］. http://rccse.whu.edu.cn/［2017-10-28］.

［13］清华大学.《清华大学一流大学建设高校建设方案（精编版）》正式发布［EB/OL］. http://news.tsinghua.edu.cn/publish/thunews/9658/2018/20180108144930424642173/20180108144930424642173_.html［2018-01-12］.

［14］北京大学.《北京大学一流大学建设高校建设方案（精编版）》正式发布［EB/OL］. http://pkunews.pku.edu.cn/xwzh/2017-12/28/content_300847.htm［2018-01-12］.

［15］浙江大学. 浙江大学公布一流大学建设高校建设方案［EB/OL］. http://www.zju.edu.cn/2018/0102/c578a773570/page.htm［2018-01-12］.

［16］中国人民大学. 我校开展一流学科建设大调研持续推进双一流建设［EB/OL］. http://ruc.cuepa.cn/show_more.php？doc_id=2357001［2017-01-12］.

［17］南方周末. 高校评价体系变革在即 争议"双一流"［EB/OL］. http://www.infzm.com/content/118323［2017-01-12］.

［18］腾讯大家."双一流"焦虑下的"集体踩踏"［EB/OL］. http://dajia.qq.com/original/category/cdb170517.html［2017-01-23］.

［19］国家统计局. 2016年中国创新指数为181.2 创新型国家建设持续推进［EB/OL］. http://www.stats.gov.cn/tjsj/zxfb/201712/t20171207_1561123.html［2017-01-15］.

［20］The Global Innovation Index. The Global Innovation Index 2017［EB/OL］. https://www.globalinnovationindex.org/gii-2017-report［2018-01-26］.

［21］科学网. 2017全球最具创新力政府研究机构25强榜单发布，中国科学院再上榜［EB/OL］. http://blog.sciencenet.cn/blog-408109-1037211.html［2017-01-22］.

［22］中国科学技术信息研究所. 2017《中国科技论文统计结果》［EB/OL］. http://www.istic.ac.cn/tabid/640/default.aspx［2018-01-16］.

［23］Nature magazine.China Declared World's Largest Producer of Scientific Articles［EB/OL］. https://www.scientificamerican.com/article/china-declared-world-rsquo-s-largest-producer-of-scientific-articles/［2018-01-26］.

［24］Web of Science 核心合集"Web of Science 类别"字段列表［EB/OL］. http://images.webofknowledge.com/WOKRS5271R1/help/zh_CN/WOS/hp_subject_category_terms_tasca.html.

［25］Nature Index.https://www.natureindex.com.

［26］SCI 数据库. http://apps.webofknowledge.com.

［27］ESI 数据库. http://esi.incites.thomsonreuters.com.

［28］世界知识产权局. World Intellectual Property Indicators 2017. http://www.wipo.int/edocs/pubdocs/en/wipo_pub_941_2017.pdf［2018-01-15］.

［29］国家科技成果网. http://www.tech110.net.

［30］国家发展和改革委员会. 关于印发国家重大科技基础设施建设"十三五"规划的通知［EB/OL］. http://

www.ndrc.gov.cn/zcfb/zcfbghwb/201701/t20170111_834860.html［2017-12-06］.

［31］科技部. 上海张江综合性国家科学中心获批复［EB/OL］. http://www.most.gov.cn/dfkj/sh/tpxw/201603/t20160307_124522.htm［2017-12-08］.

［32］安徽省发展和改革委员会. 省政府召开合肥综合性国家科学中心实施方案（2017-2020年）新闻发布会［EB/OL］. http://www.ahpc.gov.cn/pub/content.jsp？newsId=C7FE3F92-3A16-4FFC-B641-D16CAC549378［2017-12-11］.

［33］北京市怀柔区人民政府. 怀柔综合性国家科学中心获批［EB/OL］. http://www.bjhr.gov.cn/main/_133460/_137507/638741/index.html［2017-12-13］.

［34］国务院. 教育部关于印发《高等学校"十三五"科学和技术发展规划》的通知［EB/OL］. http://www.gov.cn/xinwen/2016-12/20/content_5150605.htm［2017-12-26］.

［35］科技部，财政部. 科技部 财政部 国家发展改革委关于印发《国家科技创新基地优化整合方案》的通知［EB/OL］. http://www.most.gov.cn/tztg/201708/t20170825_134601.htm［2017-12-18］.

［36］教育部科技发展中心. http://www.cutech.edu.cn.

［37］国家新闻出版广电总局. 关于中国科技期刊国际影响力提升计划项目申报的通知［EB/OL］. http://www.gapp.gov.cn/news/1663/156405.shtml［2017-12-11］.

［38］中国科学技术学会.《中国科技期刊发展蓝皮书（2017）》在京发布［EB/OL］. http://www.cast.org.cn/n200680/n202397/c57893998/content.html［2018-02-04］.

［39］JCR数据库. https://jcr.incites.thomsonreuters.com/.

［40］国家统计局. 中国科技统计年鉴2016［EB/OL］. http://www.stats.gov.cn/tjsj/tjcbw/201706/t20170621_1505832.html［2017-12-22］.

［41］中国学位与研究生教育信息网. 全国第四轮学科评估结果公布［EB/OL］. http://www.chinadegrees.cn/xwyyjsjyxx/xkpgjg/index.shtml［2018-01-22］.

［42］牛珩，周建中. 海外引进高层次人才学科领域的定量分析与国际比较——以"长江学者"、"百人计划"和"千人计划"为例［J］. 科技管理研究，2017（06）：243-249.

［43］国家科技评估中心，科睿唯安. 中国国际科研合作现状报告［EB/OL］. http://www.ncste.org/u/cms/www/201712/200927062x7h.pdf［2017-12-28］.

［44］教育部. 介绍十八大以来出国、来华留学工作有关情况［EB/OL］. http://www.moe.gov.cn/jyb_xwfb/xw_fbh/moe_2069/xwfbh_2017n/201703/t20170301_297668.html［2017-11-20］.

［45］新华网. 中共中央办公厅、国务院办公厅印发《关于做好新时期教育对外开放工作的若干意见》［EB/OL］. http://www.xinhuanet.com/politics/2016-05/04/c_128956339.htm［2017-11-20］.

［46］教育部. 教育部关于印发《推进共建"一带一路"教育行动》的通知［EB/OL］. http://www.moe.edu.cn/srcsite/A20/s7068/201608/t20160811_274679.html［2017-12-03］.

［47］Nature News & Comment.China by the numbers［EB/OL］. https://www.nature.com/news/china-by-the-numbers-1.20122?WT.mc_id=TWT_NatureNews［2017-12-15］.

［48］科学网. 中科院发布《科学结构图谱2017》［EB/OL］. http://news.sciencenet.cn/htmlnews/2017/10/390631.shtm［2017-12-20］.

［49］宣勇. 建设世界一流学科要实现"三个转变"［J］. 中国高教研究，2016（05）：1-6+13.

［50］科学报. 范剑青：数据科学的学科建设、发展和展望［EB/OL］. http://news.sciencenet.cn/htmlnews/2015/10/328181.shtm［2017-12-22］.

［51］搜狐网. 朱雪忠：我国专利数量的失控及其危害［EB/OL］. http://www.sohu.com/a/216559912_275304［2018-01-22］.

［52］国务院. 国务院关于印发实施《中华人民共和国促进科技成果转化法》若干规定的通知［EB/OL］. http://www.gov.cn/zhengce/content/2016-03/02/content_5048192.htm［2017-10-26］.

［53］科技部. 国务院办公厅关于印发促进科技成果转移转化行动方案的通知［EB/OL］. http://www.most.gov.cn/

yw/201605/t20160520_125686.htm［2017-11-26］.
［54］教育部. 教育部办公厅关于印发《促进高等学校科技成果转移转化行动计划》的通知［EB/OL］. http://www.moe.edu.cn/srcsite/A16/moe_784/201611/t20161116_288975.html［2017-11-22］.
［55］中国科学院. 中国科学院关于新时期加快促进科技成果转移转化指导意见［EB/OL］. http://www.cas.cn/tz/201608/t20160823_4572052.shtml［2017-11-22］.
［56］凤凰网. 阿里成立研究机构"达摩院"，3年投入要超千亿，还期望诞生下一个颠覆性创新技术［EB/OL］. http://tech.ifeng.com/a/20171011/44711420_0.shtml［2017-12-26］.

第二章
相关学科进展与趋势

第一节 化 学

一、引言

2015—2017年,中国化学学科取得显著进展。在各级政府和全社会的支持下,化学研究队伍越来越壮大,化学学术论文发表的数量和质量持续处在国际领先位置,中国化学正进入一个由"跟跑"向"领跑"转变的关键时期。本次化学学科发展报告涵盖了有机化学、无机化学、物理化学、分析化学、高分子化学、核化学与放射化学等6个主要分支学科,以及计算化学、流变学、环境化学、绿色化学、理论化学、晶体化学、多孔材料、化学生物学、纳米化学等交叉学科或研究领域。

二、本学科近年的最新研究进展

(一)有机化学

我国有机化学在学术论文发表方面成绩突出,在《美国化学会志》《德国应用化学》《化学科学》3个顶级杂志上的论文数量仅次于美国,居第二位。

有机反应与合成方法学方面,实现了铜催化苄基上的碳氢键不对称氰化、铜催化的对映选择性烯烃的分子间三氟甲基氰基化、炔烃的三氟甲基叠氮化等一系列反应;给出了含有螺环或桥环结构单元化合物的合成方法;发展了手性氮氧配体与不同金属络合催化的不对称反应;在有机催化及烯烃三氟甲基化引发的进一步转化方面发展了一系列方法学;解决了基于共轭烯炔类化合物为反应原料的多样性合成和选择性问题;在高活性配体开发及应用研究方面,提供了新思路、新配体、新应用。

有机合成化学方面,完成许多具有挑战的天然产物的全合成;首次全合成生物碱 hosieine A、海洋糖脂主要组分 vesparioside B 及 schilancitrilactones B、C 等;修正了对耐药性革兰阴性菌有抑制活性的天然产物 aspergillomarasmine A 的结构;设计了4种二萜生物碱统一的合成策略;首次实现了具有促进神经干细胞生长的吲哚生物碱 alstoscholarisine A 和灯台类单萜吲哚生物碱 aspidophylline A 的不对称全合成。

有机氟化学方面，自行研制了首个具有全新分子骨架的氟化试剂 CpFluors；发展了二苯基碘鎓离子催化下芳炔的氟代反应，钯催化苯乙烯的分子间高立体选择性氟-砜化反应，手性氮杂环卡宾催化下脂肪醛的氧化对映选择性 α-氟化反应等一系列反应；实现 $TMSCF_3$/$TMSCF_2Br$ 与重氮化合物这两种卡宾前体的交叉偶联，对可烯醇化酮的高效二氟甲基化反应，镍催化下（杂）芳基硼酸的二氟烷基化等。

天然产物化学方面，中国学者报道了 6000 多个新天然产物，在国际学术期刊上发表 1600 多篇论文。在高被引的 *Journal of the Natural Products*、*RSC Advances*、*Organic Letters* 等期刊中，至少 30%、多达 70% 的文章来自中国。

（二）无机化学

基础科学问题方面，首次提出了 [Zn_8^I] 和 [Mn_8^I] 等金属团簇"立方芳香性"的新概念；构筑了迄今最大共轭结构的平面型 Möbius 芳香性体系；首次确认了稀土金属元素在化学条件下的最高氧化态；用实验澄清了长期争论的"盐在离子液体中是否结为离子对"的问题；合成了首例以锑为自旋中心的阳离子自由基，并证实了自由基的自旋中心主要分布在锑原子上。

新能源与电池方面，创造了目前世界上已发表文献中纤维状太阳能电池效率的最高纪录 9.5%；首次提出了制作完全由无机材料构成的钙钛矿太阳能电池；首次合成出了带有敞开的锂离子通道、大量反应活性位点的氧化钴负极材料。

无机催化方面，成功地制备了钯负载量高达 1.5wt% 的单原子分散钯催化剂；提出了制备非晶态镍钴配合物和 1T 相 MoS_2 复合全解水催化剂的新方法；首次实现了均相超分子金属有机框架 SMOF 的构筑；首次合成出六方晶系的铂镍合金枝状纳米晶；首次发展了基于铱卟啉金属配体的金属-有机框架。

新材料方面，研制出新型碳材料所制备的超级电容器；制备出可通过外部电压来控制动作的人工肌肉器件；首次在人工分子机器体系上观察到柔性直链对整体分子运动的抑制影响和运动形式的调节作用；首次揭示了小分子在限域空间中的有序自发聚集行为；首次合成了一维有序的 MoO_3 纳米管阵列。

（三）物理化学

化学动力学方面，大大提高了构建反应体系势能面的效率和精度；首次在理论上对 H+H_2O 的初始基态、第一对称和反对称伸缩振动激发态反应进行了全维态-态量子动力学研究；揭示了地球原始大气中氧气起源的新机制；确认了镧系金属元素最高氧化态可以达到 +V 价。

电化学方面，大幅提升 M-N-C 型催化剂的电池性能至 1W/cm^2 以上；实现了燃料电池 Pt 基纳米催化剂的工程化和产业化；原创性地提出"离子筛分传导"新概念；首次实现 MXene 和单层有序介孔碳的复合材料。

离子液体研究方面，发展和完善了网上"IPE ionic liquids database"；提出了吸收分离指数、混合物的相对体和混合物的相对黏度等新概念；提出了液–液相分离的机理；利用质子型离子液体作为催化剂和反应介质，实现了常压转化 CO_2 合成恶唑烷酮类化合物。

催化方面，首次实现了从烷烃到高附加值直链烷基硅的高效转化；捕捉到了甲醇制烯烃反应中由二甲醚活化形成的类亚甲氧基物种；构建了国际上最早的自然–人工光合成杂化系统，并实现了完全水分解过程；开发了国际领先的高温浆态床费托合成油新技术；开发了具有自主知识产权的全球首套煤基乙醇工业示范装置并投产成功。

光化学方面，创造性地提出了具有更高的环保价值和经济价值放氢交叉偶联反应；制备了发红光的碳量子点，可用于对肿瘤的诊疗一体化；用荧光探针首次检测到癌细胞线粒体的极性与正常细胞的差别；获得了双波长的单模激光，极大地提高了光子学集成器件的集成度和灵活性。

生物物理化学方面，进行了残基特异性力场 RSFF 应用于环状多肽的模拟；系统研究了炎症相关花生四烯代谢网络的调控方式；得到了非平衡热力学新理论；研究了 DNA 聚合酶的工作机理；发展了活细胞超高分辨成像新方法 SIMBA；开发了扫描单分子 FCS 技术。

胶体与界面化学方面，首次阐述了构筑的柱芳烃超分子囊泡的微观结构；在光照下实现了 ATP 的高效合成；首次以一步混合法批量制备了内相液滴由互不相溶的食物油和硅油组成的 Janus 乳液；首次实现由方解石单晶构成的微透镜阵列的人工合成；率先提出了基于植物叶面界面性质和固液界面作用进行农药胶体分散体系精细化研发的理念；发展了实时单分子定位的程序包 SNSMIL。

（四）分析化学

核酸、生物活体、细胞分析及新方法开发方面，新的 DNA 探针成功地检测到了细胞膜脂类区域的事件和相互作用；建立了基于微米管整流的活体分析新原理和新方法；实现了单个碱基差异 DNA 分子的实时分辨；揭示了细胞膜包被病毒的机理；将青蒿素用于鲁米诺化学发光体系并首次应用于血红素和血迹的检测与成像。

色谱学方面，建立了目前为止关于 Hela 细胞最为全面的蛋白质组和膜蛋白质组表达谱数据集；开发了"硫醇–烯基"点击聚合的硅球表面修饰技术并商品化构建了一个在线阵列二维液相色谱分离平台；研制了液相色谱荧光检测器；首次将 SH2 超亲体与固定金属离子亲和色谱相结合，使蛋白质组学的研究达到了一个前所未有的深度和广度；发展了同时对 50 种活性物质的快速色谱质谱定量分析方法。

质谱方面，纳米发电机在大型分析仪器首次应用，实现离子源的精确控制；设计了高功率密度激光电离飞行时间质谱仪；在单细胞层次上成功完成了对神经元功能、代谢物组成及其代谢通路的研究；开发了一种基于光交联剂和质谱联用的蛋白质检测策略；开发了一种基质电喷雾扫描涂层技术；构建起新型全自动的细胞代谢研究平台。

(五)高分子化学

高分子合成方面,以有机化合物为催化剂进行丙交酯开环聚合,产物无毒性残留物;发展了新型温和条件下可见光诱导的活性自由基聚合体系;提出了"烯烃单体上的极性基团可以活化聚合反应"和"慢链行走"的概念;实现了对聚合物立构规整度的控制;发展了一种新的氨-炔"点击聚合"反应;实现了螺旋聚二乙炔组装体的可控合成。

功能高分子方面,以原位聚合方法在干扰素C残基接枝聚合聚丙烯酸(齐聚聚乙二醇)酯,抗癌效果突出;发展了一类在肿瘤细胞内活性氧或酯酶催化下的新型电荷反转高分子;证明涂有纳米粒子的细菌可以有效地传递口服DNA疫苗;构建了一种超灵敏的电荷/尺寸双响应的酸敏感基因输送系统;制备了一种能够诱导肿瘤细胞自噬、肿瘤细胞死亡的纳米组装体;制备了可以高特异性检测酶的柔性晶体管;制备了铁离子与没食子酸配位聚合物的纳米点。

高分子物理方面,发现在拉伸场下,很小的分子链变形即可导致串晶中shish部分的形成;研究了用高速薄壁注射成型方法制备PP/PE的交替多层结构;设计了一种双层结构的聚硅氧烷液晶变形材料;制备了既有很强保水能力,又有离子迁移通道的两性聚电解质凝胶;研究了异戊二烯橡胶在形变过程中出现的自增强现象;获得了一种可制取高弹性纤维的PAA/PEO纺丝液;提出了"大分子冶金学"的新概念。

(六)核化学与放射化学

合成了在高酸下对镧系和锕系元素具有很强萃取能力的萃取剂DEHDGA;制备了含偕胺肟基和羧基的PP(聚丙烯)纤维吸附剂,对海水中铀的吸附容量可达到$0.81 mg \cdot g^{-1}$;用^{131}I标记了卟啉化合物TPPOH和TPPNH2;获得了一种可用于淋巴结显像的PET/MRI双模态影像探针$^{64}Cu(DTCBP)_2$-IO-Dex;研究了pH和氧气浓度对$^{79}SeO_3^{2-}$在北山花岗岩中的迁移扩散行为;初步衍生出基于绿色选择性结晶分离理念的稀土分离及镧锕分离;发现铀金属有机骨架材料可以作为高灵敏度传感器用于准确探测低剂量电离辐射;研究了中、低放固体废物超级压缩饼在2立方米废物包装箱内的固定配方;研究了气载氚标准装置以及气载氚监测仪现场校准装置、现场校准方法、影响校准结果主要因素等,解决了国内气载氚监测仪量值溯源的难题;研发了核素的自动化快速化学分离装置。

(七)交叉学科或研究领域

1. 计算化学

发展了以吸附材料为基质的漫反射近红外光谱分析方法;提出了采用化学计量学辅助HPLC-DAD的策略;首次报道了TET蛋白对三种DNA甲基化衍生物不同催化活性的分子机制;发展了扩展广义自适应偏置力自由能计算方法;开发了能准确计算预测生物大分子-配体结构构象的方法DOX;发展了基于GPU并行与化学反应分析结合的

ReaxFF MD 新方法；研究了 GCGR 不同结构域对其活化作用的调控机制；发现了首个作用于 APC 口袋的抑制剂；首次证实了 ROR γ 作为肿瘤靶标的可能性；扩展和更新蛋白－配体复合物三维结构和亲和性数据库 PDBbind-CN；建立了国际上第一个蛋白质－配体共价结合数据库 cBindingDB；发展了药物分子设计的自动设计方法 AutoT&T；发现了蛋白质结构数据库中被明显低估的卤键作用；对高分子晶体生长微观机理的理解上升到一个新的层次。

2. 流变学

证实了"两相"模型预言的物理图像，并提供了粒子相可逆非线性流变的实验证据；引入受限"本体相"动力学参数，修正"两相"流变模型；发展了纳米复合材料线性黏弹性模型；开发了利用离子间静电能界定离聚物和聚电解质的新方法；设计了新型流体基超材料；发展了高性能石墨烯气凝胶制备技术；研制出一种可穿戴式的电子传感织物；研制了系列高性能的磁流变材料及电磁吸波材料；在表面活性剂、原油开采和原油输运领域，建立相关流变反应动力学方程。

3. 环境化学

发现了北京霾事件中颗粒物浓度的周期性变化规律以及气象条件的关键作用；通过实验室模拟与量子化学计算发现，甲苯氧化以生成甲酚为主，提示应重新评估这一大气氧化过程在臭氧和二次有机气溶胶形成中的作用；提出了碘甲烷对无机汞的光化学甲基化的两步机制，发现了新的汞光化学甲基化途径；首次发现了纳米银在自然转化过程中的稳定同位素分馏现象。

4. 绿色化学

构建了一种实现氢气的低温制备和纯化的催化剂；一步法高选择性得到了 C_{5+} 汽油烃类产物，是二氧化碳转化的一大突破；在温和的反应条件下实现了高选择性合成气直接制备烯烃；设计合成了一系列功能化的低共熔溶剂，并能够再生和回收 SO_2；创新性地实现了原生木质生物质在单一催化剂上的一步直接催化转化为液态烷烃；在烷基羧酸脱羧偶联反应中首次引入导向的概念，实现了铜催化的烷基羧酸脱羧偶联反应；实现了首例共轭双烯与环丙烷的不对称［4+3］反应；开发了浆态床过氧化氢生产技术，生产成本下降 20%，污染物排放下降 70%；发展了高效的金属有机催化方法和技术，聚乙烯废塑料降解研究取得重大进展。

5. 理论化学

建立起连续、完整的"哈密顿天梯"，提供了解决化学与物理中的相对论问题的整套框架；实现了晶格能、结构优化和振动光谱的高精度计算；提出了包括非占据轨道信息的 XYG3 型系列新一代高精度普适泛函；首次在全维水平计算了四原子反应的态-态微分截面并与实验取得了高度吻合；首次实现了三原子分子在刚性金属表面解离吸附的全维量子动力学研究；率先构建了费米子库的级联运动方程以及统一描述了声子库、电子库、激子库的耗散子运动方程；实现对化学反应自由能面的快速扫描和不需要反应坐标和预知反应机理的情况下反应动力学信息的快速获得；首次严格区分了化学激发态、生物发光态和生物荧光态及其辐射的不同；首次阐明了黄素单核苷酸在荧光素酶中荧光猝灭的机理；发现在 Ir 表面石墨烯生长中的非线性动力学行为是由点阵失配导致的生长前沿非均一性决定；发展了分子激发态衰变的速率理论。

6. 晶体化学

制备出一例兼具主体框架刚性和热刺激响应的动态 MOFs 材料；得到一种仅呈现 X 射线光致变色性质的杯状金属有机配位物的分子晶体，为 X- 射线敏感材料的研究提供了新思路；利用"一锅煮"室温超声方法合成了一种高对称性锰氧簇，是高自旋分子呈现磁有序的罕见例子；首次报道了基于 MOF 材料催化二氧化碳转化为恶唑烷酮类化合物；首次利用 MOFs 微单晶获得了高偏振度的多光子泵浦激光构筑了迄今结构最为复杂的双孔二维 COFs 并确定了其结构；设计合成了一例罕见的、多孔的、具有金刚石拓扑的五重穿插的氢键有机框架材料。

7. 多孔材料

首次发现羟基自由基（·OH）存在于沸石分子筛的水热合成体系之中，通过紫外照射或 Fenton 反应向沸石分子筛水热合成体系额外引入羟基自由基，能够显著加快沸石分子筛的成核；开发了一种全新的用以合成具有高度有序二维六方介观结构的介孔碳材料的胶束融合 – 聚集组装方法；提出了离子杂化多孔材料分离乙炔和乙烯的新方法；构建了具有高效选择性催化的反应容器，对催化 α，β- 不饱和醛的加氢反应具有高活性和高选择性；实现了 CO_2 直接加氢制取高辛烷值汽油，在接近工业生产的条件下，该催化剂实现了 CH_4 和 CO 的低选择性，烃类产物中汽油馏分烃类的选择性达到 78%。

8. 化学生物学

设计发现双功能化学探针，实现对 FTO 蛋白的"可视化"跟踪；通过多功能分子促进细胞内 Tau 蛋白的多聚泛素化修饰，实现 Tau 蛋白的降解，为阿尔兹海默症的治疗提供潜在新策略；揭示了一种全新的胆固醇转运途径；发现了过氧化物酶体细胞器的新功能；首次化学设计合成了五号真核生物染色体，开发了定制酵母环形染色体的方法，创建了一种高效定位生长缺陷靶点的方法，解决了合成型基因组导致细胞失活的难题；对铜离子伴侣蛋白的可靶性的化学生物学研究，为肿瘤的靶向治疗提供了优秀的先导化合物，同时为抗肿瘤策略的研究开辟了全新的领域。

9. 纳米催化

研制出将二氧化碳高效清洁转化为液体燃料的新型钴基电催化剂；提出用 MOFs 作为选择性加氢反应的调控器，设计和构筑了三明治结构催化剂，并证实了这种催化剂设计理念具有一定的普适性；成功实现了煤基合成气一步法高效生产烯烃，被产业界同行誉为"煤转化领域里程碑式的重大突破"；发现棱柱形状的碳化钴的纳米粒子能够实现合成气到低碳烯烃的高效直接转化，取得了超过 60% 的低碳烯烃选择性；制备了稳定的单原子分散钯 – 二氧化钛催化剂，在碳碳双键和碳氧双键加氢反应中都表现出了出色的稳定性和极高的催化活性，帮助化学家在分子层面理解非均相催化，在均相和非均相催化之间搭起桥梁。

三、本学科发展趋势和展望

近年来，与西方化学强国研究经费缩减、研究人员和学生减少、学术论文发表速度减缓不同，我国化学研究经费持续增长、研究队伍不断扩大、学术论文发表数量和质量均位居国际前列。但是，中国化学从学术论文发表大国成为学术研究强国的路途还充满挑战，如何从学术论文发表为目标的研究理念和模式转向直面解决纯化学和应用化学中问题的研究理念和模式，是实现由"跟跑"向"领跑"转变的关键。期盼中国化学早日实现学术研究强国梦，在纯化学学术研究上多出原创性成果，在应用化学研究领域为我国国计民生做出更大的贡献。让化学使中国的国防更强大，让化学使中国老百姓的生活更幸福。

第二节　昆虫学

一、引言

昆虫是地球上物种多样性最高、数量最多、生物量最大的动物群体，也是地球上进化最成功的无脊椎动物，与人类生产及生活关系密切。一方面，昆虫为人类提供食物、药物、功能食品（如蜂胶与蜂王浆等）、工业原料，有传粉、转化有机物参与物质循环等许多功能。另一方面，昆虫中的一些种类是农牧业的害虫，由此每年造成作物损失约2800亿美元，占总产量的14%，且所造成的损失有逐年增加的态势。而且部分昆虫传播疾病，影响人类健康。因此，开展昆虫学研究具有重要的理论意义和实践价值。

根据发展战略报告的要求，本报告主要借助文献计量学的方法，通过对昆虫基因组学、昆虫分子生物学、媒介昆虫学、传粉昆虫学、入侵昆虫学、全球变化昆虫学、害虫综合管理等新兴昆虫分支学科的发展研究报告的调研，旨在于了解我国2012—2016年近5年来昆虫学发展的现状，总结国内外昆虫学研究的新进展、新成果、新方法、新技术，制定我国昆虫学未来发展的规划与战略。

二、本学科近年的最新研究进展

（一）中国昆虫学在国际上的地位

昆虫学是我国重要的学科之一。目前中国昆虫学学会的会员大约 11900 人，从事昆虫学研究的机构大约 112 个。从近年昆虫学文献计量分析来看，我国在国际上发表的昆虫学研究论文排名第二，引用率也排名第二，这说明中国昆虫学研究在国际上有很强的竞争了。进一步借助于 WOS 数据库共检索 2011—2016 年，我国昆虫学研究者的 SCI 文献 6589 篇。而且各年 SCI 文献量显示，整体态势呈现显著增长趋势。在 6589 篇 SCI 文献中，被引频次总计 40649 次，篇均引用次数 6.17，H-index=57。从 2011 年的引用频次 303 次迅速增加到 2016 年被引用 14751 次，5 年的引用频次增长 47.68 倍。

（二）中国昆虫学的主要研究重点及标志性成果

1. 中国昆虫学的主要研究重点

中国昆虫学研究者通过 SCI 在国际上展示的昆虫学研究重点表现在昆虫 RNA 干扰研究、昆虫生物化学与分子生物学研究、杀虫剂及其抗性生理、昆虫系统发育及分类学研究、昆虫的发育及生理学研究、害虫防治研究等几个方面。经 CSCD 在国内展示的昆虫学研究重点，主要表现在昆虫分类学、昆虫生物化学与分子生物学、昆虫生理生态学、种群生物学、害虫生物防治、天敌与资源昆虫、昆虫超微结构、昆虫生物多样性与群落结构等方面。

2. 我国昆虫学基础研究的重大进展与标志性学术论文

近 5 年（2012—2017 年），我国昆虫学基础研究取得了很大进展，大量的研究成果发表在国内外刊物如 *Science*、*Nature*、*Nature Biotechnology*、*PNAS*、*Annual Review of Entomology* 等高影响力杂志。

（1）昆虫迁飞学的生态学效应

基于英国洛桑试验站（Rothamsted Research）位于英国南部的 2 台昆虫垂直雷达长达 10 年的监测数据，计算了所有迁飞昆虫种群的数量和生物量，发现每年约 3.5 万亿迁飞昆虫飞过英国南部，其生物量约 3200 吨。除了小型昆虫（体重 10 毫克以下）外，它们通过主动寻找和利用有利的季节性气流来实现远距离迁飞。该研究不仅揭示了昆虫迁飞的宏伟过程和对生态系统服务功能的影响，同时也展示了昆虫垂直监测雷达的应用前景。

（2）转基因抗虫作物生态风险性分析

发现种植 Bt 棉花后，由于减少喷洒杀虫剂，导致捕食性节肢动物种群数量增多，蚜虫种群下降，且周围的花生和大豆地中害虫天敌也相应增高，显示 Bt 棉花种植其后将会增加棉田及其周围作物的服务功能。而且，我国目前华北多种作物共存的多样化种植模式，可以延缓了棉铃虫对转基因棉花的抗性。

而对我国长江流域 Bt 棉花与红铃虫的互作关系表明，红铃虫对 Bt 棉花的抗性发展得到了有效控制。

（3）蝗虫型变分子与生态学机制

发现了蝗虫中 MicroRNA-276 通过上调 brm 促进卵孵化同步；蝗虫微孢子能够显著抑制飞蝗的群集行为并且可诱导群居行为的飞蝗向散居行为转变；系统阐明了蝗虫嗅觉调控分子特征、神经递质分子多巴胺的代谢通路、肉碱类小分子代谢物、免疫应答分子、small RNA 等重要的功能分子或者基因，揭示了飞蝗型变中行为调控的分子机理，为理解蝗虫型变过程中行为、体色和免疫能力的变化机理以及适应意义提供了重要基础。

（4）外来入侵害虫的入侵机制及综合防控

阐明了红脂大小蠹、本地种黑根小蠹、伴生真菌和寄主的协同入侵机制，提出并验证了入侵种与本地种之间互惠的"化学信息调控"、与伴生菌"共生入侵"两个假说；从而明确了红脂大小蠹这一在原产地是一次期性害虫为何入侵我国后能迅速爆发成灾的原因，提出了"返入侵"假说。基于过去 15 年来自全球 100 多个国家（地区）近百种物种种间竞争取代经典案例，发现大部分外来入侵害虫种间竞争取代现象发生在新入侵物种与其他物种之间，外来生物入侵明显加剧了物种竞争取代及种群地位的演化，加快了物种多样性的丧失。

（5）家蚕基因组序列解析

构建了一个可以在雌性家蚕中特异表达转录激活蛋白（tTAV）的家蚕转基因系统。其中作为效应因子的毒性蛋白 tTAV 的表达受到了四环素开关和性别开关的双重控制；系统介绍了家蚕比较基因组学、甲基化组学、转录组学和蛋白组学方面的研究成果；总结了基于家蚕基因组序列对家蚕丝腺发育与丝蛋白合成、变态发育、免疫与疾病抗性等重要性状分子基础解析研究方面的突破性进展。

（6）昆虫滞育与激素调控

发现昆虫滞育过程中基因表达受到强有力地抑制，导致血液中低水平的三羧酸（TCA）中间产物循环，在滞育解除时脂肪体激活，释放出大量的 TCA 中间产物作用于大脑刺激调节肽合成，促进生成昆虫生长激素蜕皮激素。

（7）昆虫真菌遗传进化及寄主适应机制

发现绿僵菌由专化性菌经过中间型过渡物种向广谱菌方向加速进化，并与昆虫寄主表现出协同进化的特征，伴随着其基因扩张、蛋白家族扩张、基因组结构及生殖类型变化等，尤其是与昆虫寄主识别相关的 GPCR 受体蛋白家族在广谱菌中得到了显著扩张，从而能够识别及适应更多的寄主种类。

（8）白蚁降解木质纤维素机制

通过培菌白蚁巢内和肠道内共生微生物群落结构的研究，揭示了年轻工蚁和年老工蚁肠道细菌群落结构差异，阐明了培菌白蚁不同年龄工蚁与其共生微生物协同降解木质纤维素的过程机制。

3. 在害虫可持续控制方面的重大应用、重大成果

我国昆虫学研究始终面向国家需求，解决我国重大害虫成灾与控制的关键问题。近5年（2012—2017年）来，有关"生物靶标导向的农药高效减量使用关键技术与应用""防治农作物病毒病及媒介昆虫新农药研制与应用""长江中下游稻飞虱暴发机制及可持续防控技术""水稻条纹叶枯病和黑条矮缩病灾变规律与绿色防控技术""青藏高原青稞和牧草害虫绿色防控技术研发及应用"等获得了国家科技进步奖二等奖，为保障国家粮食安全和生态安全发挥了重要作用。

三、重要分支学科发展的主要成就

近5年来，我国昆虫学面向国家需求和科学前沿，以解决我国重大害虫成灾与控制的关键基础科学问题为目标，突出宏观与微观生物学相结合的研究特色，整合分子生物学、基因组学、生理学、行为学和生态学等学科，其中在昆虫基因组学、昆虫分子生物学、媒介昆虫学、传粉昆虫学、入侵昆虫学、全球变化昆虫学、害虫综合管理等新兴昆虫分支学科发展尤为迅速，取得了很大进展。

（1）昆虫基因组学

随着国际昆虫学界发起的"i5k计划"（5000 insect genome project）和"1KITE计划"（1K Insect Transcriptome Evolution）实施，我国在多种重要昆虫种类的de novo基因组测序和分析、功能基因组、进化基因组、生物信息学和基因组编辑等领域得到了长足的发展。

（2）昆虫分子生物学

主要以重要的农业和卫生害虫为对象，在昆虫变态发育、昆虫表皮几丁质代谢、昆虫生物钟调节的分子机制，以及昆虫生殖的分子机制、昆虫免疫调控、昆虫嗅觉与味觉感受分子机理、昆虫多样性与分子识别、昆虫肠道宏DNA条形码等领域取得了很大进展。一方面从生理、生化与分子水平研究昆虫个体及其器官系统和组织的功能机制，揭示昆虫生命活动的本质，另一方面侧重运用分子标记的方法研究昆虫多样性、生态群落内种间关系及多样性。

（3）媒介昆虫学

媒介生物包括蚊虫、蜱、猎蝽、蚤、螨和白蛉等吸血传播病原的节肢动物，以及传播植物病毒的蚜虫、粉虱等，对人类健康、农业、畜牧业以及纤维制造业造成了严重的危害。主要研究进展集中在传播动物疾病媒介昆虫组学、传播动物疾病媒介昆虫生理生化特性、传播动物疾病媒介昆虫毒理学、传播动物疾病媒介昆虫防治与环境的关系以及传播动物疾病媒介昆虫与病原的互作研究，重点分析了稻飞虱–病毒–植物、烟粉虱–病毒–植物、蚜虫–病毒–植物等三者的互作关系，阐明了媒介昆虫传播病毒的作用机制。

（4）传粉昆虫学

传粉昆虫作为开花植物的重要授粉媒介，在传粉服务功能中起到非常重要的作用。目

前的主要进展在：调查和分析传粉昆虫多样性，阐明传粉昆虫与植物的相互适应关系，揭示农田景观、转基因作物、新烟碱类杀虫剂、狄斯瓦螨（Varroa destructor）、蜜蜂病毒和微孢子虫对传粉昆虫的影响，并探讨传粉昆虫的保护与利用的研究，通过蜜蜂授粉与绿色防控技术集成，增产传粉昆虫的生态服务功能。

（5）入侵昆虫学

主要以西花蓟马、烟粉虱、苹果棉蚜、扶桑绵粉蚧、稻水象甲、马铃薯甲虫、红脂大小蠹、椰心叶甲、柑橘大实蝇、苹果蠹蛾、红火蚁等入侵昆虫为对象，明确这些入侵昆虫生物学和生态学特性、扩张与扩散机制、寄主植物—入侵昆虫—伴生微生物互作以及入侵昆虫天敌和寄生蜂的生物学特性，研发了生物防治、物理防治、化学防治、RNAi技术和不育技术在捕获和防治入侵昆虫方面的效果，为入侵害虫的防控提供了技术支撑。

（6）全球变化昆虫学

作为全球变化生物学研究的重要分支，昆虫对全球变化响应研究主要集中于研究气候变化与人类活动下昆虫响应的特征、机制与规律。目前有关昆虫对全球变暖、极端温度、温室气体（大气 CO_2 浓度、O_3 浓度）升高、干旱降雨、全球氮沉降响应以及对放牧活动、种植模式和农田景观响应等方面研究进展很大，提出了全球变化下天敌昆虫保护和害虫防治的预警。

（7）害虫综合防治

作为昆虫学的一门应用科学，农业害虫综合防治主要发展了害虫防治新理念，提出了基于服务功能的昆虫生态调控理论，研发了诱集植物、驱避植物、植物源引诱剂、驱避剂、性信息素迷向防治技术、推－拉技术等害虫防治新技术，发展了抗药性治理技术、杀虫剂高效使用技术等化学防治技术，并且在棉铃虫抗性治理、水稻飞虱治理、小菜蛾治理、西藏农牧害虫治理中取得了重要成绩。

四、本学科国内外研究进展比较

从整体上来看，近年我国昆虫学已有很大发展，在国际上发表的昆虫学研究论文排名第二，引用率也排名第二，有的研究、技术或分支学科已达到国际先进水平和国际领先水平。但与美国等国际先进水平比较仍有较大的差距。主要表现在如下。

1）原创性较弱。很多研究都是跟踪模仿国外的研究，大都是验证国外提出的一些假说，很少有昆虫学理论来自于我国。

2）产学研结合不紧密。由于我国评价的体系，一些大学和科研院所主要以发表SCI论文为主，导致一些科研部门只注重发表SCI论文，不关心其潜在的应用价值。

3）缺乏研究的系统性。由于我国科学研究体制的独特性和经费资助的非连续性和相对投入不足，围绕同一昆虫学的科学主题长期深入研究的、连续发表高水平论文的实验室不多。

4）缺乏从科研到应用的桥梁。缺乏研究集成单项技术成果和大规模应用的配套技术，缺乏上规模的害虫防治、资源昆虫利用的企业。

5）交叉融合不够。如有关媒介昆虫和传粉昆虫的研究，多聚集于分子生物学技术的应用，而缺乏基础生理、生态理论研究，制约着我国媒介昆虫、传粉昆虫学的长期发展。

五、本学科发展趋势和展望

昆虫学是一门基础性的学科，同时又是一门实践性很强的学科，是指导害虫管理和有益昆虫保护的基础。未来昆虫学的发展将以现代组学和信息科学为指导，围绕害虫管理和有益昆虫保护为主线，以代表性模式昆虫、重大农林卫生害虫和主要资源昆虫为研究对象，高度重视多学科交叉与渗透，宏观与微观相结合，及时运用现代生命科学和生物技术领域原理、技术和方法武装本专业学科领域，促进昆虫学科的又好又快发展。主要的发展方向如下：

1）昆虫组学与功能解析。选择我国重大害虫和资源昆虫为对象，开展其基因组、代谢组、蛋白组学研究与功能解析，为深入研究害虫猖獗为害机制、资源昆虫利用的创新奠定基础。

2）系统调查我国重要地区的昆虫类群，特别是传粉昆虫、天敌昆虫、分解昆虫资源，提出这些昆虫的保护利用对策。

3）加强传粉昆虫研究，开展传粉昆虫多样性及其提供传粉服务关系的分析，提高传粉昆虫的生态服务功能。

4）加强宿主 – 媒介昆虫 – 病毒三者之间关系研究，切断媒介昆虫传播病毒的食物链，减少病毒的传播。

5）深入研究植物 – 害虫 – 天敌之间的三级营养互作关系的生理与分子机制和化学信息联系，揭示植物 – 害虫 – 天敌之间的协同进化，为害虫可持续控制提供新原理与方法。

6）开展天敌的保护利用、天敌昆虫人工繁育的营养学与生理学基础、天敌引进后的适应性及其与本地种的竞争，发展害虫的生物防治方法。

7）研究昆虫对全球变化（包括全球气候变化、景观格局变化、土地利用变化）响应特征和机制，提出全球变化背景下农林重要害虫预警。

8）利用遗传工程等现代生物技术手段，寻求害虫可持续控制的新途径。

第三节 心理学

一、引言

加强社会心理服务体系建设，培育自尊自信、理性平和、积极向上的社会心态，是党和国家在新时代中国特色社会主义建设目标中对心理学提出的更高要求，也是中国心理学界必须要肩负的时代使命。在这样的时代背景之下，中国心理学研究者已经开始大量关注和探究新时期中国社会所面临的心理学问题，特别是与社会治理紧密相关的心理和行为问题。"社会治理"是多元主体共同参与的过程，强调通过基层民主的方式，让不同利益群体之间开展良性互动沟通，以此来预防和化解矛盾。因此，社会治理无法脱离具体的人，更不能忽视特定的个体、群体和社会阶层的心态、诉求、社会行为及其互动过程，而心理学恰恰致力于探索这些问题及其内在机制，所以心理学在社会治理中扮演着重要的角色。

二、本学科近年的最新研究进展

1. 社会治理中的公众心态研究

近年来国内的心理学研究重视探究公众心态的现实状况，对全社会的一般心理特性做出了勾画与描述，涉及公众基本心理过程，具体关注的问题包括：公平感知、社会情绪以及社会信任。公平感知方面的研究揭示出，低阶层者不仅比高阶层者更加感到社会不公，而且即使是遭遇了同样的不公，他们也要更为敏感。从归因上来解释，高阶层者多认为低阶层者之所以处于社会底层是因为他们自身能力本来就差；而低阶层者却认为高阶层者之所以处于社会上层并非源于他们的能力与努力，而是基于某种外部因素，如关系、门路、体制漏洞。高阶层更倾向于认为社会本就应该是不平等的，有些群体就应该一直处于上层而有些群体一直处于下层也是理所当然；低阶层者则更倾向于反对这种看法，他们更强烈地呼吁社会平等，认为人与人不应该有等级与地位之别。根据社会阶层心理学的理论观点，高阶层者因为掌控的资源更为丰富，所以他们在追求自身目标的过程中，心理上相对来说并不大依赖环境的支持，因而对于客观情境要素（如是否公平）也相对不很敏感；但是低阶层自身资源有限，在目标行动中就特别依赖于环境的助力，长此以往使得他们对环境要素是否符合自己的预期变得非常敏感。比如低阶层更容易将模棱两可的事件解释为不公平事件，真正遭遇不公平时也比高阶层表现出更低的容忍程度，甚至在生理指标上（脉搏、呼吸、皮电、肌电等）表现得也更为强烈。

因此，要在一个分层的社会中更好地实现公平，就必须增强社会阶层之间的流动性，使下层民众有更多机会能够向上流动，打破阶层固化的陷阱。具体的策略有，环境干预的策略，如减少相关的社会限制，增强弱势群体的自我掌控感，限制感的减少能够明显改变低阶层者对社会分层和社会公平的态度，进而提升低阶层者的目标承诺和目标达成水平，促进其向上流动。自我干预的策略，如运用自我肯定的技术，让个体强调、确认和肯定自己最看重的价值观，能够在一定程度上克服来自于社会的歧视与威胁，重构整合的自我意识，使自我系统恢复平衡。行为干预的策略，如将决策时间调整为秋季或年终，有助于改善决策的质量。

社会情绪方面的研究主要探讨了群体情绪凝聚现象的内在机制及其影响因素。具有相似情绪体验的个体更容易聚集在一起而形成群体。群体身份认同是激活群体情绪的重要条件，当一个人把自身归属于某个群体时，他就会站在所在群体的角度考虑问题，从而体验到与其作为一个独立个体时不同的情绪。群体具有一定的强制力，它能塑造和规范其成员的情绪体验和情绪表达。群体中存在着一套共享的文化、知识体系，这一体系中包含了对事件的一致的解释策略及情绪表达规范，使得群体成员体验到一致的情绪。情绪比较可以使个体依据群体中其他成员的情绪信息调整自己的情绪，尽量与他人保持一致，最终形成情绪凝聚。群体情绪凝聚有助于促进团队效能、提升群体成员的合作意识和合作行为；有助于成员间的相互谅解，以及提升群体凝聚力和吸引力。群体情绪对群体行为有显著的预测作用。虽然目前我国民众的群体情绪总体上是积极健康的，但在诸多问题上仍存在消极情绪高于积极情绪的现象：社会治安、国际舆情和灾害事故类事件都存在较高的社会焦虑，社会治安和灾害事故类事件引发了较高的愤怒和悲伤情绪。群体的消极情绪（如愤怒）可以有效地预测群体成员对外群体的态度，甚至引发集体行动。为此，关注社会群体的消极情绪及其对后续群体行为的影响，适时地采取切实有效的干预措施对社会安定团结具有重要意义。

社会信任方面的研究主要关注影响信任的因素以及信任对社会、经济的影响。信任是个体基于认知或情感，在对自身脆弱性和风险承受力判断的前提下，对他人未来行为结果抱有的积极期待心理状态。信任通常并不是基于直接证据推理判断的结果，而是基于普遍上的一致或者赞同的情感契约。一个人的认知、家庭、人际互动情境、人格特质、情绪等因素都会对其信任产生影响，而信任对社会和谐及经济发展也有影响。社会信任影响公众的身心健康，其中，关系信任、家属信任和制度信任，对城市居民的自评身体健康和自评精神健康都有显著影响，而一般信任则只对城市居民的自评精神健康有显著影响。提高社会信任，则能够减缓阶层之间的对立，能够提高农民工与城市居民的交友意愿，从而降低农民工的社会距离。社会信任也能促进经济发展，影响家庭衣着消费、居住类消费及家庭设备和文教娱乐类消费。社会信任还能够促进创新，理论解释是社会信任降低了社会交易的成本，极大地促进信息的传播与沟通，特别是在复杂程度越来越高的创新活动中，知识交流与传播的高效性与及时性将极大地促进创新活动的开展。这方面的研究还发现，我国社会信任经历着内群信任向外群信任的转变、人际信任向制度信任的转变、低社会信任向高社会信任的转变。

2. 社会规范与价值观层面的心理研究

研究者既对于社会所倡导的道德心理展开研究，也对违背社会规范的心理和行为（如犯罪）及其矫治做出探索。道德与社会治理的关系表现为道德与社会冲突、阶层伦理与道德困境、道德滑坡与爬坡等。在西方，随着贫富分化的加剧，底层社会（特别是白人劳工阶级）的集体道德逐渐被恐惧、焦虑和愤怒所代替，进而偏好更加严苛、紧张的社会规范，并以破坏性的、民粹主义的形式爆发。我国的研究表明，低阶层在控制感较低时对贫富分化更少做内部归因，进而表现出更低的体制合理化倾向，即对社会现实不满，但又不愿意通过个人努力去实现向上流动的梦想。低层人群的失望感和个人责任推脱，反映了当下中国社会的消极工作伦理，或许意味着阶层固化的危险。

道德心理的研究主要关注道德判断的机制、道德的起源和影响等问题。近年来，神经科学家也探讨了道德行为的脑机制和基因-文化协同进化机制，暗示着道德认知、情感在一定程度上也是生理、生化反应。而道德所引发的社会冲突、由社会分层所引发的道德差异以及道德的滑坡与爬坡等问题，是与社会治理关系非常密切的道德心理学问题。中国犯罪心理学的研究既着重于犯罪行为发生的心理背景或基本问题研究，也着重于广泛的应用研究。例如学者在犯罪心理画像、心理测试、刑侦微表情、犯罪心理评估技术与循证矫正等方面，做出了非常突出的研究。

3. 社会治理中的组织与管理问题

研究主题主要涉及绩效、变革、情绪、领导力等方面。通过对这些主题的分析可以看出，组织行为学领域的研究关注组织中个体和群体的心理和行为规律，涉及个体、人际、群际等多个层次，综合特质、情境等多方面的因素，力图科学地解释、预测乃至改变、控制组织中各主体的感受和行为。组织行为的研究关注两方面的核心结果，一是"经济人"层面的绩效表现，二是"社会人"层面的满意度和幸福感，而这些，都需要组织中的"管理人"来达成。

4. 危机管理中的心理问题研究

探讨了民众风险认知、安全感与风险决策以及群体性事件这些危机心理过程，分别重点探讨了民众风险认知的社会放大框架、民众心理安全感的三元模型以及管理失误引发群体性事件的心理机制。通过这些分析，能够为良好、有序、高效地应对突发危机事件，特别是处理好突发事件中涉事群众的心理和行为反应，提供一定的启示与参考。

三、本学科国内外研究进展比较

本报告对国内外该领域的研究进行了比较,具体从概念与理论、文化差异、研究方法、应用层面4个方面进行举例分析。在概念与理论问题方面,本报告以公众心态的内容为例,发现国内外研究在概念、理论与研究范式等方面均有较高的一致性;在很多结论上,中西方被试也表现出大体相同的效应,可以互为验证。不过在一些具体的研究问题与研究取向上,国内外研究还是存在不少区别,例如,国外研究在社会信任研究领域更加重视心理过程模型的揭示。在文化差异方面,研究发现中国人在内隐和神经层面跟西方人一样表现出自尊和自我增强的动机,但在外显层面却不会像西方人那样表现出明显的自我积极性;另外,中国人在一些经典的道德认知上跟西方人很不相同,这也导致了国内外道德心理学研究中许多问题的研究立场不同。而且在德育方面,国内研究也带有明显的文化特异性。在研究方法方面,国内研究者开始跟进国外,采用多层模型分析、贝叶斯估计等新的统计方法,计算模拟这种新的研究范式,以及实验方法。当然国内研究大部分还是基于较为传统的研究方法,在这方面与国外研究相比,仍有一定的差距。在应用层面上,以危机管理这一内容的中国化应用为例,应对策略主要有两种思路:一是"亡羊补牢",属于事发后的应对;一是"曲突徙薪",属于事发前的应对。前者需快速反应、重在灭火,具体而言可采取三个策略:首先是立即响应,隔离旁观;其次是平等对话,平息情绪;第三是控制沟通,信息透明。这体现了国内学者所做的研究与我国的特殊国情的联系,在参考国外研究方法和成果的基础上有更多独特的现实思考,表现出了更注重应对现实问题的研究取向。

四、本学科发展趋势和展望

综合报告的最后部分,从各个主题对本学科的发展做了展望。在公共心态方面,为了更好地推进心理学视角下的社会分层与社会公平研究,还需要在理论的本土化、方法的多样化、结论的应用化这几个方面实现更多的突破与创新。今后研究的重要内容还包括群体共享情绪放大的发展过程、内在机制以及与群体行为之间的关系,并且有必要对网络的群体情绪的发生发展规律进行深入的研究,并借助网络实时地监控不同群体的情绪变化,迅速有效地开展群体情绪干预。未来的研究也更加需要切合社会发展的需要,着重探究如何提高社会信任,从整体上关注影响信任的各种因素,从而提出更加整合的理论,并更多地将研究成果应用于社会发展与治理中去。在社会规范与价值观这一研究领域,未来研究应该加强探讨偏差、偏见及其修复的问题。另外,应该加强更具生态性的研究,同时引入新技术探索新领域。在组织与管理方面,研究者应当更多注重从现实的组织环境、管理情境出发,深入实践,研究真命题;并且要打开视野,在交流中听取其他学科的理论和概念,从而产生更具有创造力和生命力的研究。特别值得强调的是,未来的研究视角可以提倡用助推取代直接的强制干预。所谓助推，即通过对选择环境和决策过程的设计来改变人们的选择，而不使用强制或者奖惩的方法。最后，在危机管理方面，相关研究的跟进既有必要对民众风险认知特点和心理安全感现状进行持续追踪，还需要在理论层面，通过实证研究对风险认知模型、心理安全感模型和群体事件的冰山模型进行验证和修订。

第四节　景观生态学

一、引言

景观生态学从其诞生到现在已经历 70 余年，而中国景观生态学发展相对较晚。相对于国际景观生态学研究，中国景观生态学研究与社会经济发展的结合更为紧密，将景观生态学原理与方法广泛应用于中国宏观政策制定中，如国家生态保护红线框架规划、全国生态功能区划修编、重点生态功能区调整，以及国家与省市层面生态保护规划、城市与区域发展规划、生态保护政策制定等。在这一发展过程中，中国景观生态学在诸多研究领域均取得了重要进展，集中体现在多学科交叉融合、专题领域深化探讨、典型地区特色工作、综合评估与应用实践等方面。

二、本学科近年的最新研究进展

重点探讨了景观生态学原理和方法在生态系统服务评估、全球变化和生物多样保护方面的应用，发展了具有中国特色的学科交叉体系。

1）景观生态学与生态系统服务研究：①针对生态系统服务重复计算、形成机制、系统分析和综合管理以及景观服务等问题开展了创新性研究，提出了系统性的生态系统服务决策分析框架及可操作性解决方案，深化和完善了生态系统服务评估与景观生态综合研究的理论认知；②在面向应用的理论研究方面，揭示了景观生态保护和恢复实践驱动下的生态系统服务相互作用关系及其尺度效应，从时空变异性方面提出了生态系统服务价值化评估的改进方法；③通过引入多种生态系统服务评估的生物物理模型，定量揭示了重点区域不同尺度生态系统服务的时空变异特征;④自主研发了一些单项生态系统服务物理量评价方法以及区域生态系统服务综合评价和模拟系统(SAORES)。

2)景观生态学与全球变化研究:①通过对内蒙古高原等典型地区案例研究,系统剖析了区域景观格局与生态系统过程的演变特征及其驱动力,以灌木入侵草原、湖泊面积变化等作为实例开展了景观变化的驱动机制研究;②系统揭示了林草交错带景观格局对植被动态的影响,复原了景观尺度上土壤水分的分布,提出了土壤水分决定林草植被格局的模式,发现林草交错带森林斑块大小随降水量减少而变小,主要受土壤水分调控,而森林斑块大小影响了其生长、死亡与更新特征;③通过气候变化下的土地利用变化模拟,大幅度提高了全球土地利用模拟数据的空间分辨率,扩展了气候变化情景与区域城市扩展模型的结合方式,深化了土地利用模型和社会经济/生态环境模型的耦合。

3)景观生态学与生物多样性研究:①通过系统研究中国60余座主要山地植物群落和植物多样性的分布及其与环境变量的关系,揭示了生物多样性海拔梯度模式的区域差异,及地形、土地覆被类型等景观要素对山地生物多样性格局的显著影响;②利用山地植物群落的大样地精细定位研究,揭示地形通过筛选作用驱动物种生态位分化、通过隔离效应影响物种遗传结构,成为群落Beta多样性空间格局的控制性因素,和群落构建与多样性维持的重要机制;③对于城市、农田等人为景观,关于乡土植物的受胁和外来物种入侵等方面有较多的案例研究。相关成果广泛应用于自然保护、景观建设等规划实践中,并积极介入"国家公园"规划设计与评价应用。

针对国际景观生态学关注的热点领域,如城市景观生态、乡村景观生态、森林景观生态、湿地景观生态和山地景观生态等方面,结合中国面临的实际问题,开展了系统深入研究,逐渐确立了具有中国特色的研究方向。

1)城市景观生态研究:①围绕城市景观的结构—过程—功能—服务及其动态演变,在城市景观格局与动态、城市景观格局生态环境效应的定量关系、城市景观生态规划与评价等方面,开展了系统深入的研究;②在基础理论研究方面,发展了一系列城市景观格局及动态的精细表征与量化方法,构建了城市三维景观格局指数及模型分析手段,初步明确了城市空间增长的三维模式及驱动机制;③在生态效应研究方面,揭示了城市景观结构、格局与城市热岛效应的响应关系及其尺度效应,从过程视角解析了城市景观格局的水文、水质两方面的水环境效应特征;④在应用研究方面,完善了多尺度城市生态评价技术并积累了丰富的城市景观格局优化研究案例,发展了面向生态安全与生态系统服务体系建构的城市生态用地需求与评价分析方法。

2)乡村景观生态研究:①将景观生态学原理和方法应用于乡村景观分类、评价和动态研究,提升了乡村景观分类与评价的生态功能性,揭示了传统聚落景观空间格局特征和文化内涵;②从不同时空尺度上开展了农业景观格局对生物多样性影响机制研究,提出了农田生物多样性保护的景观途径;探讨了景观异质性等空间格局特征对重要功能群生物,尤其是农业天敌生物和传粉生物多样性及其生态系统服务的影响,提出了从农业生态系统和景观镶嵌体尺度上开展农业景观综合管理，是恢复农业景观生态系统服务的重要途径；③在传统农田作物多样性和异质性对水分、养分高效利用和病虫害控制研究的基础上，开展了传统农业生产模式的景观生态特征研究，阐明了农业景观格局对生产、生态过程的影响机制，推动了传统农业文化景观的保护。

3）森林景观生态研究：①采用"空间代替时间"的方法，估算了兴安落叶松林火后不同演替阶段碳库储量，利用非线性回归方法拟合森林生态系统碳库恢复轨迹，量化碳库积累量大于释放量的转折点，评价生态系统恢复潜力，揭示了森林生态系统火后碳库由碳源转变为碳汇的时间及恢复到成熟林所需的时间；②通过空间直观森林景观模型模拟，阐明了长期灭火对北方林的生态影响以及林火干扰情势变化，提出了继续实施现行高强度灭火政策下，必须制定大范围可燃物管理措施（计划火烧、粗可燃物去除等）的建议；阐明了气候变暖会增加小兴安岭地区阔叶红松林、阔叶落叶松林和山杨白桦林的固碳速率，降低云冷杉林的固碳速率，研究结果为气候变暖条件下我国森林景观的管理提供了科学依据；③利用野外实地调查数据和ALOS/PALSAR雷达影像数据，并结合林型、地形和林分信息，构建了森林生物量的预测模型，为景观尺度模拟森林地上生物量提供了一种重要途径；通过大规模的野外系统调查和采样，结合遥感和地理信息系统技术，首次查清了我国森林植被的固碳现状和固碳速率，为中国应对气候变化的国际谈判提供重要数据。

4）湿地景观生态研究：①通过构建长江口盐沼模型（SMM-YE），模拟研究了滨海湿地结构与功能对入海泥沙减少、海平面上升的响应，揭示了互花米草入侵造成的滨海湿地景观变化及其生态环境效应；结合历史资料补充野外泥碳样品，研究发现与气候相关的净初级生产力比分解速率对长期碳积累更为重要；②综合长时间序列的多源数据，提出了水位—高程转化法，阐释了流域水利工程对河流与湖泊湿地植被变化的影响；通过改变模型条件提高模拟精度，建立了基于MODIS影像和逻辑斯蒂回归的水位—面积模型；发展了基于植被指数和地表水文指数的湿地植被变化监测方法，提高了湿地资源调查效率；③在湿地恢复过程中，提出了基于植被演替理论制定生态恢复策略；通过融合传统文化与现代景观生态学理论，提出了"调整后的池塘-耕地梯田（MPLT）"土地/水利用的水库消落带湿地恢复和综合利用模式；通过开展全国湿地景观保护成效评估、构建湿地公园研究体系，为湿地景观规划和设计的研究与实践提供新思路。

5）山地景观生态研究：①基于山地垂直带谱研究，深入探讨了山体效应对于山地垂直分异的作用机制，发现山内基面高度是评估山体效应的重要指标；②重点关注生物多样性、水循环与生物多样性，明晰了山地景观生态功能的分布格局与驱动因素，尤其是海拔分异的显著影响；③以高山林线为例，阐明了山地景观生态结构与格局对全球变化的响应机制，提出高山树线对气候变暖的响应并不是海拔的抬升，而是对种群更新的促进及树线种群的显著增大；④基于土地利用变化分析，指出滑坡、泥石流、地震等自然灾害对于山地景观生态格局产生显著影响，并加速了山区生态环境退化；水坝、公路建造等大型人类活动也会显著影响山地景观生态格局，产生新的景观类型，加大景观破碎化。

针对我国黄土高原、喀斯特地区和干旱区荒漠绿洲存在的突出生态问题，运用景观生态学原理和方法，系统探讨了这些问题形成的根源和解决方法。

1）黄土高原景观生态研究：①通过集成流域景观数量结构、空间格局和景观类型的属性特征，构建了可用于格局－过程耦合研究的景观空间负荷比指数，发展了能够反映土地利用格局与土壤侵蚀过程的多尺度土壤侵蚀评价指数。在此基础上，通过集成景观功能空间分布对养分流失的影响，水流方向、河网密度、降雨、生态系统功能，距离和水流路径、降雨、河网密度、养分投入的空间异质性，构建了流域源汇景观功能对比指数、养分截留功能指数和流域汇流累积指数，从不同角度探讨了源汇景观理论的应用价值；②总结了景观格局与生态过程的耦合方法与途径。针对野外定点直接观测研究，提出了多尺度格局－过程耦合研究的野外定点监测和遥感监测技术体系；明确了基于景观指数的格局－过程耦合和模型模拟两类景观格局－过程耦合研究方法；③通过构建生态系统服务研究范式与框架，系统开展了黄土高原地区关键生态系统服务的定量评估与权衡分析，分别从自然系统、社会系统角度梳理了社会系统对自然系统的依存关系，提出了生态系统组成－结构－过程－服务的研究范式，从样地、小流域和区域多尺度揭示了水源涵养服务、固碳服务以及土壤保持服务等对生态恢复的响应，提出了基于 GIS、生态系统模型和多目标空间优化算法的区域生态系统服务空间评估与优化工具；通过发展新的方法和模型，揭示了黄河泥沙减少的原因，确定了黄土高原水资源植被承载力阈值。

2）喀斯特地区景观生态研究：①通过喀斯特地物光谱特征分析，首次构建了基于混合地物光谱吸收特征的光谱指数来提取喀斯特裸露基岩信息；改进了喀斯特植被遥感分类方法，提高了基于传统遥感影像提取不同喀斯特植被类型的精度；提出了先进行遥感图像分割处理、再利用线性光谱混合模型来提取喀斯特地物信息，降低了喀斯特高异质性景观对石漠化信息提取的影响；②发现了人为干扰是坡面和小流域尺度喀斯特土壤养分和水分格局存在特殊的"空间倒置"现象的主要原因；揭示了喀斯特坡地存在地表流失－地下漏失特殊的水土二元流失现象，发现喀斯特坡地土壤侵蚀以小于 50 吨／平方千米·年为主，人为干扰会增加地表侵蚀产沙量；人为作用也会促进喀斯特景观结构的改善，随着生态工程的实施，喀斯特石漠化地区植被景观格局恢复特征明显；③系统地开展了生态工程背景下喀斯特地区生态系统服务变化综合评估，发现生态治理措施提高了喀斯特生态系统服务价值，提升了水土保持及固碳功能；明确了喀斯特地区作为西南生态安全屏障区的功能定位，提出了喀斯特世界自然遗产地保护及国家石漠化公园建设等喀斯特景观保护措施。

干旱区荒漠绿洲景观生态研究：①发展了历史时期土地利用与绿洲重建定量分析方法，推动了干旱区景观历史学及其应用研究；②开展了多尺度典型绿洲土地利用、绿洲化时空过程重建与影响因素分析，发展了干旱区土地利用、绿洲化时空过程及其驱动因素定量分析方法；③发展了绿洲城市景观梯度分析方法，揭示了绿洲城市化沿交通干线轴向的"团块－轴向"扩张特征；开展了一系列典型绿洲城市化时空过程与驱动因素辨析和应用研究；④开展了典型荒漠灌丛景观的空间分布、组成与结构特征研究，揭示了荒漠灌丛的

"肥岛效应"和"盐岛效应",以及荒漠景观的土壤碳聚集效应;⑤开展了典型绿洲景观格局及其生态水文效应研究,发展了干旱区生态水文过程研究方法,探讨了土壤盐分时空变化与绿洲景观格局的相互关系。

针对区域景观生态系统服务评估的复杂性,重点探讨景观多功能性及可持续性的定量评估方法,拓展了景观生态学的应用领域:①基于生态系统服务评估模型,丰富了农业景观多功能性评价指标体系,提升了景观多功能性的制图能力,为景观功能分区、多功能区热点识别和多功能性权衡提供数据支撑;②集成生态系统提供的多种供给、调节、支持和文化服务,开展了面向景观多功能性提升的生态旅游区、自然保护区等景观规划管理;③实现了景观可持续性评价模型从社会-经济系统向社会-经济-环境系统的扩展,揭示了关键生态过渡带和城市景观尤其是城市环境的可持续性变化;④将景观可持续性评价纳入关键生态过渡带和城市景观的规划设计,改进了优化算法明晰城市绿地对紧凑城市的可持续规划影响,实现了景观可持续性评价与区域规划设计的整合。

三、本学科国内外研究进展比较

近年来国际上景观生态学研究的特点突出表现在两个方面。

1)重视基础理论研究,不断有新理论提出和新范式出现。景观要素(组分)在城市生态系统中的服务及其与城市发展可持续性的关系是近年来景观生态学家研究的热点问题。当前城市的发展呈现出空间上复杂化和非线性化的特点,学界普遍认识到,实现景观服务的优化,必须充分认识到人和自然在连续的时间和空间尺度上的交互作用,必须将复杂性科学的概念和方法引入到生态系统的管理中才能取得更好的整体效应。

2)重视景观生态学与新兴领域和热点问题的结合,突出研究领域拓展。传统的景观生态学研究重视景观异质性、破碎化/片段化、连通性等对生态过程的影响,这些问题仍是当前国际景观生态学研究关注的基本科学问题。相对于这些传统研究领域,当前景观生态学的新兴领域主要包括:①生态系统服务的评估、调控与提升。2014年 *Landscape Ecology* 杂志出版了生态系统服务专刊,从生态系统碳-水服务的建模与评估,多重生态系统服务的综合分析与评估,生态保护整合分析方法等方面论述了景观要素对生态系统服务的影响;②不同生态系统服务之间的权衡。与传统景观生态学研究相比,这些研究多以与人类生活密切相关的生态系统服务为研究对象,探讨不同情景下的变化。新的理论范式的形成使得景观生态学的理论与方法与实际应用的关系更加紧密。

相对于国际景观生态学研究,国内景观生态学的发展与经济社会发展的结合更为紧密,更加偏重于对土地利用、城市化、生态系统服务等问题的探讨,这和当前中国面临的现实生态环境问题密切相关。主要特点包括两个方面。

1)重视景观生态学原理与现实问题的结合,强调景观生态学的应用价值。国内学者针对我国面临的重大生态问题,如水土流失,草地、湿地退化、保护区效果评估等,开展了一系列大尺度综合研究，这些研究为国家相关宏观政策的制定提供了重要参考依据。其中，"全国生态环境十年变化（2000—2010年）遥感调查评估"项目的实施，揭示了我国生态系统格局与构成、生态系统的质量与服务、国家层面上的主要生态环境问题特征及其变化趋势，明确了我国新时期所面临的生态环境问题，为我国国情调查和宏观生态环境管理提供了科学依据。除了将景观生态与中国现实生态问题相结合，中国学者也在景观生态学基础理论领域进行积极探索。傅伯杰等提出了将生态系统过程和生态系统服务结合起来的理论框架，这为在实践管理中通过对景观要素结构的合理配置，进而实现景观服务的最大化提供了理论基础。陈利顶等提出的"源－汇"景观概念及基于此概念而构建的"源－汇"过程的景观空间负荷比指数也有了进一步发展，该指数的构建为定量评价景观格局对某一特定生态过程的影响提供了衡量方法。

2）城市景观生态学在国内蓬勃发展，目前已经成为国际上重要的研究力量。将景观生态学原理与方法应用于城市景观格局演变与生态系统服务评估中是中国景观生态学研究的一大亮点。国内学者提出了北京市生态用地的分类与规划，分析了北京市生态系统服务重要性及其空间格局，并据此规划了保障北京市生态安全的七类生态用地，该研究为北京市土地利用规划和有效管理提供了依据。同时，结合中国城市发展面临的突出问题，针对城市化进程、城市增长模型、大气颗粒物污染的空间格局、城市热岛效应、城市生态系统服务、城市可持续发展等诸多方面都进行了深入的理论探讨与案例研究。

总体上，中国景观生态学者在把握当前国际景观生态学前沿基础上，充分结合中国自身发展所面临的生态环境问题，在水土保持、城市生态、生物多样性保护、生态系统服务评估等应用价值明显的领域表现出更为活跃的特点，并取得了一系列重要的研究成果。

四、本学科发展趋势和展望

尽管中国景观生态学研究已经取得了突破性的进展，然而制约景观生态学发展的问题依然突出，需要在理论和方法上不断创新。理论突破重点包括以下3点。

1）景观信息图谱理论的建立。景观信息图谱理论的建立将成为景观生态学发展中里程碑事件。景观格局指数的建立与发展成为推动景观格局分析与景观生态学发展的主要驱动力，但如何更好地运用景观格局指数来解读景观格局－过程－功能－服务之间的内在关系应该是景观生态学未来发展的重点，这也将成为解决景观格局分析停滞不前的关键。基因信息图谱目前在解释生物的形成过程中成为非常有用的手段，那么基因信息图谱分析的方法是否可以进而借用到景观格局分析中是景观生态学工作者需要思考的问题。

2）格局－过程－尺度－效应理论的建立。格局－过程－尺度－效应理论与景观信息图谱理论存在相似之处，是否可以借用景观信息图谱理论把格局－过程－尺度－效应的关系刻画出来？未来需要关注以下方面：①自然过程与社会经济、人文过程的综合集成；②多尺度生态过程的综合集成；③景观镶嵌体中生态流及其再分配机制；④种群、群落和生态系统过程的景观镶嵌体理论。因此，在现有格局－过程－尺度理论基础上，需要进一步与异质景观中的物质流、能量流、物种流和信息流相关联，逐步建立"格局－过程－尺度－效应"的理论。

3）景感生态学的发展。景观与感知密切相关，很大程度上感知对景观的价值及利用会产生重要影响。所以，在未来景观生态学学科发展中需要进一步关注各类主体（Agent）对景观感知的影响，从而更好地为景观生态规划和设计服务。具体地，对于面向人类需求的多功能景观，需要深入揭示人的认知能力、价值判断和文化取向对于景观功能和服务需求的意义，从而在多功能景观规划设计中真正搭建起景观格局－功能与服务－人类福祉提升的桥梁。

方法突破重点包括以下 3 点。

1）源－流－汇分析范式的建立。源－流－汇分析是基于景观中的生态过程和物质流、能量流、信息流，对传统的景观类型进行再分类，从而分析源、流、汇景观类型在流域（特定空间）的数量结构与空间配置关系，揭示景观格局对生态过程的影响。然而，由于源、流、汇景观与生态过程的特殊关系，斑块属性随着生态过程变化而变化，且随着过程发生和时间演替，斑块的源、流、汇属性也会发生变化。因此，源－流－汇的属性具有较大的不确定性，如何建立客观、科学的源－流－汇分析范式，对于定量揭示景观格局与生态过程之间的内在关系具有重要意义。

2）格局－过程－设计范式的建立。景观生态学是一门应用性很强的学科，如何将景观生态学原理和方法应用到具体的工程实践和规划设计中一直是景观生态学家追求的目标。当前亟须建立适合景观生态学的"格局－过程－设计"范式，提高景观格局－过程研究成果的应用价值。

3）景观连续体－梯度分析范式的建立。对于景观变化来说，以往研究强调景观的异质性和斑块边界的唯一性。事实上，不同尺度上景观是一个逐渐变化的连续体。空间上景观连续体对维持区域生态的完整性具有重要意义；时间上景观连续体可以体现景观的过程和功能的可持续性。现有的斑块－廊道－基质分析范式将异质景观中的要素视为各自独立的斑块、廊道和基质，忽略了景观过程和功能在时空尺度上的连续性和整体性。建立适合景观过程和功能研究的景观连续体－梯度分析范式，有助于客观识别景观演变和格局变化所带来的过程和功能变化，从而推动景观生态学的深化发展。

第五节 资源科学

一、引言

资源科学是一门横跨多个学科领域的交叉学科,主要研究自然资源的形成、演化、质量特征和时空分布及其与人类社会发展的相互关系。本报告分析了新时期我国社会经济发展面临的资源问题,系统地总结了最近几年资源研究取得的重要进展和资源学科建设成就,比较了中外资源研究的热点和重点,提出了未来学科发展的方向与重点研究领域。

进入21世纪以后,人类社会面临三大全球性资源问题:一是以气候变暖为主要特征的地球系统变化导致生物多样性损失、淡水资源减少和土地退化,农业生产的可持续性和人类食物安全受到威胁;二是人类活动诱发的水体污染、土壤污染和大气污染等环境问题造成可更新资源的可利用性改变,资源数量减少,质量下降;三是全球经济一体化导致争夺能源、矿产等战略性耗竭资源的地区冲突更加频繁,地缘政治和地缘经济风险加大。

我国自然资源的种类和数量丰富,但人均占有量仍然较少,在后工业化、快速城镇化推进和全球变化影响下,当今面临的资源问题更为复杂:①水、土资源利用强度大,有效需求增加和有效供给不足并存,资源优化配置日益复杂;②能源结构不合理,煤炭、油气等传统化石能源比重高,且资源存量、流量、开发强度与经济结构都存在较大的区域差异;资源禀赋与资源消费空间分离,资源流动成本高;③自然资源资产管理与高效利用机制不完善,资源生态补偿机制尚未健全,资源综合利用效率低,浪费严重;④资源型地区(城市)转型缓慢,社会经济发展与生态环境问题交织,成为制约全面建成小康社会的难点地区;⑤原油、铁矿石、铜矿石、粮食等大宗战略性资源的进口量逐年增加,对外依存度趋高,国家资源安全风险调控的难度加大。

二、本学科近年的最新研究进展

"十二五"期间,国家一系列重大研究计划优先安排了大批与资源环境相关的项目,特别是国家基础性工作专项,在2012—2016年间有关资源调查和资源开发利用研究的项目多达 120 多项。这些项目的实施，积累了大量的基础数据和资料，发表论文和研究报告 22 万多篇，出版专著、专题图件数百部，我国资源科学研究和应用技术创新取得长足进展。

（1）自然资源调查向国际化、专业化方向深入

自然资源调查（包括特定科学目标的考察、勘探）始终是资源科学研究的基础。在最近 5 年进行和完成的资源调查上百项，其中 2012 年完成的"中国北方及其毗邻地区综合科学考察"，由中方科学家主导，联合俄罗斯和蒙古国科学家，先后有 170 多位中外专家参加；2013 年完成的"澜沧江中下游与大香格里拉地区综合科学考察"，揭示了澜沧江流域与大香格里拉地区的自然资源、生态环境和社会经济梯度变化规律，评估了气候变化及人类活动对区域水土资源的影响，以及山地灾害的敏感性。这些跨国科学考察项目的执行，标志着我国自然资源综合科学考察已经走向国际化。

2012 年以后，国家设立"基础性工作专项"稳定支持资源调查。专项依托中科院相关研究所和教育部直属高校，调查区域涉及全国，青藏高原、新疆、西南等边远地区是重点。调查对象以生物资源居多，除了传统的动、植物资源调查，也进行特色微生物、中药资源、水产及渔业资源调查；其次是水资源、盐湖资源、沼泽湿地资源、近海海草资源等特色资源考察和非常规能源资源勘探以及全国矿产资源图集编研等。自然资源调查研究正在向着专业化、规范化、精细化方向深入。

（2）五大传统资源研究不断深化

根据自然资源的形成条件及特征，传统研究重点的土地、水、生物、能源与矿产、气候等五大类自然资源仍然是近几年资源研究的核心内容。

土地资源是保障国家和区域粮食安全的基础，在工业化、城镇化过程中，节约集约用地成为土地资源创新研究的热点，土地整理规划、土壤污染治理、生态环境影响评价等工作得到加强，遥感和 GIS 等技术方法广泛应用于土地利用变化动态监测。

在"水专项"等国家科技计划的支持下，水资源研究成果丰硕，我国学者在中外文期刊发表有关论文近 2 万篇，并攻克了一批重点行业废水深度处理关键技术，构建了我国水环境治理基础技术体系和监测预警网络。"黑河流域生态 – 水文过程集成研究"为国家内陆河流域水安全提供了基础理论；"流域水循环演变机理与水资源高效利用"建立了基于水循环的"量 – 质 – 效"全口径多尺度水资源综合评价方法。

生物资源研究范围很广，发文成果多，其中森林资源和中药资源研究突出。基于时态 GIS 技术的森林资源更新管理成为一个重要的研究方向，时空数据库构建形式和森林资源数据更新流程更加实用化，为森林资源实时监控和碳储量估算提供了技术支持。全国第四次中药资源普查，搜集并积累了大量中药材资源的标本及其种子等实体材料，建立起覆盖全国的中药资源动态监测体系。

矿产资源研究方面，中国地质科学院等国土资源部下属研究结构借鉴国内外矿产预测经验，创新性地提出了矿床模型综合地质信息矿产预测方法体系，建立了全国矿产资源潜力评价预测数据库。我国深部、复杂矿体采矿及无废开采技术进展显著，大直径深孔采矿

等共性关键技术显著提高了采矿的强度与安全性。"三深"（深地、深海、深空）地质及矿产资源勘查、绿色矿山建设示范等取得重大进展。

气候资源研究从跟踪全球变化热点，逐步向全球变化的区域响应方向转变，如气候变化对华北平原水资源和农业需水的影响研究揭示，华北平原气候总体趋向于暖干化，潜在蒸散呈下降趋势；未来气候变化情景下，区域水分盈余量下降，华北地区干旱化趋势加重。

（3）海洋资源研究异军突起

21世纪是"海洋世纪"，近几年来我国在可燃冰资源勘探开发、海洋生物资源利用、海洋装备研制、海洋资源与环境评价、海洋资源保护等领域都取得较大发展，基本摸清了我国近海海洋环境资源家底，更新了近海海洋基础数据和图件，构建了中国数字海洋信息基础框架。在海水资源综合利用、海洋可再生能源开发、海水淡化工程设计等方面实现跨越式发展。

（4）为落实领导干部自然资源资产离任审计提供科技支撑

编制"自然资源资产负债表"是党的十八届三中全会做出的重大决定，国务院于2015年9月颁发试点方案。但"自然资源资产负债表"编制涉及财政、统计、环保等诸多部门，研发负债表是一个复杂的过程，目前处于探索试编阶段，亟待建立科学范式与技术体系。中科院地理资源所以国家生态文明先行示范区湖州市和安吉县为例，实践探索了"先实物后价值、先存量后流量、先分类后综合"的编制路径，构建了由"总表 – 主表 – 辅表 – 底表"组成的报表体系，提出了由资源过耗、环境损害、生态破坏构成的表式结构，确立了土地、水、林木等几类主要自然资源的核算指标，形成一套可复制、可拓展、可推广的标准化与自动化系统，完成全国首张"市（县）自然资源资产负债表"（"湖州模式"和"安吉模式"），为后续全国范围"自然资源资产负债表"编制工作的推广，落实领导干部自然资源资产离任审计，建立生态环境损害责任追究制度提供了坚实的科技支撑。

（5）完成学科史梳理，专业性研究机构和专业队伍不断壮大

经过3年多的反复研讨、史料收集与整理，中国自然资源学会编著的《中国资源科学学科史》于2017年12月正式出版。通过学科史梳理，形成一些重要的基本认知：资源科学是人类在认识、开发、利用、配置和管理自然资源的历史过程中形成的知识体系；是中国古代资源利用经验传承和现代资源科学研究积累的结晶；资源科学属于发源于本土、问题导向型交叉学科，很多知识是从传统学科中吸收而来，并且在科研活动组织上也与传统学科有着很深的渊源，但科学目标、思想方法和科学解释是新的，从学理上看并不是某个或某几个传统学科的简单延伸和发展。

近几年来资源科学人才培养体系和学科教材体系建设有序推进，目前全国含有"资源"且与自然资源学相关的本科专业共12个，招生单位470多家；大约有300余家高校和科研院所招收研究生，本硕博三级人才培养体系现雏形，但专业设置依旧存在多源化和分散性，我国资源科学与技术人才培养专业体系的建设仍然任重而道远。

资源科学研究队伍在最近几年迅速扩大，从事资源科学与技术研究的专业机构已有90多家，约140所高等院校设置有从事资源科学研究的二级机构。

三、本学科国内外研究进展比较

根据 Science Citation Index Expanded（SCI-EXPANDED）和 Social Sciences Citation Index（SSCI）两个外文引文数据库，以及中国科学引文数据库（CSCD）进行文献分析，2012—2016 年间资源科学研究各领域高频关键词（key words）的排序是：水文水资源、土地利用与土地变化、气候变化、生物多样性、可更新能源、矿产资源等。

1）水文水资源研究。关注重点是：①全球水资源气-固-液三形态转化引发效应，尤其是蒸散、蒸腾作用对大气系统如季风的影响；②气候变化下用水规模和水短缺评估及洪水预报；③不同地形地貌条件下地下水地表水的质量和通量；④流域水质及水污染治理；⑤人类活动（如灌溉、土地利用变化）对水资源的影响等。

我国最近几年在该领域研究进展尤为突出，论文数量已经跃居世界第一，发文量较多的机构是中科院地理资源所、中国水科院、中科院新疆生地所、中科院寒旱所和北京师范大学，研究集中在水资源需求较大和矛盾较为突出的地区，如黑河流域、黄土高原、钱塘江流域（主要是治污）、青藏高原、黄河流域、西北干旱区等，研究内容包括流域水资源、水资源可持续利用、水问题分析、水政策研究、水污染治理、水足迹研究等。中国多位科学家担任有关国际学术组织重要职位，水文水资源研究在世界上的位置越来越突出。

2）土地利用与土地变化研究。2012 年联合国"未来地球计划"促进了土地资源深入发展，关注重点集中在：①土地生态系统中污染物转化，包括水质及其渗透等；②土地覆被变化；③土地生产力，包括旱地、森林、草地面积变化动态；④土壤吸收或释放气体（主要是碳、氮）的观测与模拟；⑤土地利用过程中的极端事件及影响评价。

中国土地利用和土地变化研究在最近几年发展也非常迅速，论文数量位居世界第三（前两名是英国和德国），发文量较多的机构是中科院地理资源所、西南大学、南京大学、中国农业大学和中科院新疆生态与地理研究所，在土地利用与覆被变化监测、土地生态安全、土地资源承载力等领域具有优势，中国农村空心问题研究是个特色。

3）气候变化与生物多样性研究。全球地表温度变化和从水圈寻找引起气候变化的答案以及气候变化对生态系统的影响，海洋生态系统与陆地生态系统相互作用机理等是热点。

我国气候变化与生物多样性研究正在从过去的以模仿、跟踪为主，逐步转向结合中国国情开展全球变化应对研究，农业气候资源、大气气溶胶的时空变异性和化学特征、干旱半干旱地区水循环、生物多样性保护等是研究热点。相对来说，我国气候变化研究更多地倾向于为国家服务，基础研究与发达国家比要少一些。生物多样性研究以及城市生态问题也引起中国学者的更多关注，研究水平与国际同行相比不相上下。

4）能源与矿产资源研究。生物质废物的生物炭热解、微藻生物燃料的应用、氢气智能能源系统开发，太阳能、风能、页岩气、可燃冰等新能源以及战略性新兴矿产资源研究等是热点。

能源与矿产资源多为国家战略性资源，中国在该领域的研究水平偏上，论文数量居世界第二（第一是美国），关注热点多是开发可替代传统化石能源（煤炭石油）的新能源，如风能、光能、水能、生物能等，研究机构主要有中国电力科学研究院、浙江大学、天津大学、国网能源研究院、华北电力大学等。矿产资源研究倾向于成矿预测和成矿规律，研究机构主要有中国地科院、中国国土资源经济研究院、中国地质大学、中国地质调查局、中科院地理资源所等。

四、本学科发展趋势和展望

党的十九大明确提出，要加强"国有自然资源资产管理和自然生态监管"，中科院《创新2050：科技革命与中国的现代化》战略研究报告前瞻领域一半与资源相关，可以预见：未来10年是中国资源科学发展的重要发展期，将从传统的多学科综合走向交叉、融合与集成，并随着科技创新和大数据应用技术的发展，在深度、广度和精度上达到一个新的高度，独立的理论体系、方法论和学科建制将逐步确立和完善。

水土资源耦合研究、海洋资源的综合利用、能源与矿产资源的高效开发、自然资源资产核算的理论基础研究、国家资源安全研究和资源技术方法创新将成为未来6个重点研究领域。

第六节 感光影像学

一、引言

感光影像学是感光和影像的一个交叉领域，既古老又新兴。说其古老是因为它的起源要追溯到1727年德国解剖学家舒尔茨（J. H. Schulze）发现硝酸银具有感光性，此后经过近两个世纪的不断探求、传承和发展，以卤化银为基础的银盐照相术（Silver Halide Photography）在20世纪初期成长为一个全球性产业，在20世纪中后期发展到其鼎盛时期，成为传统感光影像学的主干和内核。但是，从20世纪初开始受到来自以数字相机为代表的数字影像的冲击，逐渐退出市场。说其新兴是因为感光影像从数字技术、网络技术和新媒体的发展和进步中不断得到新的动力和发展空间，成为正在悄然兴起的数字影像学的主要组成部分，数字影像时代正在向我们走来。

二、本学科近年的最新研究进展

（一）数字影像的内核与外延

用 0 和 1 的排列组合描述的数字影像资源，广义称为数字内容（digital contents）或数字资产（digital assets）是数字影像系统的内核。当然，数字内容是更宽泛的一个概念，已经不再限于影像的范畴，包括文字、图形、符号、音频、视频等，实际上只要可以数字化的东西最终都可以成为数字内容。

图 1 概念性地描绘了数字影像系统的基本架构，所有与影像关联的应用或系统都以典型设备的方式与处于核心的数字内容关联。这个"地球"的北半球基本上都是数字影像捕获和输入设备，从大家熟悉的数字照相机开始顺时针旋转，有智能手机和平板电脑等为代表的数字移动终端（基本上都内置了照相/摄像功能）、以 X 光机为代表医用影像、以遥感卫星为代表的遥感影像、以平台和滚筒扫描仪为代表的数字化扫描设备等；南半球主要是影像的处理和输出设备，从大家熟悉的计算机开始，顺时针旋转可以看到以印前处理系统为代表的图像设计和处理、以喷墨打样机为代表的数字彩色硬拷贝打样系统、以发光显示为基础的平板显示器（包括液晶、OLED、等离子等不同的显示器，也包括基于专业显示屏的彩色软打样系统等等）、以反射阅读为基础的电子纸显示系统、印刷直接制版、直接成像印刷机、数字印刷机、数字冲印设备等。实际上还有很多影像输出难以用图标的方式表示出来，例如光刻（特别是无掩模激光束、电子束和离子束直描蚀刻）、辐射固化、激光束 3D 打印等。这些系统基本上都属于以光化学反应为基础的能量载体影像转换为物性载体影像的感光影像系统。3D 打印相对特殊，既有基于光化学反应的系统，但更多是基于光物理变化（光之热熔、烧结等）。因此，图 1 既不是数字影像系统家族组成的完整版本，也不是最终版本。

与传统影像系统形成鲜明对比，数字影像系统的内核完全是一个开放平台和纽带，绝大多数情况下属于商业化的公共平台，任何系统都可以接入并共享其中的数字资源，其核心技术分散在影像捕获/产生、处理、存储、传送、管理和呈现的各个环节。只要在某一环节甚至子环节掌握了具有竞争力的关键技术，企业无论规模和地域都可以在今天的数字影像领域中享有自己的空间和份额，协作共赢是发展的必然选择。因此，数字影像时代不再可能出现像传统影像时代那样一统天下的"大咖"，取而代之的是掌握核心技术、具有核心竞争力的高科技企业。

图 1　数字影像系统的内核和外延：数字内容是核心

（二）感光影像在数字影像中的地位和作用

从图 1 可以看出，在数字影像系统整个北半球的数字影像捕获、输入系统中，感光影像起将能量载体影像或物性载体影像转换为数字影像的作用，是数字影像资源的入口，其在捕获影像的同时已经完成了捕获影像的分色、模数转换、属性描述和存储等操作。在诸多关键技术中，传感器是核心。与传统影像或银盐照相的传感器不同，作为数字影像资源入口的感光影像系统内部发生的变化绝大多数属于物理变化的范畴，如：光导、光电子等。例如，数码相机的传感器主要采用面阵列的 CCD 或 CMOS 器件，其作用就是将进入照相机到达感光板的光子转换为电子，随后被系统读取和数字化。

在数字影像系统整个南半球的影像输出设备／系统中感光影像起将数字影像转换为可视影像的作用，即数字影像资源的出口。在所列的系统中，以发光显示为基础的平板显示器属于数字影像转换为能量载体影像（光子发射）的影像系统，但以反射阅读为基础的电子纸则（如，利用微小带电颗粒的电泳实现显示的 E-Ink 系统，利用微球旋转实现显示的 Gyricon 系统）和其他设备或／和系统（包括没有表示出来的光刻、辐射固化、激光束 3D 打印等）均属于数字影像转换为物性载体影像的感光影像系统。

三、本学科国内外研究进展比较

数字影像时代的感光影像学具有学科分支多和高度交叉的特点，每一个分支都可能隶属或对应不同的产业。这种对应行业的发散性和多学科交叉的属性是数字影像时代感光影像学的一个特点。例如，CCD、CMOS、闪存、光导体、色粉、互联网、物联网等都属于不同的行业，看似与感光影像没有关联和瓜葛，但是一旦将它们融入数字影像中或按照数字影像的要求进行设计和制造后，就成了今天感光影像不可或缺的器件或系统。因此从研究角度看，数字影像时代感光影像学的研究一定是瞄准数字影像系统的某一分支或系统或组件或单元或材料或更加细分领域的专业化研究，整体设计下的专业分工和协作成为关键。可以预测，数字影像未来覆盖的领域会越来越宽，将大家带到同样的起跑线上。这给我国在很多领域追赶、甚至超越并引领世界创造了前所未有的可能和机会。

第七节 岩石力学与岩石工程

一、引言

岩石力学与岩石工程学科是运用力学原理和方法来研究岩石力学有关现象与工程问题的一门应用科学。近年来，我国岩石力学与岩石工程学科伴随着我国社会与经济的飞速发展，获得了长足的进步。总体具有以下几个特点。

（1）我国正逐渐由岩石力学与工程发展建设的大国向强国转化

我国已建和拟建的岩石工程项目数量之多，规模之大，为世界瞩目。一大批大型水利水电工程开发与建设、矿山与能源开采、地下能源存储及核废料处置、地面与地下交通建设及其他许多岩石工程的建设陆续展开。从中国第二大水电站白鹤滩水电站，到国家九大科技基础设施之一的中国"天眼"工程（FAST）；从"12·20深圳山体滑坡事故"处理，到规划建设雄安新区、打造绿色智慧雄安等工程项目，都活跃着我国岩石力学与工程领域科学家们的身影。中国在岩石力学的理论与工程实践中已经取得的成果受到世界同行的关注与高度认可，现任国际岩石力学学会（ISRM）主席Eda Quadros在"2016年国际岩石工程安全学术大会"的一次讲话中说道,"岩石力学与工程当前的重心在中国"。

(2)坚持为经济建设主战场服务,工程实践与科技创新相互促进

随着"一带一路"国家战略推进,一带沿线区域涉及了大规模的岩体工程。我国岩石力学与岩石工程科技工作者坚持为经济建设主战场服务,紧紧围绕工程实践及实际问题展开科学研究和科技创新。交通、能源、水利水电与采矿工业等各个领域的岩石力学与工程问题,在实践中产生需求,又因为需求促进了科技创新与学科发展。

(3)学科交叉融合度越来越高,计算机网络技术对学科发展注入新活力

从近年来岩石力学与工程领域的主要科技成果来看,学科交叉融合度呈现出越来越高的趋势。另外,随着"互联网+"、云计算、大数据等信息技术迅猛发展,将传统岩石力学与工程体系和现代计算机技术相结合,推动着智慧化管理和数据互联共享,促进了我国土木、交通、水利、能源、国防等基础设施建设领域相关信息化水平的整体提高。

(4)青年学者科研创新活跃,梯队建设初见成效

我国岩石力学与工程领域相关研究论文数量与质量稳步提升,其中青年学者的贡献尤为突出。近10年来,中国在国际岩石力学与岩石工程界的影响进一步扩大,国际学术地位也得到进一步提升。除了积极参加本学科国外的大型学术会议,我国岩石力学与工程科技工作者还多次组织承担大型在华国际会议,为青年才俊提供展示平台,促进老、中、青学术梯队的壮大,使大批中青年科技工作者初露锋芒,梯队建设初见成效。

二、本学科近年的最新研究进展

近年来,由于我国经济的飞速发展而伴随的岩石工程项目的大规模兴建,在水库大坝、铁路隧道、跨江(海)桥隧等重大工程项目及地下采矿工程、人防工程及地下空间利用方面的快速发展,促进岩石力学与岩石工程学科在理论与数值分析、勘测技术、实验装备、开挖与加固技术以及灾害和预警等方面取得了大量创新性成果。

(一)岩石力学理论与数值分析

1)岩石强度和强度准则:针对传统的Hoek-Brown强度准则(简称为H-B准则)中存在的不足,结合工程实际,进行了具有针对性的完善,并提出了相应的修正准则。其中,非线性三维H-B强度准则(简称为GZZ准则)已成为国际岩石力学学会建议方法之一,该准则既考虑了中间主应力对强度的影响,又完全继承了H-B准则的优点,可直接使用试验和工程经验积累的H-B准则相同的参数。

2)岩石的变形与流变性状:以非线性蠕变损伤模型为研究重点,相关学者分别提出了长期弹模与损伤变量之间的函数表达式;建立了具有初始损伤的岩石损伤变量演化方程,形成了能够反映具有初始损伤的岩石瞬时弹–塑性变形、稳定蠕变和加速蠕变的蠕变全过程本构模型;给出了弹性损伤元件及黏性损伤元件的蠕变特征方程;建立了反映岩石

体积变化特征的统计损伤本构模型；推导了纯剪破坏时的泊松比与内摩擦角关系式。

3）岩石断裂与损伤力学：揭示了岩石断裂、损伤裂纹的萌生、扩展和贯通机制，并将岩石裂纹扩展机理的研究从二维扩展到三维，注重从岩石力学试验结果分析裂纹真实的扩展状态。

4）岩石动力响应：采用改进的霍普金森压杆（SHPB）试验装备，对不同加载率或应变率、动静组合加载、不同轴向静载、冲击速度和波形以及不同裂隙倾角等条件下的岩石动力学特性进行了系统的试验研究；针对岩石爆炸动力学特性，建立了基于CT扫描及X射线技术的三维数值模型、爆破波对含节理隧道的损伤数值模型；发展了单次波加载技术，波形整形技术和激光位移测量等动力学实验新技术，提出并发展了三套动态断裂实验方法，提出了位移与力非连续模型（DSDM）。

5）岩石多场耦合模型及应用：在理论研究方面，相关学者研究建立了基于等效渗流阻的渗流–应力耦合模型、压剪作用下考虑裂隙剪胀特性的渗流应力耦合模型、应力作用下岩石的溶解动力学模型、岩石弹塑性应力–渗流–损伤耦合模型、考虑渗流–应力耦合作用的层状盐岩截面裂缝扩展模型和泥岩渗透性演化数学模型以及低温冻融条件下岩体THMD耦合模型。在试验方面，研制了岩石应力–渗流耦合真三轴试验系统，多场多相耦合下多孔介质压裂–渗流试验系统、多场耦合煤矿动力灾害大型模拟试验系统。

6）深部岩体分区破裂化：建立了静水压力条件下动态开挖卸荷对深埋圆形洞室各向同性围岩分区破裂化影响的力学模型，确定了卸荷速率和岩体动态力学参数对深埋洞室各向同性围岩分区破裂化现象的影响；建立了非静水压力条件下深部圆形洞室围岩分区破裂化的非欧几何模型，获得了圆形洞室围岩应力场、破裂区和非破裂区的数量。

7）岩体数值方法：在岩石多尺度本构模型、蠕变本构模型等研究基础上，发展了以连续方法（有限元、边界元、FLAC、刚体界面元）和非连续方法（DEM、DDA、NMM、Key block）为代表的岩体数值模拟方法。其中，对于非连续方法的研究，相关学者重点解决了以拟变分不等式描述接触条件，以及多种分析方法相结合求解大规模或复杂岩石工程问题。在连续方法研究方面，基于模拟岩石渐进破坏过程的统计细观损伤力学分析RFPA模拟软件，结合DDA方法，形成了岩石细观损伤到宏观开裂、块体运动全过程的DDD（discontinuous deformation and displacement）多尺度模拟方法，发展了大型方程组求解和高效数据处理的并行算法；同时，广义粒子动力学（General Particle Dynamics）算法的提出，成功实现了裂隙的扩展和连接的数值模拟。

（二）岩石工程勘测、试验与设计

1）岩石工程地质勘测：基于全站仪、GPS技术以及三维激光扫描技术等的广泛应用，提出了岩体结构面参数测量、坡体表面整体变形监测、危岩块体识别以及岩体性质原位测试等勘测方法；基于钻探技术和钻孔原位勘测，提出了多种提取岩体参数的新方法；基于大型水电工程勘察，提出了基于地勘数据中心的水电工程地质勘查与分析信息一体化设计方法及其解决方案。

2）室内测试仪器设备：发展了岩石破裂过程（computerized tomography，CT）检测、软岩剪切测试与岩石蠕变测试技术、现场变形测试、现场弹性波测试与地应力测试技术等；研发了微机控制电液伺服柔性三轴实验机、中尺寸岩石真三轴试验系统、岩石三轴蠕变试验仪、节理剪切－渗流耦合试验系统、土石混合体大型土工试验装置、深部隧道框架式真三轴物理试验系统等一系列大型室内测试设备。

3）原位试验：相关机构研制了能实现高压加载、复杂应力路径伺服控制的现场岩体真三轴试验系统、真三轴流变试验系统等大型装备；研发了深孔水下空心包体三维解除法地应力测试方法与成套技术。

4）动态设计与反馈分析：提出了水工岩体开挖与支护动态反馈分析方法、复杂环境下岩石工程安全性（稳定性）的时空预测与分析评估综合集成智能方法。

（三）岩石工程开挖与加固

1）煤炭开采：基于"切顶短臂梁"理论的切顶卸压无煤柱自成巷开采110新工法[①]，作为一项煤炭行业内的重大突破，自2009年首次在四川白皎煤矿中厚煤层开采成功实施后，近年来，推广应用于薄煤层、厚煤层等不同开采条件，为煤炭资源的安全高效开采提供了一条崭新的解决方案，被称为我国矿业技术变革的第三次探索。该技术已被写入《国务院关于煤炭行业化解过剩产能实现脱困发展的意见》《能源生产和消费革命战略（2016—2030）》和《中国制造2025——能源装备实施方案》，对于保障我国能源安全具有重大战略意义和广阔的发展前景。

2）水电工程：针对大型水电工程基础开挖问题，研究揭示了大开挖方案的卸载效应，提出了一种解释大变形与边坡稳定的共轭变形机制，建立了基于岩石强度应力比的地应力折减标准，研发了基于复杂坝基开挖保护与处理全过程的监测评估、预报预警、决策支持的数字化管理系统。

3）工程加固：针对传统锚杆对节理岩体的锚固机理和特性，提出了主控裂纹的概念，研究了裂隙角度和侧向应力大小对锚固岩块的锚固效应，建立了节理锚固锚杆在剪切荷载作用下的变形模型。针对传统锚杆不能适应岩体大变形而出现的破断失效问题，研发了具有超常力学特性的恒阻大变形锚杆/索支护新材料，既具有恒阻条件下抵抗变形的功能，又具有抵抗冲击变形能量的功能，其恒阻值及恒阻运行长度等物理力学参数均居于国际领先地位。

（四）岩石工程的监测和预警

1）边坡（滑坡）监测：研发的基于边坡深部滑动面上滑动力变化的远程实时监测系

---

① 通过沿空侧切顶自动成巷实现了一个工作面只配套掘进一条巷道，同时取消了煤柱留设。

统具有集加固、监测、控制和预报多项功能的独特优点，实践中取得了能准确预报滑坡的良好效果，起到了为矿山安全开采保驾护航的重要作用。

2）地下工程监测：在岩爆机理及特征研究方面，研发了应变岩爆力学试验系统和冲击岩爆力学试验系统，代表了目前岩爆试验的先进水平；结合声发射和微震监测技术，揭示不同类型（即时型、时滞型）岩爆孕育过程中微震信息演化特征和规律及其差异性。在监测技术方面，研制了国内首台具有自主知识产权的矿区千米尺度破坏性矿震监测定位系统，研发了基于多元地球物理信息融合与联合反演理论的综合超前地质预报技术以及基于约束联合反演理论的综合预报方法。

三、本学科国内外研究进展比较

近年来，我国学者在高水平岩石力学类期刊的文章比例及发文量大幅度提升，引用率不断提高，表明中国在国际岩石工程界的影响进一步扩大，国际学术地位也得到进一步提升。但与国际研究相比较，还存在以下不足。

1）工程技术成就上升而原创性的理论研究成果较少，岩石力学的基本本构关系、通用的岩石破坏、强度准则大多仍是应用国外学者提出的研究成果，或者在国外的研究成果基础上修正发展。

2）现场和实验室试验技术原创的试验设备和试验方法不多，目前运用的主要检测、监测方法仍以引用国外成果为主，实验室仍需采购国外厂商研制的大型贵重仪器设备。

3）缺乏原创的大型数值软件平台，尽管中国也发展了实用的数值仿真软件，但是主要商用软件还是被国外大型软件所垄断。

4）发表文章数量在国际上已处于前列，但围绕岩石工程有分量的和有国际影响力的重要发明成果仍然较少。

四、本学科发展趋势和展望

岩石力学是一门与工程应用密切相关的科学，并且岩石工程问题往往具有工程规模大、建设周期长、工程地质条件复杂、边界难定等特点。近年来，我国的科技人员为推动岩石力学与岩石工程学科发展开展了卓有成效的工作，为岩体工程建设做出了突出贡献。然而，随着我国经济建设中大规模复杂岩体工程建设的不断深入，在岩石力学基本理论与模型、计算方法、监测技术与加固手段等方面，都提出了日益艰巨的挑战，也为岩石力学与岩石工程学科提供了前所未有的发展机遇。

面对岩石力学中不断涌现的新问题与新挑战，未来的研究重点主要体现在如下几方面。

1）岩石力学与岩石工程的跨尺度破坏力学问题：发展现场原位试验平台建设的同时，加快室内小尺寸平台建设，构建岩石结构微—细—宏观多尺度试验和数值研究。

2）现场原型监测、地质力学模型与数值计算综合分析的统一研究：研发具有自主知识产权的核心高性能监测设备和大型结构地质力学模型试验，推进基础理论、模型及核心技术问题的深入研究。

3）基于现代信息技术的岩石力学与岩石工程：基于大数据分析的岩体工程设计、施工与运营中的监测分析、计算模拟与工程决策，抓住人工智能带来的机遇，为未来岩石力学与岩石工程跨越式发展奠定基础。

4）深地工程将是未来岩石力学与岩石工程的战略前沿。深部地热能源与矿产资源的安全、有效开发和可靠储存关系到我国国民经济持续发展和能源战略安全，持续、安全、绿色的能源供应是我国乃至世界经济高速发展的基本保障和刚性需求。

第八节　机械工程（机械设计）

一、引言

设计作为形成机械产品的起点和先导，决定着产品的功能品质以及制造、服务的价值。产品的结构、性能、成本、可维修性、人机环境、风格等影响产品竞争力的关键因素都是由设计阶段决定的。现代制造业的竞争往往是产品设计的竞争，产品设计创新能力已成为保持企业竞争力的核心和关键。在当前数字化、信息化、智能化发展趋势下，不同学科之间相互强力融合，这给机械设计带来了极大的机遇和巨大的挑战，也将引发现代设计方法的深刻变革。本报告主要概述了近几年我国在机械设计基础理论研究以及在不同工业领域应用中取得的创新性和突破性研究进展，通过对国内外研究进展进行对比，对今后机械设计理论和应用技术的研究趋势进行了展望。

二、本学科近年的最新研究进展

（一）机械设计基础理论

在机械设计基础理论方面，主要从工业创新设计理论、数字化设计理论、结构优化设计理论以及一些新兴的设计理论与方法等几方面进行论述。

（1）工业创新设计理论

工业创新设计是由工业设计与创新设计有机结合形成的一种前沿设计，涵盖了工业产品的概念模糊前端、产品开发过程及商业化三个不同阶段。随着计算机技术的不断发展和数字化信息时代的到来，围绕计算机技术的工业创新设计也应运而生。比如工业创意设计、计算机辅助的工业创新设计、计算机支持的人机交互设计、计算机支持的协同设计和计算机辅助的创新造型设计等。工业创新设计的实现不仅依赖于企业研发人员的领域知识与创新能力，而且其涉及多学科交叉结合的新技术。它融合了多元文化设计技术、个性化配置设计技术和基于网络大数据的创新协同设计技术等，而技术创新方法的应用可以提升研发人员创新能力和创新效率，辅助研发人员快速、高质量地开发出工业产品创新设计方案。未来工业创新设计理论的发展可简要概括为三个基本属性、五种发展特点和五大趋势。三个最基本的属性为绿色、虚拟和人性化；五种发展特点是全球产品物联化、绿色低碳可持续化、机械产品智能化、适应人机工程学化和设计能力生命化；五大趋势为国际全球化趋势、低消耗低污染趋势、创新转型设计趋势、贫富兼顾民生设计趋势和共存共赢的多元设计趋势。

（2）数字化设计理论

数字化技术不仅是世界各国促进工业发展和提高经济竞争力的关键，而且是装备制造业自主创新、强化研发设计和生产制造等各环节信息化深度应用的主要手段。近年来数字化设计理论的发展主要集中在计算机辅助设计理论和基于数值模拟的设计理论等领域。其中，在计算机辅助设计方面，计算机辅助设计集成化技术、计算机辅助设计标准化技术、计算机辅助设计智能化技术和计算机辅助设计虚拟设计取得了较大的发展。但相比发达国家在三维 CAD 系统的自主研发能力、CAD 系统数据交换和可视化集成 CAD 技术等方面还有较大的提升空间。而在基于数值模拟的设计理论领域，在当前数字化、信息化和智能化发展的趋势下，信息技术与制造技术深度结合，使得机械装备的数字化向全生命周期海量数据集成方向发展，对创意、设计、加工、工艺、装配、测试、服役、维护、报废和回收等各阶段进行了复杂建模、模拟、分析和挖掘。围绕数值模拟开展的一体化建模技术、面向全生命周期的产品数字化设计表达、建模、分析技术、基于计算反求的数字化建模参量精确识别技术和建模不确定性因素及其传播的评价体系等方面的研究取得了长足的发展。数字化设计领域的成果将助力中国的设计制造业，为行业发展打下坚实的基础。

（3）结构优化设计理论

结构优化设计是用系统的、目标定向的过程及方法代替传统设计，其目的在于寻求既经济又适用的结构形式，以最少的材料、最低的造价实现结构的最佳性能。最近二十年来，以提高结构经济性、改善结构性能为目标的结构优化设计在理论和方法上都取得了显著进展，并已经广泛应用于复杂装备的设计实践中。从结构优化研究的现状和发展趋势来看，多学科和多目标结构优化设计、结构拓扑优化设计仍是目前该领域的研究热点。同时针对工程中存在不确定性等技术难点，开展基于不确定的优化设计等相关研究也日益兴起。在这些领域，许多学者做了大量工作，并获得了相应的理论与技术成果，如基于协同设计的多学科优化设计方法、基于变密度法的拓扑优化设计方法、基于可靠性的优化设计方法和基于序列优化的混合不确定性分析方法等，其中部分技术已应用到工程实际中。在复杂结构和产品数字化设计中，整体与局部的协同优化，静、动态协同优化，多学科稳健优化，多目标协同优化，多物理场下的拓扑优化设计，大规模系统的多信度近似优化技术，动态或时变下不确定性优化和面向全生命周期的产品可靠性优化等复杂系统优化设计技术仍将是结构优化领域亟待解决的问题。

（4）新兴设计方法

为了适应国内及国际机械产品的市场竞争的需求，多种新兴的机械设计理论与方法被提出并得到了广泛的研究，主要有材料—结构—功能一体化设计理论、绿色设计理论、仿生结构设计理论和面向增材制造的设计理论。目前国内的材料—结构—功能一体化设计主要从等材、减材、增材和复合制造等角度研究高性能精确成形制造新方法，而国外多致力于研发轻量化聚合物复合材料用于取代装备中重金属部件。绿色设计的发展主要体现在产品制造过程中控制并减少对环境有害物质的使用，并对产品进行轻量化设计以节约材料；仿生结构设计通过与机械工程学科、纳米科学、信息科学及生命科学进行交叉，从而仿制出具有指向性功能的新型结构；增材制造技术则不断地应用于制造业、航空业、生物医药等诸多领域，如增材制造汽车、增材制造血管、增材制造再生关节软骨等高性能的机械或结构。材料—结构—功能一体化设计将进行多尺度结构和构建性能的研究，揭示材料组织演化与结构变形的交互作用机制，探索复合制造原理，从而实现高性能复杂结构的整体制造；绿色设计将向着低碳经济和围绕高附加值产品的制造，并提供终身服务的方向转变；仿生学则伴随生物制造技术的发展，制造出仿生物体或仿生智能机器人服务于人类的生活；面向增材制造的设计则从研发走向产业化应用，并与信息网络技术融合，带来新一轮的产业变革。

（二）工业机械设计

在工业机械设计方面，主要从结构学及机器人设计、传动机械设计、机构强度与可靠性设计、机床设计和车辆工程设计这几方面展开论述。

（1）结构学及机器人设计

现代机构学是机械工程学科的重要分支之一，是设计和开发各种现代机械装备与机电一体化产品的基础。近年来，关于机构学及机器人的研究进展迅速，在多个领域进行了拓展延伸。在并联机构领域，研究内容主要涉及工作空间分析、尺度综合、动力学建模、奇异位姿分析、少自由度并联机构设计、刚柔耦合问题等。在机器人领域，提出了并联机器人型综合的GF集理论，形成了并联机器人全域定量评价和尺度综合可视化方法，奠定了我国并联机器人型装备自主研发的设计理论基础。同时，在空间欠驱动机器人动力学综合与运动控制、少自由度并联机器人（自由度数<6）构型综合、评价与参数设计以及并联

机器人机构学理论及控制和仿生机器人等方面也做了大量研究,并取得十分丰硕的成果。

(2) 传动机械设计

机械传动是装备制造业的基础,我国能源、交通、航空航天、海洋等领域装备的发展对传动部件及系统的精度、效率、功率密度、可靠性、环境适应性等提出了更高的要求。近年来,机械传动设计的研究主要集中在以关注能量转换和传递效率为主的多介质多形式高效传动设计、以关注传动精度为主的精密复合传动设计和基于功能材料的新型驱动设计。其中包括：微型齿轮尺度界定及成型原理；微齿轮的 LIGA 技术和准 LIGA 技术、微刻蚀加工技术、细微电火花线切割技术和微细切削技术；微齿轮的接触和黏附分析、摩擦分析以及多物理场耦合分析；新型锥轮牵引式无级变速传动和脉冲式无级变速传动；液力传动和液压传动；行星齿轮多级复合传动以及液压 – 机械多级复合传动；分流式传动系统改进；梯度材料、记忆合金、复合材料、智能材料等新型材料。随着材料、机械、传感和控制等学科的发展和制造技术的进步,机械传动设计将向高性能高效率传递部件与单元设计、复合传动精密调控设计以及智能结构与新型作动器的驱动机理研究等三个方向发展。如：混合动力传动的复合模数与参数匹配；精密复合驱动传动的运动、力、能量的匹配与控制策略；具有智能结构特征的机构、驱动、测试和控制集成优化和新原理传动器的多场、多相耦合作用等。

(3) 机构强度与可靠性设计

现代机械种类繁多、结构复杂,机构的强度和可靠性设计无疑是结构设计需要关注的。但现代机械系统中,广泛采用新技术、新材料和新工艺,因缺乏实际经验,很难求得适应现有条件的安全系数。同时,原本机械零件的材料强度、外载荷和加工尺寸等都是有偏差的,存在离散性,有可能出现达不到预期而失效的情况。通过传统的依靠安全系数来获得可靠性指标的设计方法越来越受到限制。因此,不能再将安全系数当作确定值处理,不能视其为可靠性设计的依据,而是需要把它们当作概率变量来处理。国内对于机械产品的可靠性研究起步相对较晚。20 世纪 80 年代,成立了机械产品可靠性设计研究协会,制定了可靠性设计标准,机械产品可靠性设计研究逐渐得到重视。虽然国内学者们给出了系统可靠性评价指标以及设计依据,但实际应用还相对较少,总体上与发达国家的差距较大。我国机构可靠性的研究虽然起步晚,但发展快,飞机起落架机构、卫星天线机构、武器操纵机构等出现的众多故障已经促使大量科研人员投入到机构可靠性的研究中。随着新技术新方法的应用,机构可靠性的研究趋势表现为对计算机仿真的应用持续深入、专业可靠性分析软件的开发有望突破、多学科理论技术交叉融合。

(4) 机床设计

随着国家科技重大专项"高档数控机床与基础制造装备"的推进,中高档机床的水平得到持续提升、高档数控系统实现关键问题的突破、功能部件的核心关键技术问题也被逐渐攻克。机床设计正朝着"高精度、高效率、复合化、智能化、绿色化、高柔性化和模块化"方向快速发展。同时,近年来的研究成果集中在数控机床动力学特性仿真与分析技术、高精度数控机床误差分析与误差补偿技术、数控机床可靠性分析与可靠性优化设计技术等方面，自主设计研发的数控机床也逐渐走向国际市场。"重大专项"成果显著，在高精类数控机床、高速类数控机床、复合类数控机床、曲面类数控机床、重载类数控机床、磨削类数控机床等机床的设计技术都取得了较大的突破。未来，面向国家重大需求，为提高我国数控机床制造与加工精度，将进一步开展数控机床误差测量 – 复合误差建模 – 误差实时补偿及实例验证一体化研究工作，这也为研究数控机床和三坐标测量机等高精度加工测试设备的误差作用机理提供试验手段，为设计、制造高性能的三轴／五轴精密加工和测量设备提供支撑，也是为提高这些设备的加工与测量精度提供重要的技术途径。

（5）车辆工程设计

车辆工程是机械设计的重要研究领域，是机械设计最新进展的集中体现。车辆工程发展和技术进步是评价机械设计学科发展的重要指标。中国品牌乘用车产销量的快速增长和轨道交通行业的高速发展反映了我国汽车行业整体的技术进步。新材料和新结构不断涌现，新的计算和分析方法可以完成海量数据的分析与设计，结构的安全性和可靠性达到了历史的最高水平。以新材料、新工艺、新结构为基础的汽车轻量化技术已广泛应用到车辆设计当中。主动和被动安全技术都已被各个厂家看作是领先竞争对手的核心技术。与此同时，为缓解能源压力，减少环境污染，混合动力、燃料电池等新能源车辆设计技术不断涌现，也已成为技术创新和产业升级的核心技术之一。随着车辆行业的高速发展，对整车综合性能的优化设计研究，将会成为学术和工程领域的热点和难点。此外，汽车安全技术的范围越来越广，并逐步朝着集成化、智能化、系统化、全员化的方向发展。新能源技术是公认的未来车辆主流发展的方向之一，同时对于促进汽车产业转型升级、提升产业国际竞争力、建设环境友好型社会也具有重大意义。

第九节 控制科学与工程

一、引言

自动化程度已经成为当今世界衡量一个国家科技发展水平和综合国力的重要标准之一，而以自动控制和信息处理为核心的智能自动化技术更已成为推动生产力发展、改善人类生活以及促进社会前进的主要动力。自动化学科经过几十年的发展，基础理论发展日趋成熟，广泛应用到工业、农业、军事、交通运输、商业、医疗等方面。近年来，计算机、网络和通信等信息技术的发展为现有的控制理论提供了广泛的发展机遇，促使控制理论自身的发展，也催生出新的学科增长点。

二、近年的最新研究进展

在我国的研究生培养体系中，自动化对应的一级学科"控制科学与工程"，下属有5个二级学科："控制理论与控制工程""模式识别与智能系统""系统工程""导航、制导与控制""检测技术与自动装置"。

（1）控制理论与控制工程

控制理论与控制工程是控制学科的基础，从控制理论诞生，到如今智能控制理论和针对复杂系统的控制理论正在得到越来越广泛的关注和研究。国内对经典PID控制的研究成果颇为显著，不仅对PID控制为何能如此广泛而又有效应用于工程系统给出了理论回答，而且可以对工程师设计PID参数提供理论指导。

智能控制是在解决具有高度复杂与不确定性以及控制性能要求越来越高的背景下产生的。在大数据时代的背景下，数据驱动控制、学习及优化的研究和发展，符合科技发展大趋势的潮流，以及国家所提出的大数据、人工智能等前沿领域的科技战略。数据驱动控制摆脱了对受控系统数学模型的依赖。自适应动态规划方法以经典的动态规划方法为理论基础，融合了人工智能先进技术，为解决大规模复杂系统优化控制问题提供了新的途径，我国在自适应动态规划方法的研究上得到了很大的发展，在智能电网、智能汽车、工业过程等方面实现了应用。

平行控制是解决复杂系统优化与决策的强有力的方法。近年来，平行控制与平行管理在理论框架、核心技术、应用示范等方面取得了丰硕的研究成果。

理论上的丰硕成果推动了自动化技术在各行各业的应用和实践。

近年来，国内学者针对环境保护自动化从理论基础与实践应用方面开展了较为深入的研究。在城市固废处理自动化方面，我国在建模、优化控制、参数估计和预测、故障诊断等方面都有新的研究成果。自动化技术在空气质量预测、大数据和源解析、综合决策平台中的应用推动了我国大气污染治理的进程。

机器人是实现复杂系统智能控制的重要载体，机器人技术也在现代社会发展中起到了日益明显的作用。仿人机器人技术不断突破，进入了国际领先行列。智能化工程机器人引领了工程机械行业技术升级。水下机器人取得重大成就并在海洋研究等领域获得成功应用。特种机器人实现了从无到有，品种和应用不断壮大。特别是在建筑智能化方面，我国的优势集中在桥梁、隧道等大型基础设施建设中对自动化和机器人技术的应用。结合计算机、通信和人工智能技术的发展，仿真科学与技术在控制科学、系统科学、计算机科学等学科中孕育发展，尤其在我国国防和军工领域得到广泛重视与应用。

在《中国制造 2025》、制造业信息化工程等国家重大战略项目的指引下，我国制造系统控制技术在设备、管理、模式及信息等几个方面取得了一定的研究成果，研制了先进的智能控制器及控制软件，打造了一批数字化样板车间和智能工厂标杆企业。随着互联网技术的飞速发展，"互联网+"协同制造将网络与制造紧密相连，催生了制造业发展新模式。

流程工业自动化方面，随着两化融合程度不断加深，生产过程检测、建模、控制、优化、决策支持等新理论、新方法、新技术不断取得突破，逐步形成了企业资源计划、生产执行系统、生产过程系统等多层次的集成自动化系统。

智能交通系统方面，在轨道交通、地面公共交通、城市停车、交通流理论与交通信号控制、交通规划与设计、交通大数据等方向都取得了长足的进步，取得了一批国际领先的研究成果，我国在智能交通领域的研究水平已经实现了与世界同步。我国城市轨道交通的自动化水平得到了很大的提高，已经从人工驾驶发展为自动驾驶，进一步发展为无人驾驶。高速铁路得到了快速发展，目前已建成"四纵四横"为骨架的高速铁路网络。网络信息服务方面，P2P 服务、网格服务、面向服务的构架、云计算、物联网、大数据等信息服务技术，为网络信息服务带来了巨大的机遇。

（2）模式识别与智能系统

在计算机视觉方面，我国科研人员在前沿问题上力求占领理论与技术的制高点。生物信息学把控制科学与工程的研究对象从机械、电子、物理、化学等系统扩展到了以分子和细胞为基本单元的生命系统。多媒体分析领域中，地理分布、社会事件、描述生成、信息对抗、跨社区多源异构等特征分析获得了我国学者的青睐。脑机接口技术得到了长足的发展，研究成果覆盖肢体辅助、神经康复、军事及娱乐等应用。在智能楼宇领域，采用信息技术和智能化控制手段，使得楼宇管理日趋简单、高效。

（3）系统工程

2016 年，中国科学院数学与系统科学研究所给出了系统学的历史来源、定义、内涵与外沿，在系统科学与工程领域产生了巨大反响，为系统科学的发展指明了发展方向。系统理论与方法方面，我国在复杂网络、多个体系统、系统工程等方向都取得了诸多研究成果。复杂系统作为基础研究也已经列入《国家中长期科学与技术发展规划纲要（2006—2020 年）》中。

（4）导航、制导与控制

随着定位导航技术在过去 20 年中日趋成熟，我国研究学者围绕强非线性、高不确定性、参数快速变化、强耦合等航空器/航天器的系统特点，逐渐将研究重心转移到了复杂环境中的航空器/航天器控制上，分析和探讨航空器/航天器的智能控制方法，推进我国航空航天事业加速迈入智能自主时代。

（5）检测技术与自动装置

对于动态系统故障诊断与容错控制，在微小故障诊断、间歇故障诊断、闭环系统的故障诊断方面我国已取得了系统性和关键性的研究成果。在中国经济转型能源消费升级的背景下，新能源适应性控制技术、先进测量及智能控制技术、探索机器学习和人工智能应用等热点问题引领了发电自动化技术的发展。发电自动化技术适应新能源的发展，提高了特高压互联电网的可靠性，推动智能电厂建设。分布式能源并网相关进展主要体现在控制策略、无缝切换技术，自动频率、电压及黑启动技术、能源管理技术、协同控制技术、含分布式电源的能源系统自愈控制和故障诊断研究等几方面。电力公司和很多高校、研究机构都开展了微网体系、能源互联网等关键技术研究工作并取得一定进展。对于智慧城市研究，自2013年以来，围绕智慧城市的发展，我国学者从智慧城市的起源与发展历程、体系结构设计、关键技术、标准化与评价体系、典型案例等多个方面展开了广泛的研究。

自动化领域除了在以上的传统方向上有了重要发展之外，还在交叉学科和新兴应用方面具有旺盛的生命力。近年来，互联网的全面普及以及移动智能设备的广泛应用同时促进了社会媒体的爆发式增长，促进了社会计算研究理论的迅猛发展。信息物理融合系统是改变人类生活最有前景的前沿技术之一，经过十余年的快速发展，已广泛应用于制造业、智能电网、智能交通、智慧医疗、智能网络等多个领域，正以史无前例的方式重塑着更多研究领域。

无人机、无人车、无人船是网络化自主无人系统的重要节点，对提高军事、民用、国防科技水平具有重要的意义。无人机是一种完全在电子设备的监控下可以自动完成全部飞行过程的飞行器，在导航、结构设计、建模与飞行控制以及组网和多机协同等方面已经取得了一定的成果。无人车是指自动规划行车路线并控制车辆实现预定驾驶目标的智能汽车。当前无人车领域的研究热点主要集中于无人车的环境感知、决策和控制三个方面。无人船是指在水面进行自主或半自主方式航行的智能化平台。目前，我国在无人机、无人车、无人船等技术领域都取得了一定的技术突破，对军事、民用等方面都会产生深远的影响。

随着物联网、互联网、大数据与云计算等现代信息技术的发展，为实现具有创新力和可持续发展的现代新型农业，智慧农业应运而生。在科技部、农业部、地方政府的支持下，已实施多个示范项目，在各个方向都有技术积累。随着国家"三权分离"等农业政策的实施，智慧农业将有更广阔的前景，深刻改变目前的农业生产方式。

三、国内外研究进展比较

纵观国内外发展现状，我国学者已成为国际控制理论界不可忽视的一支力量。不过，在结合重大实际需求提炼理论研究课题和在多学科交叉基础上开辟新的研究方向上，与国外同行相比，仍有一定差距。

（1）控制理论与控制工程

在非线性控制和鲁棒控制领域，国内研究水平无论在理论上还是工业应用上都属于国际先进水平。平行控制与平行管理在中国进入快速发展时期。在国外，对平行控制与平行管理的研究，目前还较为初步和分散。我国可视化技术的水平和国外还有一定的差距，特别是与在制造领域上普及应用还有差距。但是我国在飞行器设计仿真、汽车新能源动力系统建模和状态估计方面已经取得了有目共睹的进步。

中国机器人学者研究内容和范围与国外相差不多。当前国内外的工业机器人产品存在显著差异，国内企业尚未形成规模化生产的局面，国外机器人在高端应用领域占据着绝对优势。从机器人使用密度上看，目前我国工业机器人使用密度仍然远远低于全球平均水平。我国建筑用机器人的发展总体上落后于国外先进水平，优势则集中在隧道、桥梁等大型基础设施的建设方面。新技术与新理念在建筑领域的应用拓展不足。

我国智能楼宇的起步较晚，总体上我国智能楼宇的发展紧跟世界潮流。

在流程工业自动化方面，国内在工艺过程模型和先进控制、生产优化、企业生产经营决策支持方面取得了一些具有国际先进水平的研究成果，但在全流程角度实现集成建模和优化方面还需要加强。在产学研合作方面，我国流程自动化与国外同行相比，仍有一定差距。

在智能交通方面，我国的研究起步相对较晚，但呈现强劲的发展势头，相关研究和应用都取得了突破性的进展，涌现了一批国际领先的研究成果。我国在智能交通领域的研究水平已经实现了与世界同步。在网络信息服务方面，国内学者和企业在网络信息服务应用方面的研究已取得了一系列的成果，但仍需加强在基础理论以及具有重要应用前景的技术问题上提出原创性的主流方法和核心算法。

（2）模式识别与智能系统

在基础理论方面，我国在低秩学习、分类器集成、多示例多标签学习等处于国际前沿水平。在深度学习应用方面，国内学者也取得了可喜的成绩。我国在多媒体社会事件分析起步较晚，近年来取得了丰硕的成果。在多媒体描述生成方面，国内整体上处于跟跑阶段。在脑机接口理论研究方面，国内与国际完全接轨。在脑机接口应用方面，国内与国外的差距较大。

在计算机视觉方面，国内研究在各个方面都具有先进性的工作，但在近两年的热点问题上，具有指导意义的开拓性工作较少。我国生物信息学在组学数据处理和分析、多类型数据整合、定量系统生物学与合成生物学等处于世界先进行列，在中医药系统生物学与网络药理学独树一帜。但是多项核心新技术源自国外，总体水平较国际先进有滞后。

（3）系统工程

在系统理论与方法方面，复杂网络的研究工作已经从最初的迅猛发展进入到平稳上升的阶段。

（4）导航、制导与控制

我国在航天航空领域持续和大量的投入，极大地促进了该领域的发展。但是，在先进控制方法的应用广度和深度上，还停留在初级阶段。

（5）检测技术与自动化装置

我国的研究开发有了显著的进展，在动态系统的容错控制、常规电源的清洁燃烧及电网适应性、微电网、基于信息与通信技术的智慧城市构造等研究方向均取得了一定的研究成果，处于国际领先水平。在发电自动化研究方向，国内外均未满足对可再生能源、燃煤发电、核能发电等发展的需求，并面临着将火力发电厂的控制融入智能电网的巨大挑战；我国在新型清洁发电技术的应用上发展较为缓慢，同时也欠缺储能辅助的发电灵活性控制方面的研究。在分布式能源并网方向，国际上对分布式发电、能源互联网技术等问题较为关注。

在智慧城市的研究方向，我国目前正着力于从信息与通信技术基础、城市的"智慧成长"、知识经济与创造力经济等角度实现智慧城市的构建；而国际上对智慧城市的目标型、动力型、应用型等构成要素的调查研究已经有了长足进展。

（6）无人系统

与先进国家，尤其是美国相比，仍存在较大差距。需要开展自主感知和复杂环境下的无人系统研究，提出更加有效的控制方法和策略，并进行无人系统的试验，使无人系统能够在国防和民用更好的使用。

四、本学科发展趋势和展望

为促进控制科学与工程发展，提出以下几点战略需求。

1）要关注交叉学科。科学上的新理论、新发明的产生，新的工程技术的出现，经常是在学科的边缘或交叉点上。重视交叉学科将使科学本身向着更深层次和更高水平发展，这是符合自然界存在的客观规律的。

2）产学研一体化。由易到难逐步推进，形成比较完整的项目研究、生产制造、使用管理的产业链。

3）基础人才的培养和支持。各年龄段的人才培养至关重要。目前我国的中小学教育缺少科技实践教育相关课程，动手能力匮乏。需要加大国内外交流，学习科技较发达国家的人才培养机制。

4）基础和重点研究的资金支持。目前我国对科技的经济投入已经超过了很多国家，但是要赶超发达国家甚至引领世界先进水平需要在基础研究和重大应用方法上给予更大的资金投入。

5）提高科研人员的社会地位和经济收入。科研现在已经成为一种职业，在纯粹的兴趣驱动之外，如果科研职位不够吸引人，那么优秀的年轻人很难选择科研作为终身职业。如果国家导向能够提高科研人员的社会地位，提升科研人员的经济收入，会有更多人投入科研事业，形成一种良性循环，促进我国科技的发展。

第十节 标准化

一、引言

20世纪末,信息技术飞速发展,标准必要专利的出现,标准在市场竞争、产业发展以及公共管理中发挥着越来越大的重要性。WTO/TBT协议把标准和标准化在国际贸易和全球治理中的地位进一步提高。标准学作为一门介于技术科学、经济学、管理学、法学和社会科学之间的综合性边缘学科,呈现出综合和分化的趋势,与其他学科的交叉渗透趋势越来越明显。近年来,随着标准学研究与实践的不断深入,新的理论与方法不断涌现,标准化学科领域不断丰富与发展。除了传统经济学、管理学以外,网络经济学、法学、产业创新理论、公共治理理论、社会学等领域全面介入,对标准、标准化的内涵和外延开展跨学科研究,形成标准化学科发展的新特点。由于国际标准化组织、部分国家标准化组织和政府的共同努力,全球的标准化学校教育和职业培训水平也有了不同程度的提高。

我国从20世纪末开始实施标准化战略,大力支持自主创新技术形成国家标准,推动我国自主创新技术转化为国际标准。进入21世纪之后,我国政府开始启动标准化改革,鼓励产业联盟制定标准,进而推动社会团体组织制定标准,并于2015年启动修改《中华人民共和国标准化法》(以下简称《标准化法》)。我国的改革开放、加入WTO、实施标准化战略以及标准化体制改革的实践,为开展标准化科学研究提供了广阔的天地,我国的学术界也逐渐形成自己独有的学术特点和研究方法。

二、本学科近年的最新研究进展

本部分包括标准化知识体系研究、概念和分类、标准经济学、标准与技术创新、标准必要专利、正式/非正式标准化组织研究、标准化管理体系研究与实践进展。

(1)标准化知识体系研究

21世纪之前,标准化学科知识体系研究尚处于起步阶段,标准化学科更多是围绕理论基础、学科地位等问题展开研究。20世纪80年代，钱学森提出"标准学"的概念，认为标准学"介于自然科学和社会科学之间，社会科学成分更大一些"。标准学的研究方法主要是运筹学、控制论等系统工程的方法。钱学森指出"标准学还是尚在研究的东西，尚未建立自己的知识体系"。进入21世纪以来，国内外针对标准化学科知识体系的研讨逐渐兴起。国内许多学者从不同的角度提出了标准学的知识体系框架。

（2）概念和分类

分别从社会学维度、经济学维度、哲学和术语学维度进行了分析研究。

（3）标准经济学

从微观经济和网络外部性、标准的经济效益两个方面进行了分析和研究。例如，英国商业、创新和技能管理部门（BIS）2010年发布了题为《标准经济学》的报告，对标准和经济增长及生产力增长之间的关系、标准和贸易之间的关系进行了深入细致的分析。

（4）标准与技术创新

标准对技术创新的作用一直是学术界争论的焦点。欧盟2008年发出的《标准化对创新的贡献日益增加》政府公函指出，标准化对创新和竞争力的贡献日益明显，提出利用其创新性的市场来强化欧盟利用其知识经济优势的竞争地位。Choi，Lee和Sung对1995—2008年间科学网（Web of Science）上发表的528篇标准化和创新的相关研究进行了系统分析，发现标准化与创新正在引起越来越多学者的关注，这些研究主要分布在管理、经济、商业、工程管理、计算机和通信等研究领域。Choi，Lee和Sung对这528篇文献的标题、关键词和摘要进行汇总，并利用摘要信息对这些文献进行聚类，发现这些研究主要分布在三大领域：标准化对创新的功能作用，这一领域的研究文献合计165篇；标准化战略和影响，这一领域的研究文献合计181篇；不同类别的标准研究，这一领域的研究文献合计182篇。

（5）标准必要专利

学术界对标准必要专利（SEP）的研究可以分为两类，即产业创新和战略理论的研究以及法学界的研究。因为专利是创新知识的重要载体，而标准必要专利问题大量出现在标准联盟（standard consortia）当中，其中还体现出专利持有人的竞争战略，所以标准必要专利成为产业创新和战略理论研究的重点。而法学界对标准必要专利的研究主要关注专利在标准化组织中的披露，"公平、合理和无歧视"承诺的实际意义以及在实际当中的不公平、专利劫持、专利陷阱、诉讼当中的救济问题等。

（6）正式／非正式标准化组织研究

对标准化组织的考察涉及管理学、社会学、经济学、社会治理理论等领域，考察对象包括各类传统标准化组织、标准联盟等组织。这类研究主题可分成两个类别：第一类是站在标准化组织（传统标准化组织、标准协会、标准联盟等）的角度，研究标准化组织的性质、标准化机制、标准竞争战略以及标准化治理等；另一类是站在公共治理的角度，研究标准化组织的公共治理问题，将标准与区域经济发展和国家创新体系、体制改革等主题联系起来。国外对标准化组织的研究主要包括墨菲和耶茨对 ISO 的考察以及 Simcoe 对标准制定组织的考察等几个方面。国内对标准化组织的研究说明我国对于团体标准化研究与国外的研究有很大不同，涌现了一批基于我国特有的标准化体系、特色产业集群的团体标准研究主题，还包括标准化体系改革问题研究、团体标准案例研究等。

（7）标准化管理体系研究与实践

标准化技术方法应用于质量管理的巨大成功源自于 1987 年 ISO 9001《质量管理体系》标准的发布，自此，"管理体系方法"的概念已被应用于环境、信息安全、食品安全和职业健康与安全等许多其他领域或组织的业务活动。近年来，随着管理体系标准数量的增多以及应用多个管理体系标准的组织数量的增加，ISO 意识到这些标准需要具备更强的一致性，但也意识到有必要根据这些组织的职能及其目标以及所处的不断变化的业务环境修订这些标准。这使得 ISO 管理体系具备了新的体系结构和通用术语、核心定义的基础。因此，ISO 制定了《ISO/IEC 导则，第 1 部分，ISO 综合补充规定——ISO 的特定程序，2016》（ISO/IEC Directives, Part 1, Consolidted ISO Supplement—Procedures specific to ISO, 2016）附录 SL（规范性）《管理体系标准提案》，规定了有关管理体系标准项目合理性论证的过程和准则、管理体系标准起草过程指南以及管理体系高层结构、通用术语和核心定义等内容，要求所有新的管理体系标准项目或修订现行的管理体系标准项目都应遵循该导则。

通过分析可以看出，经济学的贡献主要是在标准的经济效益、标准的必要专利与技术创新等方面，法学的贡献主要是在标准的必要专利与反垄断、专利政策等方面，社会学的主要贡献主要是在标准的基本概念和分类，标准化组织的考察等方面。

三、本学科国内外研究进展比较

我国这几年标准化学术研究发展迅速，从原来的相对比较落后，到现在为止有些领域的研究成果已经达到国际水平。我国早期的改革开放，引进国外资本和先进的技术政策让我国的市场成了全球技术标准竞争最重要的主战场之一。进入 21 世纪之后我国的自主创新战略和标准化战略大大促进我国企业的竞争力，形成我国后发技术和标准也加入了全球技术标准竞争的态势。这为我国的学术界开展相关研究提供了广阔天地，形成了我国标准化学术研究相对比较独特的研究、路径和特点。突出体现在标准化的理论研究和标准化教育等方面。

通过分析可以看出，国际层面的理论研究已经发展到经济学、产业创新、法学、公共管理、社会学等领域全面介入，各领域的研究不断深入的阶段。我国的标准化研究的特点主要是应对我国在经济全球化以及市场化改革过程中所面临的问题和挑战，主要包括对标准化知识体系的研究、在宏观、中观、微观三个层面的标准经济贡献率的研究，不同国家标准化体制对比研究和我国的标准化体制改革研究等。本部分还包括对国内外标准化教育最新发展的归纳总结和对比研究。

四、本学科发展趋势和展望

由于经济学、工商管理、公共管理、法学、社会学等不同学术界跨学科研究的进展，标准学科领域的知识体系已经不断完善，理论研究将更加深入，标准理论的跨学科特性将会更加凸显。在经济学、法学、工商管理、公共管理、社会学等各个学科将全面介入的基础上，在深度和广度方面向纵深发展。标准的跨学科混合研究是今后发展必然的趋势。例如，经济学、社会学、公共管理都在探讨标准的学术意义，都试图回答什么是标准，标准如何分类等，但是所观察的视角不同；又如，目前经济学和法学在分别研究标准必要专利所产生的垄断现象，经济学的研究往往借用法学的研究结果，法学的研究也会借用经济学的研究结果。可以预料，这种混合式的研究所产生的结果将会具有非常重要的学术意义和社会指导意义。

我国标准化教育今后的发展方向是：必须加强标准化人才培养，推进标准化学科建设，支持更多高校、研究机构开设标准化课程和开展学历教育，设立标准化专业学位，推动标准化普及教育。加大国际标准化高端人才队伍建设力度，加强标准化专业人才、管理人才培养和企业标准化人员培训，满足不同层次、不同领域的标准化人才需求。

在实施标准化教育过程中，我国可以借鉴日本和韩国的经验，由政府推动标准化知识的初中和高中教育。成人的标准化职业教育和培训要充分发挥我国现有各级标准化管理部门、国家和省市标准化研究院（所）、各级标准化协会的资源优势，提高向社会提供在职的标准化教育和培训课程的能力；也可以与大学的标准化教育相结合，共同开展标准化成人教育和培训。培训对象可以是专业标准化组织的工作人员，也可以是企业的专业技术人员和标准化人员；要根据不同培训对象的具体需求设置不同的标准化培训课程。

我国标准化理论研究已经取得了一定的成果。但是，由于标准化学科是一门新兴的交叉学科，需要填补的空白还很多。从整体看，更多学科研究的加入，正在使基础理论研究逐渐深入，对标准化学科本身的定义更加清晰，也有了更多对标准和标准化内涵和外延的学术讨论。我们相信，在政府的支持下，在广大标准化工作者和教育工作者的共同努力下，标准化学科理论和知识体系建设将得到进一步发展，为我国的国家创新体系建设、标准化体制改革以及国民经济的发展做出更大贡献。

第十一节 测绘科学技术

伴随着大数据、云计算、物联网、智能机器人等新技术的快速发展，测绘与地理信息科技的发展也储备了源源不断的新动力，与这些新技术的融合，测绘地理信息正成为大众创业、万众创新的重要领域，测绘与地理信息相关学科发展迅猛。当前，我国测绘与地理信息学科发展正进入全面构建智慧中国的关键期、测绘产品服务需求的旺盛期、地理信息产业发展的机遇期、加快建设测绘强国的攻坚期，其内涵已从传统测绘技术条件下的数据生产型测绘转型升级到信息服务型测绘与地理信息。2016—2017 年，随着测绘与地理信息的科技手段与应用已从传统的测量制图转变为包含 3S 技术、信息与网络、通信等多种手段的地球空间信息科学，近年来更与移动互联网、云计算、大数据物联网、人工智能等高新技术紧密融合，在此转型升级过程中涌现了大量的新观点、新理论、新方法、新技术、新成果，若干关键技术取得重大进展，转型升级后测绘与地理信息技术的社会应用与服务取得重大研究成果，测绘与地理信息获取技术、快速处理分析技术、关键技术升级的攻关水平和装备水平大大得到了提升。

一、测绘与地理信息科技关键技术的转型升级

1. 大地测量与导航

大地测量与导航作为前沿性、基础性、创新性、引领性极强的战略科技领域，在国家创新驱动发展的进程中发挥越来越重要的作用。大地测量利用各种大地测量手段获取地球空间信息和重力场信息，监测和研究地壳运动与形变、地质环境变化、地震、火山灾害等现象和规律以及相关的地球动力学过程和机制，在合理利用空间资源、社会经济发展战略布局、防灾减灾等方面发挥着重要作用。

（1）北斗全球卫星导航系统

近两年，北斗全球卫星导航系统（以下简称"北斗系统"）着力发展地面基准站布网、地面数据处理中心等相关建设，并推广"北斗系统"创新应用，与多个国家开展卫星导航领域的国际合作。继续建设"北斗系统"，2015 年开始实施"北斗系统"的组网任务。到

2016年3月底，北斗全球卫星导航系统已成功发射22颗在轨卫星。BDS/GNSS精密定轨定位与数据处理及应用的理论、算法、模型、软件与服务系统等取得若干进展。目前，我国已经建成具备北斗信号接收和数据产生能力的360个CORS站，初步建成了国家广域差分服务系统，结合大型建筑物复杂环境室内定位关键技术的重大突破，实现了全国范围室外优于1米，室内优于3米的定位精度。

（2）大地基准与参考框架维护

国家现代测绘基准体系已获取了全国统一空间基准下的高精度、地心坐标成果，解决了各省级基准站网坐标框架不统一和导航定位基准不一致的问题，并利用410个国家基准站CORS开展国家基准框架维护和兼容北斗的国家基准站数据分析及监测评估，我国现代大地测量基准体系已逐渐具备高精度、涵盖全部陆海国土、三维、动态的能力。2000国家大地坐标系下的国家级测绘成果在国土资源部等17个部委得到了推广应用，已完成《大地测量控制点坐标转换技术规范》行业标准，用于规范各类控制点包括GNSS点向2000国家大地控制点转换方法和技术要求。空天一体化基准相关方面的研究进一步加强，主要研究了利用甚长基线干涉测量（VLBI）、卫星激光测距（SLR）和全球卫星导航定位（GNSS）等技术确定高精度的EOP。

（3）大地测量数据处理与地球动力学研究

大地测量数据处理的进展主要体现在GNSS数据处理、大地测量反演和地球动力学算法3个方面。GNSS数据处理借助于云计算、云存储和分布式技术，解决了地球参考框架和大地网高效处理、CORS领域、大规模GNSS网平差以及大规模GNSS基准站网数据处理等相关复杂问题的求解，大地测量反演在数据获取、模型构建、反演算法设计及地球物理解释4个部分领域均取得了长足进展，地球动力学研究主要是在现今地球动力学过程的地壳运动学特征进行高精度和高时空分辨的观测和反演等方面取得了若干成果。

2. 重力测量与地球重力场

（1）卫星重力测量

我国大地测量和地球物理学界的专家学者紧跟国际卫星重力测量研究热点和动态，在低轨卫星精密定轨、卫星重力场建模、新一代卫星重力计划GRACE-Follow-On和E.MOTION等领域开展了多方面的研究工作，积极推动我国下一代的卫星重力测量计划，提出了我国双星串飞编队的卫星重力测量模式。

（2）海洋重力场与陆海统一大地水准面精化

我国海洋重力场和海洋潮汐等海洋模型精度不断提高，现已构建了中国近海及领海海域 $2'\times2'$ 的重力异常、全球海域 $2'\times2'$ 平均海平面高、$15'\times15'$ 全球海洋潮汐和海底地形等一系列数值模型，并已完成综合多源重力场资料研制我国全海域大地水准面模型。此外，开展了联合重力位模型、重力异常、地形数据，顾及完全球面布格异常梯度改正项的确定垂线偏差的理论与方法研究，并此基础上构建了我国 $1'\times1'$ 的垂线偏差数值模型。

3. 摄影测量与航天测绘

近年来，随着航天航空技术、计算机技术、网络通信技术和信息技术的飞速发展，摄影测量与遥感正在继续朝"三多"（多传感器、多平台、多角度）和"四高"（高空间分辨率、高光谱分辨率、高时相分辨率、高辐射分辨率）方向发展，近两年摄影测量与遥感专业技术进展体现在以下几方面。

（1）光学高分遥感卫星

高分四号卫星作为我国第一颗地球同步轨道遥感卫星，与此前发射的运行于低轨的高分一号、高分二号卫星组成星座，大大提高了遥感影像获取的时间分辨率和空间分辨率。天绘系列卫星是我国第一个传输型立体测绘卫星任务，已成功发射01星、02星和03星。目前三颗星组网摄影，在轨运行状态良好。截至2017年2月，天绘一号影像全球有效影像覆盖率已达81.2%，全国覆盖率达99.9%，可望实现无地面控制点条件下1:5万比例尺地形图（20米等高距）的测制。资源三号测绘卫星作为我国第一颗民用高分辨率光学传输型测绘卫星，已实现中国境内陆地国土面积98.47%的影像有效覆盖，全球范围内有效覆盖率接近14%。2016年5月30日发射的资源三号02星，实现了两颗资源三号测绘卫星组网运行，资源三号卫星全国数字正射影像库及数字表面模型数据库由全色和多光谱影像融合生成，几何分辨率为2米。

（2）航空遥感

近年来，由于无人机具备全天时、实时化、高分辨率、灵活机动、高性价比等优势，航空遥感的主要进展体现在无人机遥感技术的发展，主要研究了微型无人机、固定翼无人机、无人直升机、滑翔机等遥感技术。

（3）高光谱遥感

在高光谱遥感传感器研制与信息获取技术方面，机载成像光谱仪商业化水平不断推进，应用领域持续拓展，无人机高光谱遥感受到了高度重视，表现出良好的技术优势和发展潜力。

（4）合成孔径雷达（SAR）

SAR正向多平台、多波段、多极化、多模式、高空间分辨率和高时间分辨率方向快速发展，目前的研究主要集中在典型的SAR系统，包括星载（spaceborne）、机载（airborne）和地基（ground based）系统。近年来，地基SAR成像系统通过合成孔径技术和步进频率技术实现方位向、距离向同步高空间分辨率成像，测量精度能够达到0.1mm。

（5）激光雷达（LiDAR）

国内在星载激光雷达、机载激光雷达、车载激光雷达和地面激光雷达系统的研制与开发取得了较大的进展，如研制了用于月球探测的CE-1、CE-2和CE-3激光成像雷达，车载城市信息采集与三维建模、车载数据采集3DRMS和全景激光MMS等系统，无人机LiDAR系统的发展迅猛。

（6）数据处理

卫星影像数据平差处理的研究进展表现在提出了区域网平差方法和联合平差方法等，

在高分辨率影像处理中，高分辨率遥感影像分类方法、高分辨率遥感图像信息提取和高分辨率遥感图像场景的机器理解等方面取得了一定的进展。此外，在移动测量和三维 GIS 方面发展了一系列新原理和新方法。

4. 地图制图与地理信息工程

地图学与地理信息技术正经历从数字化向信息化的发展变化，产品内容和产品形式向社会化、三维化、动态化、泛在化和智能化发展，基于云架构的地理信息数据网络化采集、自动化成图、智能化分析与泛在化服务正在成为热点。近年来这一学科领域的研究主要集中在地图学与地理信息理论、数字地图制图与制图综合技术、地理信息系统、移动地图与网络地图等 4 个方面。

（1）地图学与地理信息理论

在以云计算、大数据和智慧地球等新概念、新架构和新方法的推动下，地图和地图学本身的概念内涵和外延在不断地演化中，出现了全息位置地图、智慧地图和新媒体地图等衍生新概念，经典地图学理论不断演进，地图学与地理信息理论体系更加健全。

（2）数字地图制图与制图综合技术

在数字地图制图方面，多比例尺地图数据库动态更新、增量更新、级联更新、要素更新以及实体化数据模型建立正在实现，地理信息更新和地图符号化出版一体化已经实现，在基础地理信息的持续更新和地图制图综合方面，多项新方法新技术得到推广和应用。

（3）地理信息系统

地理信息系统的研究进展主要表现在空间数据感知、获取与集成，时空数据组织与管理，数据编码，数据压缩，离散格网及地址信息编码等取得一系列研究成果；地理表达与可视化方面的研究集中于自动制图与矢量数据可视化、三维建模可视化和经济社会事件可视化方面。

（4）移动地图与网络地图

随着网络地图应用的普及和新媒体地图的发展，产生了智慧地图（或称智能地图）、公众参与地图、全息地图等地图新概念，提出了混搭地图、众包地图、个性化地图等在线地图服务的新模式，探索了面向地图的多模态人机交互模式。需要特别指出的是，众包地图采用"众包模式"建立大众参与的地图服务，得到快速发展，已进入制图、救灾和规划服务等领域。此外，在网络地图设计与表达、移动地图设计与表达、在线地图中多尺度可视化方法的研究中提出了一系列新方法和新算法。

5. 工程测量与变形监测

当前，我国的"一带一路"国家战略、航天强国战略、海洋强国战略以及建设雄安新区等一系列强国战略实施推动了新技术、新装备和新方法在工程测量领域的应用，创建了新的技术体系，促进了工程测量学科的发展。近几年，工程测量学科在理论、方法与技术上的进展主要有以下几个方面。

（1）重大工程几何测控关键技术

利用有限元数值模拟、GB-SAR、地面三维激光扫描系统、数字垂线仪、GNSS 和测量机器人等先进技术与理论方法，取得了一系列创新成果，包括预变形分析和测量监控相结合的方法、基于地面三维激光扫描的构筑物高频振动测量的数据处理方法、基于测量机器人多连接点的基准网组建方法、GNSS和测量机器人相结合的桥梁施工实时几何监测系统等。

（2）建筑遗产数字化保护技术

我国在建筑遗产的全生命周期信息采集、精细化重构、健康监测、虚拟修复及数据储存组织管理等研究都取得若干新的突破。建筑遗产精细重构与表达方面，通过高保真建筑遗产数字采集技术与装备，形成了空地联合遥感遥测的建筑遗产形貌结构信息提取的关键技术；以激光三维扫描技术、高光谱技术等为代表的高新测绘技术及物联网与大数据技术的发展，实现了建筑遗产全生命周期信息的全模态高精度采集以及文物本体状态主动精准感知。

（3）大比例尺测图新技术

综合利用无人驾驶飞行器技术、测绘传感器技术、无线传感器网络技术、遥测遥控技术、通信技术、POS定位定姿技术和GPS差分定位技术等，以各种成像与非成像测绘传感器为主要载荷，快速获取国土、资源、环境、事件等多源空间信息，并可进行快速处理、多维建模和高效分析。

（4）变形监测与安全检测技术

以计算机技术、网络技术、电子测量仪器技术、传感器技术、通信技术为一体的变形监测系统发展迅速，基本取代了传统的变形监测方法，变形监测已进入了自动化、智能化和信息化时代。主要进展包括核安全壳内外观缺陷检查/监测系统的建立和物联网安全监测云服务等。

（5）移动测量技术

在移动测量方面，研制出了一系列的不同用途的移动测量设备，从传统的室外移动测量到室内移动测量都有不同突破。主要研究了车载移动测量系统中定位定姿系统误差校正与补偿；通过3D SLAM技术，将激光扫描设备放在机器人上，可以进行室内激光扫描和定位，并支持在线三维实景空间浏览、人机交互、定位导航、协同信息交互以及高精度实景测量等；解决了激光雷达系统原始数据由初始极坐标向当地坐标系统的坐标转换等问题。

（6）高精度定位技术

近年来，iGPS、伪卫星和超宽带等高精度定位技术发展迅速，在航空航天制造、大型部件精确定位、飞行导航、室内定位等方面被广泛研究与应用。iGPS目前在柔性装配、大型部件精确定位以及自动钻铆机位置标定等过程中开展了大量应用研究；伪卫星技术结合GPS开始用于大坝形变监测，以及采用伪卫星技术实现厘米甚至是亚厘米精度的室内定位系统；当前已开发出了基于UWB技术的iLocateTM无缝定位系统，实现了2D/3D的实时定位。

6. 矿山测量与地下空间测量技术

当前，矿山测量的内涵已经发生了深刻的变化，且仍在不断发展。矿山测量已不仅限于几何量测与分析，还包括矿山采场环境、矿区地表环境等非几何量的观测与分析，测量手段也由常规的测绘仪器设备发展到包括地球物理仪器、环境参数物联网传感器等。近年来，矿山测量研究进展主要体现在以下几个方面。

（1）矿山测量中的测绘新技术

连续运行卫星定位综合服务系统（CORS）、无人机航空摄影测量、地面三维激光扫描技术在露天金属矿山控制测量、地形测量、三维建模等方面得到了推广应用，惯性测量由于其自主式、全天候、机动灵活的特点，给矿山测量提供了一种新型的测量方式。

（2）基于InSAR时序形变的矿区全盆地沉降研究

利用SBAS、永久散射体技术等时序InSAR技术，研究了InSAR技术在矿区开采沉陷中的应用，发展了矿区地表沉降监测以及老采空区稳定性监测技术；将DInSAR与时序SAR（PS-InSAR、SBAS、SqueeSAR、CR-InSAR）技术相结合，系统研究了大气、噪声、形变相位的精确分离方法，此外，在煤矿区老采空区地基稳定性评价体系、矿区地表三维形变和非法开采监测预警等方面取得了若干研究成果。

（3）感知物联网在矿山测量中的应用技术

从技术和服务模式两个方面研究了矿山物联网的发展趋势，阐述提出并研究了雾计算技术、网络分形结构、矿山物体的本体描述与知识化、云计算与大数据技术等概念及其在矿山物联网中的应用与发展。

（4）基于GIS一张图的煤矿生产技术在线协同管理系统

建立了基于Web GIS的煤矿生产技术在线协同管理系统，采用统一GIS平台、统一数据库、统一管理平台的方式，通过地测、通防、生产、机电等专业的分布式在线协同工作，实现了煤矿生产技术各个专业基于采掘工程平面图的"一张图"管理。

7. 海洋与江河湖泊测绘

近两年来，海洋与江河湖泊测绘在海洋测量调查平台、海洋测绘基准与导航定位、海岛礁与海岸地形测量、海底地形与地貌测量、海洋重磁测量、海图制图与海洋地理信息工程等6个技术领域取得较大进展。

（1）海洋测量调查平台

现已建成各种吨位和不同类型的海洋调查船160多艘，实现了从引进改装到自主研发制造各类专用或综合型调查船的技术突破，使得海洋调查的范围从近海领域扩展到远海、大洋乃至极地地区。其中，国内自主研发的无人智能测量船，在智能巡航、躲避风浪和稳定性等方面都取得了重大技术突破；国内自主研发一系列无缆水下机器人AUV产品，已成功突破AUV下潜万米大关。

（2）海洋测绘基准与导航定位

在海洋测绘基准方面，开展了深度基准模型构建、垂直基准转换、验潮站深度基准确定、我国海平面系统偏差及高程基准偏差、海底大地测量基准建立、陆海基准的无缝连接、水下参考框架点建设与维护和陆海大地水准面无缝连接等的理论与方法研究。在海洋导航定位方面，开展了星站差分GNSS、惯性导航系统与超短基线声学定位系统相结合的高精度水下定位检验测试，完成了中国沿海无线电指向标全球差分系统 RBN-DGPS 差分台站的升级改造和沿海北斗连续运行服务参考站 BD-CORS 的建设。

（3）海岛礁与海岸地形测量

提出了基于移动测量技术的海空地一体化海岸带机动测量方案，开展了无人机航空摄影测量和三维激光扫描技术在海岸地形测量方面的应用研究，利用机载 LiDAR 实现了海岛城市高精度 DEM 数据获取和滩涂地形 4D 产品快速制作。

（4）海底地形与地貌测量

开展了深水多波束测深系统和全海深多波束测深系统的测试工作，研究了 3D 声呐图像和实时图像拼接技术，提出了多波束测量声速剖面反演、精细海底地形的线性化恢复和浅水水深测量等一系列新方法。

（5）海洋重磁测量

国产自主知识产权海洋航空重力测量系统的研发实现了关键技术的重大突破，研究并试验验证了基于差分定位模式与基于 GPS 精密单点定位模式的航空测量成果精度的一致性。在海洋重力测量数据处理方面，研究进展体现在四个方面：一是海洋重力测量数据采集与处理实现全过程自动化与智能化；二是重力仪性能评价实现技术流程标准化和评价指标的系统化与定量化；三是精细化海洋重力测量数据处理方法体系更趋科学严密；四是构建了多源海洋重力数据融合处理理论。

（6）海图制图与海洋地理信息工程

研究了海图与航海通告一体化生产技术实现方案以及数据库模型，提出了一种基于 Web GIS 的海洋地理信息共享平台，深入开展了 e-航海航保信息标准化研究和应用技术研究，推广了多功能标、MSP 服务、e-航海船台及手机服务等应用。

二、转型升级后的测绘与地理信息技术的社会应用与服务

1. 地理信息基础框架建设与服务

地理信息基础框架建设与服务关键技术获得突破，基础地理信息数据库规模化动态更新，制定和形成了一系列的技术方案与标准规范，研发了相应的生产和管理软件系统，建立了一套适用于规模化动态更新工程的技术体系，解决了跨尺度和跨类型数据库之间的基于增量更新技术的工程化应用推广的难题，实现跨数据库联动更新技术工程化应用，构建了基础地理信息的要素级多时态数据库模型，实现不同版本之间同名要素的自动关联，以及自动变化提取和统计分析，基于增量式入库模式，实现了三个尺度、四种类型、多个现势性版本的国家基础地理信息集成建库，以及基于 C/S 架构的动态管理和基于 B/S 架构的在线服务。当前，我国已实现了全国 1∶5 万基础地理信息的全面覆盖和全面更新，形成全国"一张基础图"；全国各省、市、自治区测绘地理信息部门继续扩大 1∶1 万基础地理信息的覆盖范围，加快 1∶1 万数据库的建设和更新。到 2016 年年底，全国已有约 70% 陆

地国土面积实现1∶1万基础地理信息（含地形图）的覆盖。

2. 地理国情普查与监测

第一次全国地理国情普查采用航空航天遥感、全球导航卫星系统、地理信息系统等测绘地理信息先进技术，以优于1米的高分辨率航空航天遥感影像数据为主要数据源，通过多源遥感影像快速获取与处理、现场调查、信息提取、海量数据集成建库、地理统计分析等技术手段，在获取海量地理国情空间数据，搭建基于先进的数据库运行云平台的全国高分辨率遥感影像、地表覆盖、重要地理国情要素、多尺度精细化数字高程模型等7个大型空间数据库的基础上，首次查清了反映地表特征、地理现象和人类活动的基本地理环境要素的范围、位置、基本属性和数量特征，并形成了这些基本地理环境要素的空间分布及其相互关系的普查结果，包括10个一级类、58个二级类和135个三级类共2.6亿个图斑构成的全覆盖、无缝隙、高精度的海量地理国情数据。当前，已在战略规划、土地覆盖和土地利用、国土疆域、自然灾害等方面开展了大量地理国情监测工作，地理国情监测数据获取技术、获取手段多样，涵盖了航天、航空、低空、地面等多个层面和光学、雷达、LiDAR等多种方式，能及时获取不同空间、时间、光谱分辨率的地理国情监测遥感影像数据和地面调查数据，相关研究主要集中在全球变化、土地覆盖、土地利用、生态环境、自然灾害、地表沉降等领域。

3. "天地图"地理信息公共服务平台

"天地图"地理信息公共服务平台网站经过近两年的建设及省市级节点不断接入，已集成了全球范围的1∶100万矢量地形数据、500米分辨率卫星遥感影像，全国范围的1∶25万公众版地图数据、导航电子地图数据、15米分辨率卫星遥感影像、2.5米分辨率卫星遥感影像、全国300多个地级以上城市的0.6米分辨率卫星遥感影像。目前，天地图形成了国家、省、市（县）三级互联互通的架构体系，全国已有30个省、区、市完成省级节点建设，145个市（含县级市）完成市级节点建设，并实现了与主节点的服务聚合。整体服务性能比此前版本提升4～5倍。新版天地图还开通了英文频道、综合信息服务频道和三维城市服务频道，并更新了手机地图。

4. 北斗二号卫星工程

北斗二号卫星工程于2016年完成研制建设，现已建成了由14颗组网卫星和32个地面站天地协同组网运行的北斗二号卫星导航系统，该系统是国际上第一个导航定位、短报文通信、差分增强三种服务融为一体的区域卫星导航系统，也是我国第一个卫星与地面站星地一体组网运行的航天系统，该系统作为我国面向大众和国际用户服务的空间信息基础设施，与美GPS、俄格洛纳斯等国外先进系统同台竞技，面临数以亿计的各类用户长期连续稳定使用的严苛考验，现已正式向我国及亚太地区提供导航、定位、授时和短报文通信服务，服务区内系统性能与国外同类系统相当，达到同期国际先进水平。北斗二号卫星工程建设的圆满完成，使我国拥有了完全自主的高性能北斗二号卫星导航系统，从根本上摆脱了对国外卫星导航系统的依赖，彻底掌握了时空基准控制权、卫星导航产业发展主动权、国际规则制定话语权，为我国经济建设和国防安全提供了有力保障，已取得显著的政治、军事、经济和社会效益。

5. 航天重大工程的遥感空间信息可信度理论与关键技术

针对遥感空间信息可信度理论在航天重大工程实际应用中需要解决的关键难题，以遥感空间数据获取、处理和应用全过程的质量控制为主线，突破了遥感空间数据可信度量、可信处理和可信评估等核心技术难题，首次设计并构建了航天重大工程的遥感空间信息可信度理论方法，建立了多波束激光虚焦点成像模型和多法向平面控制几何检校技术，破解了振颤"探、分、补"技术难题，建立了海量空间数据产品通用二级抽样评估优化技术，形成了自主知识产权的面向重大航天工程和相关行业空间数据质量控制技术新体系，实现了嫦娥探月、载人航天和测绘卫星等航天工程中遥感空间信息可信度保障的重大创新。

6. 省级信息化测绘体系关键技术及应用

围绕信息化体系建设的需求，结合转型发展的实际，采用产学研用协同创新的方式，利用科技创新驱动及装备升级改造，目前已建立了现代测绘基准、航空航天遥感资料快速获取、多源数据快速处理与地理信息动态更新、重大基础测绘生产业务信息化等四大技术支撑体系和现代测绘基准、应急测绘保障、网络化服务三大地理信息综合服务体系，实现基础地理信息获取立体化实时化、处理自动化智能化、服务网络化社会化，率先建成了业务化运行的省级信息化测绘技术体系。

7. 测量工程空间信息获取理论方法及软件

以空天地海测绘技术大框架下的测量工程为研究对象，在测量工程数据处理理论、空间信息获取技术、测量信息表达方法、数据处理软件等方面取得了多项创新性成果，研制了涵盖地面测量、海洋测量、卫星定位、变形监测与灾害预警4个方面的测量工程立体空间信息获取与数据处理软件系统，实现了基于测量机器人、三维激光扫描仪、卫星定位接收机、海洋测量传感器和虚拟现实等技术的测量工程空间信息自动获取、快速处理与沉浸式交互。在国内外工程测量、测深测流、卫星定位、变形监测等领域得到了广泛的应用，极大地推动了测量工程技术的进步，取得了显著的社会和经济效益。

8. 地籍与房产测绘

随着国家《不动产统一登记暂行条例》的出台以及现代测绘技术、新型测绘仪器和测绘手段的不断发展，包含在不动产范畴的地籍测量和房产测绘从理论到实践发生了较大的变化。截至2016年年底，我国已完成了第一、二次全国土地调查以及5次土地变更调查，当前已开展"一张图"工程，实现土地利用遥感动态监测，土地调查数据库的实时更新，以及土地监管机制的创新；地籍和房产权属调查工作正在有序进行，并逐步实现房地挂接，已建立了标准统一、分布合理、数据关联、更新及时、互通共享的不动产登记数据库体系，在地籍与房产测量方面，开发了集几何面积计算、分摊模型建立、属性数据入库于一体的专业软件，制定绘图、计算、生成报告一站式解决方案；随着不动产统一登记工作的开展，不动产统一登记材料数字化、数据整合关联、数据库建设、信息系统开发等研究取得了较大的进展。

当前，测绘地理信息事业正处于转型升级的战略机遇期，新型基础测绘、地理国情监测、航空航天遥感测绘、全球地理信息资源开发、应急测绘（以下简称"五大业务"）与地理信息产业发展都迫切需要测绘地理信息科技提供有力支撑，切实解决制约传统基础测绘向新型基础测绘转型中遇到的科技问题，突破地理国情监测以及航空航天遥感测绘的技术难关，解决全球测绘和应急测绘的前沿问题，破解地理信息产业发展中遇到的技术障碍，全面推动事业改革创新发展。未来5年，测绘与地理信息学科将紧密围绕"加强基础测绘、监测地理国情、强化公共服务、壮大地信产业、维护国家安全、建设测绘强国"发展战略，落实《测绘地理信息事业"十三五"规划》，以支撑"五大业务"为抓手，以创新为动力，以需求为牵引，以问题为导向，以项目为纽带，着力健全创新体制机制，提升科技自主创新能力，形成一批具有国际竞争力的民族品牌软硬件产品，进一步缩小与国际领先水平的差距。

第十二节　矿物加工工程

一、引言

矿物加工工程，在以往相当长的时期内称为选矿工程，也称矿物工程。在我国现今学科分类中，矿物加工工程属于矿业工程的二级学科。国外有些国家往往将矿物加工工程纳入冶金工程、化学工程或材料科学与工程。

矿物加工是用物理或化学方法将矿石中的有用矿物与无用矿物（通常称脉石）或有害矿物分开，或将多种有用矿物分离的过程。在矿产资源开发和综合利用产业链中，矿物加工是介于地质、采矿与冶金或化工之间的重要流程作业环节。矿石通过矿物加工（选矿）处理，可将有用矿物品位显著提高，从而大幅减少运输费用；而且可减轻后续工序如冶金、化工处理或材料制备的困难，降低后续工序的处理成本，对国民经济的发展具有重要支撑作用。

二、本学科近年的最新研究进展

近年来，中国矿物加工学科建设取得了长足进步，以矿物加工理论、工艺、装备、药剂和选矿厂自动化及过程控制为代表的选矿技术发展迅速，为中国矿物加工事业发展提供了人才保障及理论和技术支撑，在世界矿物加工技术发展过程中起到了重要作用。为了促进矿物加工学科的进一步发展，提高矿物加工学科在资源开发及环境生态相关领域中的地位和作用，对中国矿物加工学科建设、人才培养、科学研究情况进行了调查评估，系统展示了中国矿物加工学科发展的成就和学科发展的支撑条件，对比分析了国内外矿物加工学科的发展态势及未来发展趋势，指出了学科发展面临的问题与挑战。

（一）学科建设

中国矿物加工学科建设取得重大成就，形成了完备的教学、科研、设计与工程应用学科体系，拥有高水平的人才队伍和研究团队，建立了完备的基础研究、技术开发与成果转化平台。

中国矿物加工学科经过全国高校和研究机构几十年的建设，不断得到完善，具备较强的国际竞争力。本学科的本科生教育和研究生培养快速发展，截至2017年9月，中国开设矿物加工工程本科专业的高等学校共计37所，全国拥有2个二级学科的重点学科。大学矿物加工学科共有专任教师600多名，在校本科生15000多名，硕士研究生3600多名，博士生650多名。

中国拥有矿物加工学科的研究和设计机构20余家，这些专业研究和设计机构拥有高水平的研究和设计队伍，完备的科研条件与先进的技术平台，是中国开展矿物加工科学基础理论与应用基础研究、实现科研成果转化、促进科研合作与学术交流的重要基地。

中国拥有矿物加工学科相关的国家重点实验室、工程技术研究中心等国家级研究平台22个，各类省部级实验室和研究中心共计58个。这些高端研究平台既是学科发展的实力体现，也是学科发展的支撑硬件之一，对促进学科快速发展起到了重要作用。

目前中国矿物加工学科领域从业人员共40多万人，其中院士6名，国家有突出贡献的中青年专家和政府特殊津贴专家350多名，长江学者特聘教授/讲座教授10多名，国家杰出青年科学基金获得者10多名，跨世纪、新世纪百千万人才工程国家级人选20多名。中国拥有几十支各具特色的矿物加工科学与技术研究团队，这些团队为中国矿物加工科学与技术的发展做出了重要贡献。近年来，在国家纵向和矿业企业横向科研项目的支持下，国内矿物加工学科产出了大批学术著作和科技成果，所取得的研究成果为国家矿产资源开发提供了有力的技术支持。

（二）理论研究与开发应用

矿物加工科技进步为中国矿产资源开发提供了坚实的技术支撑，形成了有特色的基础理论研究方向，部分技术与装备已居国际领先水平，建成了一批技术先进、装备优良的大型现代化选矿厂。

基础研究方面，形成了基因矿物加工工程、矿物表面断裂键与各向异性理论、流体包裹体浮选效应、金属离子配位调控分子组装、复杂难选铁矿磁化焙烧与深度还原、矿物加工计算化学及药剂分子设计、矿物分选过程的数值化计算与仿真等矿物加工领域新方向与新观点，在国际选矿界产生了较大的影响。①提出了"基因矿物加工工程"的理念，以矿床成因、矿石性质、矿物物性等矿物加工的"基因"特性研究与测试为基础，将现代信息技术与矿物加工技术深度融合，经过智能推荐、模拟仿真和有限的选矿验证试验，快捷、高效、精准地选择选矿工艺技术和装备；②开展了矿物表面断裂键与矿物各向异性的理论研究，提出了利用矿物表面断裂键性质准确预测矿物的解理性质和常见暴露面及矿物表面质点反应活性的思想，为实现含钙矿物、铝硅酸盐矿物的选择性浮选分离奠定了理论基础；③深入研究了矿物晶体解离时流体包裹体组分向溶液中释放、矿物表面原子沿法向的位移及矿物表面对溶液中流体包裹体组分及捕收剂分子的吸附，形成了以"流体包裹体组分释放 – 矿物表面弛豫 – 流体包裹体组分吸附"为核心的硫化矿物流体包裹体浮选理论；④形成了基于"金属离子配位调控分子组装"的浮选界面调控新理论，通过金属离子与配体的配位形成具有特定捕收能力的配位捕收剂，或通过金属离子的配位调控形成具有特定结构的胶束，为新型浮选药剂的开发提供了新的方向；⑤围绕菱铁矿、褐铁矿、极微细粒铁矿、鲕状赤铁矿等劣质铁矿资源的高效开发与利用，开展了大量的基础研究和技术开发工作，基本达成了采用选冶联合工艺才能实现劣质铁矿资源高效利用的共识，形成了流态化磁化焙烧和深度还原等新技术理论；⑥在矿物加工计算化学领域，开展了浮选药剂分子模拟计算和设计、矿物浮选界面结构及相互作用的分子动力学模拟计算、药剂分子与矿物表面的对接技术等研究；⑦借助数值试验方法进行选矿设备研发，能够显著减少设备物理模型的制作，大大降低了研发成本，提高了研发效率，同时利用数值试验方法进行矿物分选理论的研究也有所进展；⑧开展了非均衡柱式分选过程及强化理论研究，从颗粒气泡的碰撞出发，研究了气 – 固 – 液三相体系中影响颗粒气泡碰撞矿化及分离的流体特征及颗粒界面特性，提出了从微观"涡"强化、界面调控到宏观流场强化的非均衡强化过程。

在技术开发和成果转化方面，针对矿产资源开发利用过程中的突出问题开展重点研究，在选矿工艺技术、选矿药剂、选矿装备研发及应用、矿产资源综合利用、选矿自动化、矿物材料研发和深加工等方面均取得了突破性进展，形成了一批具有重大影响力的新技术、新成果。①选矿工艺方面，开发了超大规模微细粒复杂难选红磁混合铁矿选矿技术、复杂难处理钨矿高效分离关键技术、基于页岩钒行业全过程污染防治的短流程清洁生产关键技术、有色金属共伴生硫铁矿资源综合利用关键技术、多流态梯级强化浮选技术等高效选矿工艺技术，显著提高了我国有色金属、黑色金属、煤炭与非金属资源的开发利用水平；②选矿药剂方面，成功开发了一系列新型环保绿色特效药剂，利用硫化矿新型高效捕收剂合成技术发明了新型硫氨酯等高效铜捕收剂并在大型铜矿山应用，针对各类型铁矿与不同脉石矿物研发了系列新型铁矿浮选药剂；③选矿设备方面，碎磨与分选设备的研究设计和制造取得重大突破，Φ12.19 米 × 10.97 米大型半自磨机、Φ7.9 米 × 13.6 米球磨机、320 立方米大型浮选机在国内外矿山推广应用，超大型粗粒磁选机成功应用于贫磁铁矿抛尾；④选矿自动化方面，目前国内大型先进选矿厂碎磨作业的工艺参数基本实现了自动控制，选别作业中浓度、粒度和品位的在线检测系统逐步应用，全流程优化控制技术的研究也取得了一定进展；⑤依靠工艺技术进步与装备水平提升，我国近年建成了鹿鸣钼矿、乌奴格吐山铜钼矿、甲玛铜多金属矿、袁家村铁矿、鞍千铁矿等技术先进、装备优良的大型、特大型现代化选矿厂，整体水平显著提升。

三、本学科国内外研究进展比较

我国与世界其他矿物加工强国在研究重点上有所区别，我国更重视先进技术的开发与应用，国外更多进行微观基础研究；在全球范围内，我国矿物加工学科总体上处于"工艺技术领跑、装备自动化并跑、基础研究跟跑"的地位。

面对以贫细杂为特点的资源劣势，中国矿物加工领域面临着许多国外并不存在或尚未涉及的复杂难题，并从政府的宏观引导层面和企业的实际需求方面，投入了大量技术研发力量与研发经费。在多年的努力下，成功开发了如流态化磁化焙烧 – 磁选技术与装备、低品位铜矿绿色循环生物提铜关键技术、复杂铅锌硫化矿高浓度分速浮选新技术、复杂难选低品位黑白钨共生多金属矿高效选矿新技术、适合处理中低品位硫化 – 氧化矿资源的选冶联合流程等特色技术，选矿工艺总体水平已达到了世界先进水平，部分选矿工艺技术水平处于世界领先水平。

国外注重矿物加工基础理论的研究，在矿物颗粒碎磨过程中的力学分析，分选过程中运动轨迹及力场分析，气泡 – 颗粒相互作用，药剂与矿物作用机理分析都有更加科学和深入的研究。国外也重视新型矿物加工方法，如湿法浸出、微生物选矿、细颗粒三维浮选技术工艺的机理研究，为此研制新型选别设备，强化选别过程，提高分选效果。中国在矿物加工基础理论方面同样进行了大量研究，但与国外侧重点不同，一般针对具体矿种进行应用基础研究，与资源开发实践联系相对紧密，但对微观层面研究较少；在计算机数值仿真与模拟等研究应用方面明显不足。

国外矿物加工设备发展较早，已经形成较大的规模，设备的技术性能和自动化水平较高，在众多大型矿山获得广泛应用，这些技术主要掌握在美国、俄罗斯、英国、德国、日本、澳大利亚、瑞典等发达国家。中国矿物加工设备起步较晚，但是发展迅速，与国外发达国家的差距正在缩小，如大型浮选机和磁选机已达到国际领先水平。但中国整体技术水平与国外相比还有较大差距，创新能力有待提高。

四、本学科发展趋势和展望

未来矿物加工技术的发展，需要更好地满足社会和行业需求，更加注重科研和教学过程的智能化，开展更深入的微观领域基础研究，应用领域更加多样化，同时需要知识背景更为丰富的复合型人才形成支撑。主要有以下发展趋势。

1）矿物加工技术发展更好地满足社会和行业需求。社会和行业发展对矿物加工技术的需要包括：尽可能提高资源回收率和综合利用率，降低生产成本，降低水资源和能源消耗，减少污染物排放，提高过程可靠性等。基于以上需求，矿物加工技术总体发展趋势可概括为：绿色、高效、节能、低耗、低排放、自动化和智能化。此外，矿物加工技术的发展需要从采、选、冶、化工、材料全流程考虑能量和资源消耗，建立"资源开采－选矿－冶金一体化"流程思想，加强行业沟通，获取经济效益与资源利用率最大化，是矿物加工技术发展的重要方向。

2）矿物加工科研和教学过程的智能化。当前，CFD、FEM、DEM等数值试验方法已广泛应用于矿物分选理论研究和矿物加工设备研发过程，信息技术在矿物加工科研和教学过程中的作用日益凸显，矿物加工科研和教学过程的智能化将是未来矿物加工学科发展的重要趋势之一。"互联网＋教育"和"互联网＋矿物加工"将在矿物加工科研和教学中发挥重要作用，大数据技术将为矿物加工科学研究工作提供更多便利，虚拟选矿厂将在矿物加工教研和生产管理中扮演重要角色。

3）先进检测技术和试验方法的发展促进微观领域基础研究的深入开展。矿物加工学科近百年的发展历程中，技术进步带动理论的发展是长久以来的一大特点，由于缺乏有效的研究手段，许多矿物加工分选过程中的机理尚未得到统一认识和完美解释。随着科学技术的进步，先进检测技术和试验方法的发展，将促进矿物加工科技工作者深入开展矿物加工基础研究，解析矿物加工分选过程，实现矿物加工领域的理论指导实践。

4）矿物加工技术应用空间更加广阔。循环发展是未来社会发展的重要模式，党的十八大报告中首次将绿色发展、循环发展、低碳发展并列提出。欧洲、日本等发达国家已经建立起了具有循环经济发展特色的经济模式，中国循环经济还处于起步阶段。矿物加工技术可根据待处理物料的物理、化学性质的不同，采用不同的方法进行物料分离与富集，为循环经济的发展提供技术支撑。此外，矿物加工原理可推广应用于其他领域，从而拓宽矿物加工学科领域。如高梯度磁选用于医学上红细胞的分离，生物学中离子的分离，核工业中核原料放射性固体的分离，超导磁选机分离液态氧、氧，浮选法从纸浆废液中回收纤维素，从废纸上脱油墨、脱炭黑，废旧塑料的回收，医药微生物方面分选结核杆菌与大肠杆菌等。

5）矿物加工学科未来发展需要大量复合型人才。长期以来，由于受到培养理念和传统教育模式的影响,中国矿物加工高校在专业设置、课程安排、教学计划等方面过于强调专业型人才的培养,人才专业面偏窄、适应能力不强等问题已无法满足行业发展对复合型专业人才的需求。矿物加工行业未来绿色化、高效化和信息化的发展趋势对行业人才提出更高要求。随着经济全球化、中国"一带一路"战略的实施及国内矿业集团公司国际化发展的需求,急需一批具有国际化视野的复合型高端人才。

6)中国矿产资源开发领域面临资源保障不足和需求持续增长的矛盾、资源过度开发和环境有限承载的矛盾、产业升级滞后和科技高速发展的矛盾,矿物加工学科发展机遇与挑战并存。

人类对矿产资源需求量继续增加,而优质资源的持续消耗、环境硬约束日益强化及现有技术的局限性,矿物加工学科的发展面临诸多挑战。一是国内矿产资源供需矛盾日益加剧,对资源高效节约开发利用技术的需要更加迫切;二是矿产资源禀赋差的情况不可逆转甚至更趋恶化,对复杂难处理资源综合开发利用技术的需求更加迫切;三是建设美丽国家对生态环境保护提出了新的更高要求,对矿产资源的绿色、低碳开发的需求更加迫切;四是供给侧结构性改革和产业转型升级,对发展节能低耗的矿物加工技术和装备的需求更加迫切;五是信息时代的行业转型升级,对矿物加工智能化发展的需求更加迫切;六是选矿下游产业的升级换代对更高品质矿物加工产品的需求更加迫切。

数字化、智能化是 21 世纪最重要的新兴技术,发展智能矿业有利于提升产品质量、提高生产效率和降低生产及运营成本,有利于在未来国际竞争中求得更大的生存和发展空间。以数字化、智能化技术应用为切入点,加快信息技术与传统矿物加工技术的深度融合;通过科技创新,推进矿物加工行业技术进步及产业升级,抢占国际矿业行业竞争的制高点,谋求未来发展的主动权是中国矿物加工学发展的当务之急,也是促进中国从矿业大国走向矿业强国的战略要求。

在"中华民族伟大复兴"的目标指引下,国家相继提出了"一带一路""新型城镇化建设""中国制造 2025""互联网 +"等重要战略部署,对国内外资源开发、产业布局、技术装备能力、数字化与智能化水平等提出了新的要求,作为矿产资源开发和为众多行业提供各类矿产品的重要学科,矿物加工学科肩负着使命,在资源高效节约开发利用技术、绿色矿物加工技术、智能矿物加工技术的持续创新方面,在推动行业科技进步、促进行业转型升级、为国家重大战略部署的实施做好服务方面必将会有更大的作为。

第十三节 稀土科学技术

一、引言

稀土是世界公认的发展高新技术、国防尖端技术、改造传统产业不可或缺的战略资源。我国稀土储量居世界第一，60多年来，针对国内稀土资源特点，开发了一系列先进的稀土采选冶工艺技术，并在工业上广泛应用，建立了较完整的稀土工业体系，成为世界稀土生产、出口和消费大国，在世界上具有举足轻重的地位。稀土磁性材料、发光材料、催化材料及储氢材料等稀土功能材料随着稀土在高科技领域的应用研究不断取得重大突破。稀土材料的应用越来越广，产业发展日益蓬勃，稀土材料科学技术取得了长足的进步，部分技术已达到或超过世界先进水平，有力推动了相关行业或产业的快速发展。

二、本学科近年的最新研究进展

离子吸附型（风化壳淋积型）稀土矿是宝贵的中重稀土资源，其中稀土以水合离子或羟基水合离子吸附在黏土矿物上。近期研究主要集中在风化体系稀土迁移富集的赋存状态、矿床矿石特征和稀土配分变化的成矿规律、稀土矿浸取过程动力学和传质机理，研究开发出生态环境友好型浸取剂和浸取技术。

我国在稀土资源提取、稀土分离提纯技术及产业化方面在世界上具有绝对优势。近几年来，主要针对稀土冶金过程存在的三废污染、资源综合回收利用率较低等问题，在稀土绿色分离化学基础研究、稀土冶炼分离新方法、新技术、新工艺及装备等方面开展了大量研究并取得突破性进展。研发出新型萃取剂及萃取分离体系；研发成功离子型稀土原矿浸萃一体化新技术、碳酸氢镁法分离提纯稀土原创技术，并获得工业应用，从源头解决氨氮、高盐废水及放射性废渣的污染问题；突破超高纯稀土金属关键制备技术及装备，获得了绝对纯度＞99.99%的Gd、Tb和Dy等10多种超高纯稀土金属。

稀土分析化学学科在稀土冶金分析中新型质谱技术的应用、稀土在线检测、稀土固体废物分析、分析标准、生物环境样品的前处理方法和稀土元素分析、稀土元素与精神疾病的关系、疾病生物标志物的稀土元素标记等方面的研究有了较大进展。

稀土永磁产业不仅是稀土应用领域发展最快、规模最大的产业，也是最大的稀土应用领域。近年来，我国在高性能稀土永磁材料、高丰度稀土磁体、重稀土减量化技术、新型热压/热变形技术等方面均取得较大进展。高性能稀土永磁材料综合磁性能（最大磁能积MGOe加矫顽力kOe之和）达到75，高丰度稀土（Ce）磁体年产量超过3万吨，La/Ce添加量27.1wt.%的磁体磁能积达48.9MGOe；DyCu压力扩散法制备的热变形磁体磁能积可达53MGOe。

特种稀土磁性材料主要分为磁致伸缩材料、磁制冷材料和微波磁性材料等。在超磁致伸缩材料方面，主要是开发低场下大磁致伸缩，同时具有高的机械强度、窄滞后的新型磁致伸缩材料。其中TbDyFe合金的磁致伸缩性能，与准同型相界、材料的取向密切相关，而在准同型相界处具有的高磁致伸缩性能与其微米畴内存在大量自发磁化强度方向一致的纳米畴有关。磁有序伴随的晶胞负热膨胀是导致$LaFe_{13-x}Si_x$化合物呈现大磁熵变的原因，其最大磁熵变大于先前报道的$Gd_5(Si_xGe_{1-x})_4$系化合物，$LaFe_{11.4}Si_{1.6}$化合物在相变温度TC附近晶胞参数呈现约0.4%的负热膨胀。随着信息技术的飞速发展，对高频软磁材料的要求也从传统块体延伸到薄片化、结构化及磁性/非磁性复合材料，其中具有平面各向异性的稀土–3d金属间化合物，饱和磁化强度可以达到1.5特斯拉。

催化技术可以产生巨大的经济效益和社会效益，据估计催化直接或间接贡献了世界GDP的20%~30%。加强稀土催化作用机理研究，拓展镧、铈等轻稀土在石油化工、环境治理中的应用，不仅有利于我国稀土资源的高效利用和稀土产业结构的优化调整，而且还有利于减缓日益突出的贵金属资源供需矛盾，给新型催化材料带来了转机。针对现有VOCs净化催化剂制备成本高、普适性和抗中毒能力较低等方面的不足，通过催化剂组分调节、载体优化及形貌设计，降低干扰组分在催化剂表面的沉积速率，提升催化剂的耐受性；通过不同催化模块的组合，优化配置C–H和C–Cl断链初级氧化及高效氧化等功能模块，降低贵金属使用量、提高催化剂的使用效率和对多污染物体系的适应性。

稀土光功能材料是稀土应用的主要领域之一。近年来，新研究成果、新技术和新应用不断涌现，带动了传统稀土光功能材料及其终端产业的更替与升级，推动了新的稀土光功能材料及其新产业的出现。目前，白光LED已经成为市场主流照明方式。在新材料研发方面，氮化物荧光粉仍然是基础和应用研究的主要热点和前沿。大功率LED芯片发展，带动了稀土荧光陶瓷成为发光材料研究的另一个热点，研发出系列光色可调的材料体系。稀土激光晶体领域在生长技术、功能复合、激光输出性能、器件设计等研究方向均取得了可喜进展，其中，研制了口径Φ150毫米以上Nd，Y：SrF_2激光晶体，峰值发射截面5.3×（10~20）平方厘米，小信号增益系数达到2.1倍。

稀土储氢材料经过多年的体系创制、成分优化、结构调制和表面改性研究，性能逐年提高，在电化学储氢和气固相储氢等方面得到大量应用，具有易活化、储氢容量较高（理论容量为372mAh/g）、动力学性能好等优点的AB5型稀土储氢合金应用已非常成熟，在新能源、新能源汽车和节能环保领域受到人们的日益关注。AB 3~3.8 新型稀土镁基储氢合金的研究突破了国外专利限制，克服低循环寿命、高粉化率、高自放电等问题。

现代稀土陶瓷材料体系是以稀土离子的光学和磁学为主的各种性质作为出发点，利用材料基因设计、材料计算、新型表征技术等先进手段，其内涵既包括结构－性能关系，还涉及陶瓷制备工艺以及陶瓷表征技术。随着高端稀土陶瓷性能的不断提升和新型功能稀土陶瓷的开发力度不断加强，对稀土陶瓷粒径分布、晶粒尺寸、产品一致性差提出了更高的要求。

在第二代高温超导工业化长带制备方面，以单根长度、77 开尔文下的单位宽度临界电流值为生产水平的主要标志，已有多家公司先后成功研制出长度超过 100 米，最长达到 1000 米且能够传输数百安培以上超导电流的第二代高温超导带材。采用顶部籽晶熔融织构生长法（TSMTG）制备的 30 毫米直径 YBCO 单畴块材，77 开尔文下的磁通俘获场达到 0.7 ~ 0.9 太（T），少数可以达到 1.2 太（T）。

稀土具有一定的球化作用和良好的抗微量反球化元素的干扰作用，是球墨铸铁生产使用的球化剂中的重要组成元素；以稀土作为主要成分的蠕化剂，利用稀土元素的弱球化作用原理，拓宽了获得蠕墨铸铁的工艺范围，使之在工程条件下也能稳定地获得蠕墨铸铁件。我国研究成功具有高强度（$\sigma b \geq 1000$ 兆帕，最高可达 ≥ 1600 兆帕）、高韧性（$A \geq 11\%$）的奥铁体球铁（ADI），具有动载性能高（弯曲疲劳强度达 420 ~ 500 兆帕、接触疲劳强度达 1600 ~ 2100 兆帕），已在汽车、柴油机、拖拉机和工程机械的齿轮、曲轴和各种结构件中应用。

稀土在有色合金中的应用分散在金属加工与制造的各个领域。可降解医用镁合金被誉为"革命性的金属生物材料"，是许多发达国家竞相发展的研究方向，稀土在其中发挥着重要的作用。近些年快速发展起来的能源汽车产业，也将是稀土镁合金产品的重要应用领域。稀土在铝合金中主要用于微合金化和变质处理，应用在新型的电工铝合金和过共晶型铝硅合金中。我国研制含稀土的钛合金包括 Ti-55，7715C，Ti633G，Ti-60，Ti600 等，已经形成一系列具有中国特色的高温钛合金。

新型稀土功能助剂在聚氯乙烯、天然橡胶、聚氨酯橡胶、尼龙、涂料等高分子材料中获得多项自主知识产权，部分研究成果已经实现产业化，打破了高端化工助剂由国外垄断的局面。高纯稀土基靶材在大规模集成电路、信息记录与存储、表面防护、平板显示及材料的表面包覆改性等领域具有广泛应用前景，其中 Zr-Co-RE 吸气薄膜用靶材、铝－高熔点稀土合金（Al-Sc、Al-Y、Al-Er 等）靶材、300 毫米晶圆制造用高纯 La_2O_3 靶材等实现规模应用。

三、本学科国内外研究进展比较

我国稀土矿种矿床多，稀土矿物加工水平居世界第一。风化型稀土矿是我国最早工业利用的离子型稀土矿，近几年国外相继在缅甸、巴西、智利和马达加斯加发现，但国外尚无可应用的开采技术。

我国在稀土湿法冶炼分离工艺技术优势明显，围绕包头混合型稀土矿、氟碳铈矿、南方离子吸附型稀土矿和伴生独居石矿，形成了特色湿法冶炼分离生产工艺，特别是近几年在稀土资源高效绿色提取分离新技术方面取得较大进展。国外在冶炼分离工艺方面的研究投入相应很少，缺乏冶炼分离方面的人才队伍和开发能力，主要从中国寻求技术支持。

近年来，国内质谱技术在稀土分析领域的应用增长迅速，我国政府对高端仪器设备研发加大投入力度，已有多个厂商推出国产 ICP-MS 仪，但性能有待进一步提升。在现行的 ISO 国际标准和美国 ASTM 标准中，尚未建立针对稀土产品及检测的文字标准或实物标准。目前在国际贸易中，贸易双方多采用我国现行国家或行业标准分析方法。

稀土永磁材料，国内外相关的工作均集中在高综合性能烧结钕铁硼以及重稀土减量化、高性能粘接磁体、新型热压/热变形磁体等面向产业应用。新型的热变形技术由于具有获得高性能高矫顽力磁体的潜力，已成为国内外关注的焦点。高丰度稀土的平衡利用技术结合了我国独特的稀土资源优势，也是我国的重点方向。

目前，国外研究主要围绕以 TbDyFe 为代表的稀土超磁致伸缩材料的应用研究，国内对 TbDyFe 材料的磁致伸缩机理及取向进行了深入探索，并在新材料体系开发方面，走到了国际前沿；La-Fe-Si 系列是国际上普遍认为具有重要应用前景的磁制冷工质之一，也是我国具有自主知识产权的稀土功能材料；我国最先开始研究的稀土高频软磁材料具有自主知识产权。

国际汽车催化剂市场长期被巴斯夫、庄信万丰、优美科等国际公司把控，中国汽车催化剂产业经过十余年的努力取得了长足的进步，大大缩短了与国际的差距，但在技术积淀、品牌知名度及前瞻性技术开发等方面与国际水平还有较大差距。我国在重质油裂解用 FCC 催化剂的研究开发方面已处于世界领先地位，Grace-Davison、BASF 等公司在降低汽油硫含量 FCC 催化剂方面具有一定的优势。国外用于 VOCs 净化的催化剂技术相对成熟，国内在制备催化剂所需的关键材料、装置自动化和精确控制等方面与国外存在较大差距。

围绕白光 LED 用发光材料、大功率及激光用光功能材料等领域不断推出市场化产品，然而在具有良好应用前景的 OLED 材料、稀土激光晶体等领域，发展相对缓慢，部分原始专利仍然被发达国家所掌控。而围绕稀土上转换发光和长余辉发光纳米材料及其生物应用，稀土配合物及其在太阳能电池应用等领域，相关基础研究工作与国外同步。

国内对于 AB5 型稀土储氢合金和 AB 3~3.8 型稀土镁基储氢合金，主要向高性能化和低成本化方向发展。近两年，稀土镁基储氢合金的研发方面虽然取得较大进展，合金性能优异，但鉴于日本专利的限制在生产和应用上受到了阻碍。日本公开的专利多涉及材料结构，而我国研究领域还主要局限在合金组分的控制上，研发进度落后于日本。

我国在闪烁晶体及陶瓷制备和产业化已在国际上占有一席之地，可通过第一性原理对闪烁材料进行能级设计与导带结构性能调控，并获得了能实现闪烁性能提升的最佳稀土掺杂的方案，但新材料体系创新设计与国际先进水平仍有相当大的差距。

二代高温超导带材发展可分为 MOD、MOCVD 和 PLD 三种技术路线，并且都得到了较快发展。国内 PLD 技术路线的典型代表企业是上海超导科技，是世界上具有年生产 100 公里以上高性能二代高温超导带材能力的五家生产商之一，也是世界上唯一能够自主设计和制造全部二代高温超导带材千米级生产装备的企业。而在稀土超导块材研究方面，国内超导单畴的应用研究相对滞后。

稀土铸铁材料方面，国外已解决了大断面（壁厚 ≥ 120 毫米）球铁件中心部位的石墨畸变和组织疏松等问题，制作了包括重达百吨的大型球墨铸铁核燃料储运器在内的各种重、大、特型球铁件。我国应用钇基重稀土与计算机凝固模拟技术已生产多种大型球墨铸铁件，但大型核燃料储运器尚未问世。

用作航空航天材料的高性能稀土镁合金，其开发模式正在发生转变，研发被整合进入更大的研究发展框架。我国正加大资源整合，把研究、生产、市场调研和使用部门联合起来，完成镁合金从初级产品到应用，从废品回收到循环再利用，实现全周期绿色开发。

国内关于稀土助剂的技术成果和产品应用成果不断涌现，处于国际先进或领先的地位。全球高端稀土基靶材制造商主要集中于美国、日本、德国等地，占领着全球相关行业用稀土基靶材的主要市场。国内稀土基靶材开发滞后于产业发展需求，迫切需要开发稀土基靶材的精细化加工技术。

四、本学科发展趋势和展望

随着《中国制造 2025》、战略性新兴产业发展战略的实施，对稀土材料提出了更高的要求，稀土材料发展也面临新的机遇和挑战，重点发展趋势和发展重点如下：①新型稀土材料研发：挖掘稀土本征特性、探索新功能材料、开拓新的应用领域；②先进稀土功能材料工程化、产业化制备技术研发，实现连续化、智能化、大规模稳定生产，不断提高材料性能，满足高端应用器件需求；③稀土基础材料绿色高效制备工艺研发，实现高纯化、高值化、绿色环保、平衡利用。

第十四节 材料腐蚀

一、本学科近年来的最新研究进展

（一）腐蚀电化学最新研究进展

腐蚀电化学是腐蚀与防护学科的理论灵魂，是腐蚀与防护学科的主体部分。我国从事腐蚀电化学研究的主要单位有北京科技大学、浙江大学、厦门大学、武汉大学、天津大学、中国海洋大学、南昌航空大学、中科院金属研究所、哈尔滨工业大学、华中科技大学和哈尔滨工程大学所属的腐蚀研究团队。近年来的研究热点主要集中在具有空间分辨的微区腐蚀电化学、腐蚀电化学噪声解析与腐蚀行为关系、薄液膜腐蚀电化学、纳米尺度腐蚀电化学和新型、极端环境腐蚀电化学等 5 个方面。

（二）局部腐蚀机理研究最新进展

局部腐蚀机理研究是腐蚀与防护学科研究前沿关注的另一个重要方面。厦门大学林昌健团队发展了国产扫描电化学腐蚀探针测试技术；左禹团队对点蚀的起源与生长机理和动力学有系统研究；乔利杰课题组发现氢对应力腐蚀的关系研究贡献很大；李晓刚团队和杜楠团队对点蚀的生长和发展进行了模拟仿真与激光在线测试。我国关于局部腐蚀的研究发展很快，在国际学术界发表了较多论文，逐渐产生了影响，但在研究的理论深度和研究方法的创新性方面仍存在进步空间。

（三）高温腐蚀研究最新进展

近年来，我国学者在高温腐蚀的机理研究和防护涂层的设计、研制和性能研究方面都卓有成效地开展了工作，取得了有特色、有国际影响并被国际同行关注的一系列成果。代表性研究群体为王福会研究员团队、何业东教授团队、徐惠彬和宫声凯教授团队、陆峰研究员团队和彭晓团队，有针对性设计并研制了不同组成和结构的纳米晶和超细晶涂层、开展了相关涂层的高温腐蚀性能研究，发现细结构利于生长黏附性强的保护性氧化膜，这项研究在国际上具有特色创新性。

(四) 自然环境腐蚀研究最新进展

在自然环境腐蚀硬件设施建设方面，近年来取得了长足进步。李晓刚团队领导的国家材料环境腐蚀野外科学观测研究平台完成了对我国材料环境腐蚀试验站的整合，完善了大气、土壤、水环境28个国家级腐蚀野外试验站的建设，并进行了中华人民共和国成立以来最大规模的材料投试，共投试试样5万多片，为材料环境适应性研究工作的开展奠定了坚实的基础。在数据库建设和数据共享服务方面，搭建公益性网络平台"国家材料环境腐蚀数据共享与服务网"，提升了资源的保存和利用水平。建设的材料环境腐蚀数据库系统包括：由43种环境因素和材料腐蚀（老化）数据表所构成的原始数据库、面向数据共享与应用服务的环境因素和材料腐蚀（老化）数据库、材料环境腐蚀图谱数据库及材料腐蚀信息资源数据库。这一大型材料环境腐蚀数据库的建立也标志着材料环境腐蚀站网数据共享体系的初步形成，也是到目前为止我国材料腐蚀领域种类最全，数据量最大的共享数据库。近5年来共计获国家科技进步奖二等奖1项，省部级科技进步奖一等奖6项，国际大奖2项；在国外发表论文200多篇，这些都极大地提高了我国在自然环境腐蚀方面的研究地位，标志着我国已经进入自然环境腐蚀研究强国之列，而且这方面还保持着很好的发展势头。他们在 *Nature* 杂志上提出了"腐蚀大数据"概念和技术流程，产生了广泛的影响力。

(五) 防腐蚀技术研究最新进展

近年来，国内在材料耐蚀性和新型耐蚀材料方面均做出了一些新的研究成果，具有一定的科学性和广泛的应用性。主要体现了新型不锈钢研发、新型耐蚀结构钢的研发、高耐磨耐腐蚀改性超高分子量聚乙烯特种管材的研发、铝合金表面制备没有任何疏松层的氧化铝膜、镁合金表面制备的微弧氧化膜、发展出高性能的空间技术用润滑薄膜、利用激光熔覆技术在铜合金表面制备出了高硬度、高耐磨、抗热震、与基体冶金结合的金属陶瓷复合涂层、采用热分解法制备了IrTa混合金属氧化物（MMO）涂层钛阳极、设计出了以富含氧、氮、磷杂环化合物和醇/胺缩合物为主的新型阻锈剂等技术并实现产业化。我国不锈钢产量2017年已达2500万吨，尤其是高品质不锈钢品种开始厚积爆发；耐蚀耐候低合金结构钢品种日益增加，产量接近低合金结构钢总量的20%。

(六) 腐蚀试验研究方法与设备最新进展

近年来研究者们开发出了多种类型的扫描探针显微技术，如扫描离子电导显微镜SICM、扫描热显微镜STHM、扫描声学显微镜SAM、光子扫描隧道显微镜PSTM和扫描近场光学显微镜SNOM等技术；国家材料环境腐蚀平台实现了基于大数据和无线传输的连续动态监视材料腐蚀状态，发展了标准化的技术；电化学噪声技术、丝束电极技术、原子断层分析技术、激光电子散斑干涉技术也在腐蚀研究领域得到广泛应用。这些新技术的涌现并在腐蚀科学领域应用发展，极大地拓展了腐蚀研究的技术手段，发掘了大量腐蚀新机制新理论。

（七）腐蚀设备开发进展

腐蚀评价环境试验设备由传统的盐雾试验箱、湿热试验箱的开发，转向模拟飞溅区腐蚀、深海腐蚀、潮差区腐蚀等更加复杂的腐蚀环境及多种腐蚀因子耦合作用的腐蚀环境。李晓刚教授开发了系列化的自然环境模拟装置，包括大气腐蚀模拟与加速、土壤腐蚀模拟与加速、海洋腐蚀模拟与加速和深海环境模拟与加速装置；石油化工原油腐蚀模拟与加速试验装置应用于石化企业科研工作中；他们与苏州热工研究院共同开发的核电腐蚀大型流程装置已经成功进行了设计工作，进入实际组装阶段。北京科技大学、浙江大学和青岛理工大学利用干湿循环的原理，专门研制了模拟潮差区环境的试验设备。路民旭教授、中国石油大学陈长风教授研制了大量各种类型的高温高压腐蚀设备，用于石油工业硫化氢和二氧化碳腐蚀研究。北京航空材料研究所研制了 8 因子加速试验装置，已经成功应用于 18 种标准腐蚀试验中。中国兵装第 59 研究所研制了与国外水平相当的大气腐蚀室外加速试验装置。北京科技大学负责承担的国家大科学工程计划项目，建设流体类环境结构材料试验装置、大气类环境结构材料试验装置、多相流环境结构材料试验装置、高温高压水汽环境结构材料试验装置、极端 – 多因素耦合环境结构材料试验装置、自然大气环境结构材料试验装置、特殊地域环境结构材料试验装置、力学 – 化学多场耦合环境结构材料试验装置等 8 套大型装置，开展大尺寸构件和设备的环境失效试验，这些装备已经完成设计，进入了实施制造安装阶段。以上工作将大大提升我国腐蚀与防护测试和试验的能力。

二、学科在学术建制、人才培养、基础研究平台等方面的进展

经过 30 余年的建设和发展，本学科在学术建制、人才培养、基础研究平台建设方面已经基本形成了高等学校人才培养，国家重点实验室、一批基础研究和应用基础研究的高水平实验平台及高校科研平台开展基础研究，企业研发中心进行应用研究的多层次互相交叉的框架模式。在教育部的学科培养建制上，原来是隶属于材料科学与工程的二级学科，可以进行本科专业招生。从 20 世纪 90 年代调整为隶属于二级学科材料学的三级学科，这一调整对腐蚀与防护学科产生了巨大的负面影响。首先，冲击了已经形成腐蚀与防护学科的本科教学体系，以北京科技大学腐蚀与防护学科为例，这一专业 20 世纪 80 年代是唯一的高校博士学位授予点，以每年 60 人的规模连续培养了 15 届本科毕业生，这些学生大部分已经成为我国各行各业腐蚀与防护工作的顶梁柱，甚至成为国内外有重要影响力的学术带头人。可惜的是，由一代学者努力建设的教学学科体系，随着这一调整，一夜之间付之东流，全国将近 20 所高校的情况基本相同，以湖南大学的情况最为惨重，湖南大学的腐蚀专业曾经培养了一大批国内外有影响力的学者，但是腐蚀与防护专业在湖南大学短期内

无法恢复到原来的水平，影响力逐渐下降，可见学科调整对腐蚀与防护学科的冲击是巨大的且极其负面。其次，使专业人才培养水平明显下降。据不完全统计，我国目前有 30 多所院校按照材料学专业或应用化学专业的腐蚀与防护专门化培养本科生，每年在 1500 人以上，近几年由于对腐蚀与防护专业学生的大量需求，腐蚀与防护专业学生供不应求，以西南石油大学为例，该校每年培养腐蚀与防护专门化的学生 200 多人，但是，由于是三级学科，许多专业性极强的课程被裁减，使得学生专业知识水平明显下降。事实证明，在教育部的学科目录上，将腐蚀与防护学科由二级学科调整为三级学科是有较大负面影响的，目前这种本科生培养模式无法满足社会对腐蚀与防护人才的需要。由于这种学科调整没有直接影响到研究生培养模式，我国腐蚀与防护方向的硕士研究生、博士研究生的培养一直呈平稳增长态势，目前规模大致是每年毕业硕士生在 300 人左右，博士研究生在 50 人左右。应该指出，腐蚀与防护二级学科取消后，由于社会存在大量的需求，除了大部分原有腐蚀专业的学校继续培养这方面人才外，实际上还有不少高校新开办了培养腐蚀防护人才的专业，如南昌航空大学、西南石油学院、长春工业大学、重庆理工大学、燕山大学、中国民航大学、上海电力学院等。但由于都依托于其他专业，学生专业知识不足。因此，专业名称调整与国家经济、社会、国防等对防腐蚀人才的需求是不适应的。

在腐蚀科学研究平台方面，国家腐蚀与防护重点实验室的停办，对学科发展产生了极大的负面作用，目前国内的腐蚀学科带头人，乃至国际上部分有影响力的腐蚀学者，多为该国家重点实验室所培养。好在以北京科技大学腐蚀与防护中心（含教育部腐蚀与防护重点实验室、教育部环境断裂重点实验室、北京市重点实验室）、中科院沈阳金属所、中船重工七二五所海洋腐蚀与防护国防科技重点实验室和国家材料环境腐蚀野外科学观测研究平台等一系列高水平实验研究平台不断发展，加之近年来武汉材料保护研究所的特种表面处理企业国家重点实验室和中国电器科学研究院的环境适应性企业国家重点实验室的设立，承担和完成了大量的国家科技研究计划项目，产出了一批创新性研究成果，解决了经济建设、社会发展和国防建设的许多重大关键技术问题，取得了显著的社会经济效益，在人才培养和队伍建设中发挥了不可替代的作用。

三、本学科的最新进展在产业发展中的重大应用、重大成果

近年来，腐蚀学科的重点实验室、高校基础研究平台及腐蚀科技工作者围绕国家目标和着眼于国家需求，紧密围绕"海洋开发""一带一路""中国制造 2015"等国家重大战略中耐蚀材料和技术的需求，解决了国民经济建设、社会发展和国家安全建设中的许多基础性、关键性技术难题，在我国社会经济发展和国防建设中做出了重要贡献。在耐蚀材料、表面涂装、表面处理、防锈油、缓蚀剂、电化学保护等各个领域都取得了丰硕的研究成果，防腐材料产业已经有了相当规模。

材料（制品）的腐蚀数据积累和材料腐蚀规律与机理研究满足了我国经济建设的迫

切要求。例如，在海洋开发战略中，海洋的基础设施建设，特别是深远海发展的需求，已被列为首要任务，根据工业发达国家的经验，基础设施建设中材料的选用要以材料（制品）在海洋典型环境中的腐蚀与老化数据作为重要依据。另外，西部沙漠（戈壁）与海拔3000米以上的高原环境占西部总面积的75%，青海、新疆有大片内陆盐湖干涸后的盐渍土，南疆铁路建设中由于没有混凝土在该地区土壤的腐蚀数据，曾被迫停工数月，后来根据实验室的短期试验结果，确定混凝土的选用方案和防护措施，才恢复施工。国家重大工程建设，如高铁建设、电网建设、大飞机、航天工程、南海建设、石油石化、核电、西气东输、青藏铁路、青海钾肥厂二期工程、南水北调、跨海大桥等也大量使用材料（制品）的腐蚀数据和在这些环境下的材料腐蚀规律与机理研究成果。近年来，我国工业形成了完善的门类体系，规模日益扩大，边界环境因素越来越复杂，不仅原有的工业环境出现了大量的腐蚀问题，而且新的具有强烈腐蚀性的工业介质环境不断出现（例如核电、航空、载人航天、石油、化工、高铁、海洋工程等），有的腐蚀问题制约了该工业体系的发展。腐蚀科技工作者为这些问题的解决提供了材料腐蚀学科的有力支撑。在近年的某载人航天发射工程中，出现了设备的腐蚀情况，腐蚀科技工作者与航天人协同攻关，短期内解决了问题，保证了按时发射，这是我国腐蚀与防护学科首次直接处理载人航天中的材料腐蚀失效问题。

材料（制品）的环境腐蚀数据积累和材料腐蚀规律与机理研究为节能、节材、减少因腐蚀造成的经济损失，指导材料的合理使用提供了科学依据。根据统计与调查，我国年材料腐蚀总损失高达20000亿元人民币。根据材料的环境腐蚀数据和腐蚀规律与机理研究结果，发展各种防护技术，指导和保证材料的科学使用，并采取相应的防护措施，可节约材料，节省能源消耗，减少腐蚀经济损失25%~30%。根据目前我国的防腐蚀水平，能达到上述范围的下限，则我国是每年约5000亿人民币的效益。同时，避免和减少腐蚀事故发生，延长设备与构件的使用寿命，有很好的社会效益和经济效益。

材料（制品）的环境腐蚀数据积累和材料腐蚀规律与机理研究为提高材料（制品）的使用性能和质量，促进新材料的研究开发提供了依据。材料（制品）耐各种典型环境腐蚀的数据是研究开发各种耐蚀材料的重要依据。我国建立的材料海洋大气与海水腐蚀试验站对海洋用钢的研究开发与发展做出了卓越的贡献。我们目前的耐蚀材料生产水平与国外相比，存在明显差距，严重制约了我国制造业的水平，各类设备与构件都存在服役寿命低、安全性差的问题，原因之一就是建立在微纳米层次上的腐蚀机理与规律研究不足。目前，腐蚀科技工作者正在努力工作，为材料生产企业提供这方面的研究成果直接用于生产。

近20年来，在我国由于冶金、化工、能源、交通、造纸等工业的发展，各种企业，特别是乡镇企业对自然环境的污染，不仅导致生态环境的破坏，同时使材料的腐蚀速率迅速增加，设备、构件、建筑物等的使用寿命大大缩短。我国局部地区雨水pH值已降低到3.2~3.8，导致普碳钢的腐蚀速率增大5~10倍；混凝土建筑物的腐蚀破坏大大加速。腐蚀科技工作者提供了大量的典型材料在不同污染自然环境中的腐蚀数据积累和主要影响因素的研究成果，为国家制定材料保护政策和环境污染控制标准提供了依据。

在国防建设方面，材料腐蚀与防护是为提高武器装备与型号的环境适应性而进行的材料腐蚀机理、材料腐蚀失效预测与评价、武器装备防护技术和工程化应用的研究与实践活动，是武器装备和型号发展所必需的共性技术，是保证和提高武器装备质量的重要环节，是武器装备全寿命周期工作的重要内容。现代武器装备和型号的发展，迫切要求从"事后补救"的"试验考核"模式，向开展"材料寿命预测、腐蚀防护主动性设计"的"事前预防"模式转变，"武器装备及其材料腐蚀与失效行为、武器装备环境效应预测与评价、武器装备防护技术"应成为与武器装备同步发展的重点。在我国，腐蚀与防护科技工作者在这方面发挥了重要作用，成为提高我国武器装备水平不可或缺的一支力量。

综上所述，开展材料腐蚀与防护学科的研究是国家经济建设和国防建设，科技进步和经济与社会可持续发展的迫切需要。持续深入开展本学科的基础性工作，有利于提高我国的材料与基础设施的整体水平，促进我国材料腐蚀实验研究基础性工作体系和防护技术工程体系的形成与发展，对国家建设、科技进步、技术创新，以及学科的进一步发展具有重要意义。

四、本学科国内外研究进展比较

本部分仅采用文献计量学的方法，基于前期统计的数据，进一步对 2011—2015 年间发表的腐蚀学科相关的 SCI 论文展开分析。过去五年间，中国取代美国成为发表 SCI 论文最多的国家，达到了 21.41%；美国从之前的第一掉落到第二，占总数的 15.42%，这说明随着我国经济的快速发展，我国的腐蚀学科也相应地步入了发展的快车道，进入世界第一方阵，开始了领跑与并跑。在过去五年间，腐蚀科学相关 SCI 论文发表数量逐年增加，2015 年发表的腐蚀科学相关 SCI 论文数达到 1500 篇。国内腐蚀科学相关 SCI 论文数的增长比国际腐蚀科学发展更加明显，表明近年来中国在国际腐蚀科学的发展中起到了越来越重要的作用。

目前，我国在腐蚀与防护领域的差距主要表现为：政府和企业各级领导重视不够；基础研究跟踪模仿为主，防护技术自主创新品种较少；某些关键材料和绝大部分腐蚀测试设备不能满足需求，仍然依赖进口；耐蚀材料质量稳定性不足，寿命不能保障；材料服役数据积累不足，缺乏完善的评价体系和标准规范。造成这些差距的原因包括：缺乏有影响力的基础性研究方向，基础研究与生产脱节较严重；关键共性技术壁垒严重；高精尖的测试设备研发严重不足；企业拔尖人才匮乏；协同创新体系亟待完善。对此我们提出了相应的解决对策：构建腐蚀与防护学科基础研究协同研发平台；构建腐蚀与防护应用技术协同研发平台；构建防护技术工程协同研发平台；构建以上三个平台大协作大交流的新机制；加强腐蚀与防护科普工作，推动腐蚀与防护的立法工作。

五、本学科发展趋势及展望

总之,在加强基础研究、保持领跑和并跑的同时,不断培植我国具有影响力的基础研究方向,是发展我国腐蚀与防护学科的基本原则。在此基础上,发展具有国际一流水平的耐蚀材料(特别是耐蚀钢铁材料)工程、表面处理与涂装防腐蚀工程、电化学保护防腐蚀工程、环境介质处理或工艺防腐蚀工程等领域中的新材料新工艺新技术新设备,不断提升我国腐蚀与防护学科的原始创新和应用能力,才是发展我国腐蚀与防护学科的根本目的。

第十五节 核技术应用

一、引言

核技术是研究如何将核科学研究中所揭示的原子核变化规律及其固有和伴随产生的物理现象加以实际应用的科学。核技术一般分为核武器技术、动力核技术、非动力军用核技术和非动力民用核技术。本报告中所述核技术应用主要指非动力民用核技术,分为加速器技术、放射性同位素技术、核探测与核电子学技术、核仪器仪表以及核技术在医学、农学、工业、社会安全领域的应用。

核技术应用作为综合性战略产业与制造业中的大部分行业有关,以其知识密集性、交叉渗透性、不可取代性、应用广泛性和高效益性受到各国高度重视,通过与其他产业交叉融合发展实现产业转型升级,不仅可产生巨大的经济效益,更具有不可估量的社会效益。当前,国际核技术及其应用日益呈现产业化、规模化和高科技化的特征,并仍然处于快速发展阶段,其增长速度始终高于世界经济的平均增长水平。特别是核药物和核医学的增长速度是近几十年未见的,年增长率达15%。发达国家核技术应用产业已具备相当规模并对经济社会发展产生显著的推动作用。

二、本学科近年的最新研究进展

(一)加速器

经过多年的发展,医用电子治疗加速器、同位素生产加速器、辐照加速器、无损检测加速器等应用型加速器技术日趋成熟,产业化水平也在逐步提高。2014年全球回旋加速器市值约1.52亿美元,2015年达到1.65亿美元,预计在2030年达到2.45亿美元。医用回旋加速器维护及升级市场收益逐步走高,预计在2025—2026年回旋加速器维护维修及升级的市场收益会高于出售回旋加速器的收益。质子和重离子治疗加速器成为国际放疗领域的研究热点,目前世界范围内共有约50台质子和重离子治疗装置,主要集中在美国、欧洲和日本等发达国家。国际上密封中子管的技术及指标已达到了较高的水平,并实现了小批量生产。离子注入机技术已日趋成熟并高度产业化。

我国医用直线加速器从1986年的71台到2015年的1931台,增长26倍,2006年以来数量翻了一番,而且在医用加速器研制生产上发展迅猛,能力比较全面,不仅能够开发所有医用能量档的电子直线加速器,在磁控管和新加速器研制上也取得了明显的进步。强流质子回旋加速器等指标和性能优越,我国成为少数拥有新一代放射性核素加速器的国家,是我国医疗发展特别是肿瘤治疗水平的重要标志。我国工业用电子辐照加速器的研发和生产以1~5兆电子伏特的中能辐照电子加速器为主,能量大于5兆电子伏特的高能辐照电子直线加速器技术也取得了突破性进展。为我国辐射材料改性、新材料研发、环境保护等领域的发展提供了更为有效的工具。专业化、规模化的电子辐照加速器生产基地初具规模,装备数量迅猛增长。加速器数量从2008年年底的140台/套发展到2016年的500余台/套,占全球电子辐照加速器总量2000台的近1/4,总功率超过3.5万千瓦,占全球总功率8.0万千瓦的43.8%,年增长率在20%以上,国产加速器的国内市场占有率达到80%以上。无损检测加速器性能得到国际认可,研制技术在一定程度上紧跟并引领国际上技术的发展,不仅满足了国内的需求,甚至远销世界各地,在国际市场上占有相当大的份额。

(二)同位素

由于国际上生产裂变^{99}Mo的反应堆基本都接近寿期,且对核不扩散的限制越来越重视,高浓铀生产裂变^{99}Mo的替代技术受到广泛关注,美国、荷兰、德国、澳大利亚、比利时、韩国、加拿大都在开展相关工作。韩国宣称正在设计建造专用生产放射性同位素的反应堆,拟在2017年建成,主要用于裂变^{99}Mo生产,计划年产量为20万居里。荷兰核研究与咨询公司(NRG)计划在2020年建成PALLAS反应堆用于医用放射性同位素的生产。放射性药物进展喜人,截至2016年,美国食品药品监督管理局(FDA)批准的核素药物就有49种之多,以^{18}F为代表的PET显像诊断药物以每年大约30%的速度增长,特别在一些新兴国家增长更快。

我国早在20世纪80年代末就建立了较完整的同位素科研、生产供应体系。经过近30年的发展，技术能力和水平大幅度提升，建立了包括反应堆、加速器、放射性同位素研究生产设施等在内的完备的放射性同位素研制与生产平台，拥有近万人的科研生产队伍。掌握了多种同位素的制备技术，个别放射性同位素具备相应的生产能力。已掌握了利用CANDU堆制备百万居里级 ^{60}Co同位素技术，建立了相应生产能力。依托CMRR建成了高浓铀制备裂变同位素 ^{99}Mo的平台，形成从靶制备、辐照、溶解和提取分离的全流程能力。掌握了采用间歇循环回路制备十居里级 ^{125}I的技术，在游泳池式反应堆上建成了相应的生产系统。在居里AE-Cyclone-30加速器上建立了气体靶制备满足放射性药物研制要求的 ^{123}I系统，形成批产3居里的能力。依托引进的近百台医用回旋加速器为满足临床需要大量制备了 ^{18}F、^{11}C同位素。开始研究用低浓铀制备千居里级医用裂变 ^{99}Mo和用循环回路制备百居里级 ^{125}I的工艺，以满足国内对主要医用同位素的需求。依托反应堆和加速器建立了制备 ^{177}Lu、^{64}Cu和 ^{68}Ge以及 ^{211}At的工艺，形成一定批量生产能力，质量满足后续使用需要。放射源的制备技术进步显著，掌握了万居里级 ^{60}Co辐照源的设计、制靶、辐照、运输、制源和检验技术，形成了年产600万居里 ^{60}Co辐照源的能力，产品的各项技术指标和质量水平达到甚至超过了国外同类产品水平，可满足我国约75%的市场需求；2016年，国产钴源实现销售收入2000万美元，我国同辐股份有限公司一举跃升为全球第三大工业钴源供应商。建立了医用 ^{125}I种子源制备工艺，生产的产品在临床上得到广泛应用。开展了 ^{68}Ge医用校正源的源芯制备、密封、表观活度和剂量分布测定等一系列关键技术攻关，完善了制备工艺，并建立了生产线，技术指标达到国外同类产品水平。放射性药物制备技术取得多项突破，在国家"新药创制"重大专项的支持下，恢复了 ^{3}H、^{14}C标记化合物的研究平台，重新开始为国内科研提供所需的标记物并进行标记服务。

（三）核探测与核电子学

近年来，随着核辐射探测器材料的技术突破，涌现出一大批性能优异的探测器，如CZT、GaAs、溴化镧、溴化铊探测器等。特别是平面硅工艺探测器技术的重大突破，派生出一系列先进的硅半导体探测器。这些探测器技术在现代高分辨率成像、防爆安检、空间探测、科学研究、武器装备、国民经济等领域发挥了重要作用。微电子工艺技术在核电子学线路设计上得到广泛应用，专用集成电路（ASIC）的研制成为核电子学中前端电路、信号调理等方面的主要内容。器件的抗辐射加固性能取得了长足的进步，越来越多新器件、新结构得到了关注，有些新工艺在抗辐射方面具有独特优势。

我国探测器经过多年的发展，取得了一定的成绩，国内NaI闪烁体、BGO、硫化锌是早期研发的探测器，目前已经实现批量化产品。堆外探测器、堆芯探测器均已经实现了国产化，国内实验堆、军用堆均使用我国自行研制的产品，在商业核电站的堆外探测器、堆芯探测器部分实现国产化。但结构复杂的先进探测器依赖进口，制约了技术进步。我国在专用核电子学方面器件还处于起步阶段，尤其是大规模数据获取等应用依赖国外进口，核电子学专用器件与国外差距较大。

（四）核技术在工业、农业、医学等领域的应用

国际上，通用核仪器仪表逐渐实现了系列化、智能化、数字化、网络化及现场化；堆用核仪器向在线测量、事故后监测以及全数字化仪控方向发展。材料辐照改性、食品辐照、医疗卫生用品的辐照消毒实现了工业化和商业化，目前全球有美国、欧洲、日本等50多个国家和地区批准了8个大类共计230多个辐照食品品种上市。根据国际原子能机构报告，2015年全球约有70万吨辐照食品。辐照处理废气、废水、废物技术已相对成熟。射线工业无损检测技术在工业设备和装备制造、运行质量保证、提高生产率、降低成本等领域正发挥越来越大的作用；辐照装置呈现大型化、专业化和高性能化。辐射诱变育种技术几乎在全球所有国家主要作物的改良中得到应用，目前全球已经有100多个国家利用该技术来改良粮食作物、经济作物和花卉苗木等。据不完全统计，运用此项技术，全球范围内共培育了3218个突变植物品种。精准医疗分子影像技术成为核医学发展的重要方向，分子影像设备，多模态、精准放疗设备成为主要发展趋势，PET/CT是发展较为成熟而公认的技术。

三、本学射行业科国内外研究进展比较

我国核技术应用经历了科研开发、产业化和快速发展三个阶段，现已步入高速发展期。截至2016年12月底，我国从事核技术应用研发的科研机构、高校和应用单位达400多家。中国同位素与辐射行业协会统计数据显示，截至2015年，我国核技术应用产业的产值约为3000亿元，是2010年的3倍，年增长率保持在20%左右，总体呈现加速增长态势，有望成为当前经济环境下新的增长热点。为全社会提供近10万个就业岗位，在提高人民生活水平、促进社会经济发展中发挥了不可替代的重要作用。随着产业规模不断扩大和国家重视程度的加深，相关产品的性能和技术水平提升很快，得到国际同行越来越广泛的认可。通用核仪器的发展已初具规模，逐渐跟上国际发展步伐。核医学的科研和生产水平逐年提高，医学领域应用全面拓展，核医学显像诊断和治疗技术发展迅速，实现了中国第一台国产PET/MR在医院装机的突破，性能达到国际领先水平，核心部件全部自主研发；肿瘤治疗机构在过去的30年里大幅增长，从1986年的264家增长到2015年的1413家；放疗市场由2008年的58.3亿元增长至2015年的269亿人民币，且呈加速迹象。辐射诱变育种成绩显著，据国际原子能机构的统计数据，在全世界利用辐射诱变育种技术育成的2316个作物新品种当中的27%由中国科学家育成；在全国作物耕种面积中，辐射诱变育成的品种占到二成，育成的突变品种最高年种植面积在1.3亿亩以上，每年为国家增产粮食35亿~40亿千克，辐射诱变育成的油料作物每年产量10亿多千克，年产值约为14亿~17亿元。我国辐照加工研究与产业化在国际上已处于领先地位，截至2015年年底，产值规模已达到144多亿美元，年增长率保持在15%左右，辐照材料改性、医疗卫生用品辐照灭菌等辐射加工

服务以及辐射技术装备领域实现了产业化发展,产业规模分别达到72亿、58亿和14亿美元;辐射技术在环境污染领域的应用全面拓展;射线无损检测技术已接近国际先进水平;大型集装箱/车辆检查系统等安检设备多次服务于首脑峰会、奥运会、世博会、博鳌亚洲论坛年会等多项国际重大活动的安保工作,装备到全球140多个国家的海关和边防口岸。

四、本学科发展趋势和展望

近年来,我国核技术应用产业取得了丰硕的成果,产业化进程逐渐加快,市场机遇不断增多。但我国人口基数大,基层核技术应用还很薄弱,一些先进设备和检查项目人均占有量与发达国家相比还有较大差距,而且在一些领域还存在技术种类少、产业规模小、技术水平不高、基础设施不足等问题,核技术应用产业的发展状况与核大国的地位不相适应。我国核技术应用产业年总产值占GDP的比例仅为4‰,与欧美日等发达国家的水平相差一个数量级。以全球核技术应用产业产值占GDP的平均占比2%衡量,我国核技术应用产业未来有着每年万亿元的巨大发展空间。而我国正在大力策划推动的"一带一路"沿线人口约44亿,经济总量超过20万亿美元,如果这一区域核技术应用发展达到发达国家水平,将是一个巨大的市场。

目前,我国正面临着发展核技术应用产业的大好时机:①发展新型产业,改造传统产业,促进经济结构转型升级;②加速农作物品种更新换代,促进现代农业发展;③防治污染,改善生态环境;④提高医疗水平,保障食品安全,改善人民身体健康;⑤建立良好秩序,保障社会和公众安全。所有这些挑战和机遇都为我国核技术应用产业提供了强劲的市场需求。

第十六节 油气储运工程

一、引言

油气储运是石油、天然气工业的重要组成部分。它处于油气工业产业链的中游,包括油气集输与处理、长距离输送、储存与装卸以及输配供应等生产系统,是连接油气开采、加工、销售诸环节的纽带。

油气储运在油气工业、国民经济、国家安全中具有非常重要的作用。其中，油气集输与处理是油气田开发的重要组成部分；油气管网是国家能源大动脉和重要基础设施，是油气工业上下游衔接协调发展的关键环节，是现代能源体系和现代综合交通运输体系的重要组成部分；油气储库是油气生产、运输、加工、供应系统的枢纽，油气战略储备基地是国家能源安全的重要保障设施；油气的终端配送系统则是面向用户的民生工程。

油气储运工程学科是多学科交叉的综合性、应用型学科。它以系统的安全、高效与环保为目标，综合运用数学、物理学、化学、经济学以及系统工程、化学工程、机械工程、材料工程、安全工程、信息工程等学科的思想与方法，阐明油气生产与供应中的收集、分离、运输、储存、配送等各个过程的规律，提出并实施相关问题的技术和工程管理解决方案，为相关新装备、新材料的研发提出目标和要求。

进入 21 世纪以来，诞生于 20 世纪中叶的中国现代油气储运工业迅猛发展，已进入国际先进行列。目前，已基本形成覆盖全国、联通海外的油气储运网络系统，基本满足了油气生产和消费的要求。截至 2015 年年底，我国原油、成品油、天然气主干管道里程分别达到 2.7 万千米、2.1 万千米和 6.4 万千米；我国一次运输中，原油管道运量达到 5×10^8 吨，约占原油加工量 95%；成品油管道运量达到 1.4×10^8 吨，约占成品油消费的 45%。天然气管道覆盖率不断提高，用气人口从 2010 年的 1.9 亿人增加到 2015 年的 2.9 亿人。截至 2016 年，我国已建成 9 个国家石油储备基地，储备能力达到 3530 万立方米；建成液化天然气（LNG）接收站 14 座，年接收能力超过 4000 万吨 / 年，接收站数量及总接收能力已跃居世界 LNG 进口国前列。与此同时，我国油气储运高等教育迅速发展，目前已有 62 所本、专科高校设立了油气储运工程学科，其中 5 所高校的油气储运工程学科为国家重点学科。最近，3 所高校的油气储运工程学科进入国家"双一流"建设计划。

二、本学科近年的最新研究进展

"十二五"以来，我国油气储运工程学科快速发展，取得了一系列的理论研究成果和技术突破，并有效转化为生产力，有力支撑了油气储运产业的大发展。

（一）油气集输与处理技术全面提升

油气多相流研究取得新突破，解决了段塞流捕集与控制等关键技术问题，开发了我国首款大型油气混输工艺计算软件，实现了单条混输管道最大输油量 220 万吨 / 年，最大输气量 35×10^8 立方米 / 年，最大输送距离 68 千米；针对不同类型机采油井，创新研发了单井软件量油技术并大规模应用，大大简化了集输流程；含蜡原油不加热集油机理和技术研究再上新台阶，转油站系统节气 40% 以上；化学驱油田采出液处理技术取得重大突破，大庆油田建成了世界上规模最大、年处理原油 1300 万吨的聚合物驱采出液处理系统，以及年处理原油400万吨的三元复合驱采出液处理系统；创新了低渗透和非常规天然气集输处理工艺，实现了经济、有效开发；形成了高酸性气田地面工程系列技术，以及高压、高产、高酸性"三高"气田地面工程安全设计系列技术，保障了"三高"气田的安全开发；创新了地面工程标准化建设技术体系，形成了地面建设新模式。

（二）油气管道建设技术和装备研发取得突破性进展

针对重大工程建设需要，管道设计理论与方法取得系列创新，攻克了基于应变的设计、基于可靠性的天然气管道设计和评价等一批设计技术难题，形成了一批新的技术标准；大口径、高压力管道输送用管及制管技术再上新台阶，X80钢级已成为长距离输气干线管道的首选管材；建成亚洲最大的全尺寸钢管爆破试验场，填补我国在高压、高钢级天然气管道全尺寸断裂行为和管道爆炸环境影响研究领域的空白；复杂地质条件下管道施工技术进一步配套完善，满足了中缅油气管道等各种复杂环境下管道施工的需要；油气管道大型装备国产化取得重大进展，30兆瓦级燃驱机组、20兆瓦级高速直连电驱机组、Class900 48英寸全焊接球阀、2500千瓦级10兆帕输油泵机组等一批大型装备研制成功并开始应用。

（三）油气管网运行技术不断提升

含蜡原油流变性机理和黏弹-触变性本构模型研究取得重要突破；创新建立了基于可靠性的含蜡原油管道停输再启动安全性评价方法；发展了易凝高黏原油管道流动与传热的模型和求解算法，开发了适用于多管同沟敷设、冷热油交替输送、间歇输送等复杂输送过程的仿真软件包，指导了西部原油管道等重要管道的建设和运行；形成了超大型复杂油气管网集中调控技术，开发出国内首套商业化管道仿真软件，中国石油所辖5万多千米管道实现了统一集中监控、调度和管理；天然气管网系统可靠性研究取得重要阶段性成果，油气计量技术及标准体系建设稳步推进。

（四）地下油气储存技术取得长足进步

攻克了地下水封洞库水封性评价与控制、岩体稳定性综合判识、地下水封库地质勘查与动态设计等技术难题，首座300万立方米地下水封战略储备油库建成投产；气藏型和盐穴型储气库技术进一步完善，形成了储气库地质与气藏技术、钻完井技术、注采工艺、地面工艺和完整性管理等系列关键技术，保障了11座地下储气库的安全有效运行。

（五）液化天然气技术跨越发展

历经引进、消化、吸收到自主创新过程，我国现已完全掌握天然气液化、LNG接收及再气化成套技术、大型LNG运输船的建造技术、小型LNG运输船的设计和建造技术，并已具备了较强的国际竞争实力；LNG储罐材料、设计和建造技术取得重大突破，形成了具有自主知识产权的大型 LNG 储罐技术；浮式液化天然气生产储卸设施（FLNG）的液化工艺、设备和船体技术取得突破，具备了工程实施的技术能力；冷剂压缩机、低温蒸发气（BOG）压缩机、海水开架式气化器（ORV）、LNG 潜液泵等关键装备国产化取得突破。

（六）海洋油气储运技术再上新台阶

海底混输管道多相流动、降黏减阻、水合物防治、蜡沉积预测等理论和技术研究取得重要进展，形成了深水油气田流动监测与管理技术；建立了大壁厚高钢级直缝埋弧焊钢管（UOE）的设计、制造、检验技术体系，提升了海底管道风险评价、稳定性分析、安全预警等能力，形成了具有自主知识产权的柔性管、立管设计技术；形成了水下生产系统总体方案研究与设计能力，装备国产化取得进展。以水深1500米的荔湾3-1深水气田开发工程建成投产为标志，我国海洋油气储运技术发展上了一个新台阶。

（七）油气管道完整性管理全面实施

管道完整性管理已从运营期延伸到勘察、设计、施工和废弃等全生命周期，场站完整性管理已进入推广实施阶段；管道内检测技术取得突破，直接评估方法全面应用，对管道本体和防腐层隐患的识别能力显著提高；研制成功光纤预警、地质灾害监测等技术装备，提高了管道安全预警技术水平；建立了适合我国国情的管道风险评价方法，形成了系统化的风险管控机制；发展了全系列油气管道维抢修技术与装备，建立了突发重大事故应急机制；形成了较为完备的管道安全技术法规及标准体系，为管道安全运营提供了法律法规保障。

三、本学科国内外研究进展比较

目前，我国油气储运技术总体上处于国际先进行列，其中在含蜡原油管道输送技术、大口径 X80 级钢管应用技术、油田化学驱采出液处理技术等方面处于国际领先地位。

我国油气集输与处理技术总体上处于国际先进水平。其中，聚合物驱和三元复合驱采出液处理技术、含蜡原油低温集输与处理工艺、单井软件量油技术等处于国际领先地位；油气混输的部分单项技术达到国际先进水平，但大型油气混输泵等部分装备仍有差距；已全面掌握非常规天然气、高酸性天然气的集输与处理技术，但在特高酸性天然气集输处理、天然气脱硫脱碳工艺及装置大型化方面与国外先进水平尚有一定的差距。

我国主要的管道建设技术已达到国际先进水平，有些已处于国际前列，如 1219 毫米和 1422 毫米大口径 X80 级钢管应用技术、X90/X100 级钢管研制、X80/X90 的全尺寸爆破试验研究等。西气东输二线 4800 千米采用 X80 钢管，如此大规模的 X80 钢应用属全球首例。但在管道系统可靠性设计、冻土及低温地区管道设计、长输管道阴极保护等技术方面与国外先进水平相比有一定差距；高性能管道无损检测设备仍是短板，管道建设智能化处于初始阶段。

我国油气管道运行技术总体处于国际先进水平。其中，含蜡原油管道输送技术处于国际领先地位；大型油气管网调控技术和油气计量技术总体达到了国际先进水平，但运行智能高效和供应精准匹配方面需进一步加强。我国已建成西北、东北、西南、海上四大油气进口战略通道。截至 2016 年年底，我国在役陆上油气管道总里程 12.6 万千米，但与美国 68.8 万千米、俄罗斯 26 万千米的油气干线管道总里程相比，我国油气管网规模与美国和俄罗斯还存在明显差距。

在油气储存领域，我国地下水封石洞油库建库技术已取得突破，但技术体系的完备性亟待加强；基本形成了气藏型地下储气库技术体系，但库容参数设计和运行优化、储气库钻完井、储气库完整性管理等专项技术有待进一步提升；基本掌握了盐穴型地下储气库的建库配套技术，但技术成熟度有待进一步提升。

我国天然气液化、LNG 接收与再气化技术已取得全面突破，并成功应用于一批 LNG 接收站项目的工程建设中，整体技术水平与国际相当，已形成国际竞争力，但装置产能和技术成熟度尚有差距；LNG 工程设计和建造的国际竞争力日益增强，但智能化工程设计技术尚处于起步阶段；LNG 相关核心材料和装备制造已实现国产化，但大型装备制造能力还有差距。

在海洋油气储运领域，浅水海上稠油开发与输送工艺处于国际领先水平，流动保障技术已从之前的适应 300 米水深提高到适应 1500 米水深，并形成了相应的技术体系。但是，技术成熟度还有待进一步提高，深水海底混输管道技术与国际先进水平尚有差距；在超过 1500 米水深油气田的流动保障技术、水下生产系统和装备方面与国外尚有较大的差距。

在油气储运安全领域，我国管道完整性管理已全面落地，但与国外先进水平相比仍存在不足；已具备全系列管道内检测技术与装备，但新型检测器研发和管道评价理论尚需加强；管道安全监测技术装备已基本国产化，但关键硬件设备成熟度还需提高；管道维抢修关键技术装备需进一步配套完善，油气管道事故应急协同响应机制需进一步加强；管道安全技术标准体系需进一步完善。

四、本学科发展趋势和展望

当前，我国油气储运工程学科的发展正面临重大战略机遇。习近平总书记在十九大报告中指出，要"加强水利、铁路、公路、水运、航空、管道、电网、信息、物流等基础设施网络建设"，明确了管道作为现代经济体系组成部分和国家基础设施的重要地位。2017年 5 月，国家发展改革委、国家能源局发布了《中长期油气管网规划》（本段简称《规划》）。这是我国首次在国家层面制定系统性的油气管网发展规划。该《规划》对今后十年我国油气管网的发展做出了全面战略部署，明确了油气管网发展的重大意义、指导思想、基本原则、发展目标和保障措施。按照该《规划》，到 2025 年，我国油气管网规模将达到 24 万千米，与 2015 年相比翻一番。

展望未来，油气集输与处理技术将需适应油气资源品位劣质化、油气物性多样化、建设环境复杂化和不断提高的安全环保要求，向装备多功能化、一体化、智能化、工艺简化、安全清洁方向发展；油气管道将继续向大口径、高压力、高钢级、网络化、智能化方向发展；油气储库库容大型化、区域储存量超大型化、地下储存将成为未来的发展趋势，大型油气储存设施的安全和环保问题因此而愈加凸显；液化天然气技术和装备将向安全可靠、节能高效、大型化和更强适应性方向发展，国产化率将进一步提高；海洋油气田开发正向深水、超深水发展，海洋油气储运技术需适应这一要求、形成配套技术并实现国产化；油气储运设施安全将继续成为各级政府和公众关注的热点，为此需要在理论、技术、管理各方面加强研发与推广应用，不断提高安全保障水平。

鉴于油气储运工程学科鲜明的综合性、应用型学科特点，为了适应油气储运产业发展、油气储运技术发展和安全及环保要求，油气储运工程学科需要努力推进理论和技术的原始创新、产学研结合的多学科联合攻关、重大装备国产化以及新技术新材料新装备推广应用。

第十七节　煤矿区土地复垦与生态修复

一、引言

（一）学科的产生背景与概念

任何一门学科的产生，都有一定的历史背景和历史环境，都是社会和生产力发展到一定阶段，为人类的某种需要而产生的，煤矿土地复垦与生态修复学科的产生也不例外。我国土地复垦与生态修复工作则起步较晚，1978年我国开始改革开放，使得国民经济迅速发展，采矿等各种生产活动带来的土地和环境问题日益严重，导致我国在1980年以后才正式提出和开始重视土地复垦与生态修复。

尽管煤矿土地复垦与生态修复产生于20世纪20年代，但国外系统研究则开始于20世纪70年代，我国则是80年代末，因此，该学科是一门较新的、多学科交叉的综合性应用学科。它有其独特的研究对象、研究任务，具有学科发展必然和不可替代性。其独特的研究对象就是煤矿区损毁的土地和生态环境。其研究任务是损毁土地和环境的特征及其产生机理、修复治理原理与方法及其保障修复治理的管理技术。因此，我们将研究损毁土地和环境的产生机制及演变规律及其修复治理的政策、理论、方法、技术工艺的学问，称之为"土地复垦与生态修复学"。由于学科的交叉性特征，土壤学、生态学、环境科学、土地科学等多个学科都对该学科的发展起着促进作用。

煤矿土地复垦与生态修复学科的研究对象是煤矿区因采矿活动损毁的土地和生态环境，其与煤矿生产紧密相连，具有时空动态性、涉及问题复杂和涉及学科多样等特点。

（二）学科的作用

我国煤矿土地复垦与生态修复历经 30 多年研究与实践，取得了很大发展，使我国的土地复垦率提高到了 25% 以上。煤矿土地复垦与生态修复的主要作用表现在：

1）矿区土地复垦与生态修复是保护耕地的一种有效措施；
2）矿区土地复垦与生态修复是保护生态环境、提高人民生活质量的有效措施；
3）矿区土地复垦与生态修复是缓解工农矛盾、解决就业和维护社会和谐发展的有效途径；
4）矿区土地复垦与生态修复是矿业可持续发展的有力保障。

（三）学科的主要研究领域

由于煤矿土地复垦与生态修复学科的研究任务是损毁土地和环境的特征及其产生机理、修复治理原理与方法及其保障修复治理的管理技术，因此，其主要研究领域和内容是：①煤炭开采对土地与环境影响的监测与诊断；②矿区土地复垦与生态修复的工程技术（土壤重构与地貌重塑）；③矿区土地复垦与生态修复的生物技术（植被恢复与生物修复）；④矿区土地复垦与生态修复的监管与政策；⑤矿区土地与环境修复的基础理论与原理。

（四）学科定位

由于目前我国学科目录中，还没有土地复垦与生态修复学科，所以该学科的归属一直在讨论中。

从煤矿土地复垦和生态修复的缘起和发展分析，煤矿土地复垦与生态修复学科属于矿业工程一级学科。主要理由是：①土地复垦和生态修复是采矿工程的重要组成部分；②土地复垦和生态修复的研究对象是采矿活动导致的；③绿色矿山建设中，土地复垦和生态修复是重要的任务和目标；④矿业工程学科的发展将越来越和资源、环境学科相融合。因此，建议在矿业工程一级学科下增设"矿区土地复垦与生态修复工程"二级学科。

鉴于土地复垦与生态修复涉及多个学科，尤其是涉及土地、环境等学科，因此，在"环境科学与工程""土地科学与工程"等学科下设立"土地复垦与生态修复"二级学科也是可行的。

二、本学科近年的最新研究进展

我国煤矿区土地复垦与生态修复的发展历史，形成了 4 个阶段的发展特征：①萌芽阶段（1980—1989 年）；②初创阶段（1990—2000 年）；③发展阶段（2001—2007 年）；④高速发展阶段（2008 年至现在）。

经查询检索不完全统计，从 1988—2017 年共计毕业矿区土地复垦与生态修复方面研究生为 766 位，其中博士研究生 173 位。关于土地复垦与生态重建的相关专著共 72 本，其中国内学者自己撰写的专著有 69 本，编译国外书籍 3 本。截至 2016 年，关于矿山土地复垦与生态修复的相关专利（包括已申请未授权专利）共 117 项。其中发明专利 97 项，占所有申请专利的 83%，实用新型专利 18 项，软件著作权 2 项。统计分析煤矿土地复垦与生态修复有关的国家科技奖和煤炭、教育、国土、环保等省部级科技奖二等奖以上，共检索收集到 52 项，其中国家科技进步奖 7 项，煤炭行业科学技术奖 26 项，国土资源科学技术奖 8 项，环保科技进步奖 3 项，教育系统科学技术奖 3 项，省级科学技术奖 4 项。获奖内容涉及矿区生态重建、生态修复、土地损伤诊断及生态修复、生态综合治理技术、生态环境损害治理、土地复垦与监管关键技术等。

（一）基础理论领域的进展分析

1）边采边复（采复一体化）理念与技术原理：我国学者结合露天采矿工艺和我国的实际情况，研究了露天煤矿剥离—采矿—回填—复垦的一体化工艺过程，提出了横跨采场倒堆内排工艺的露天煤矿采复一体化的基本原理及定量化表达。针对井工煤矿创新性地提出了采复一体化的"边开采边复垦"的概念和技术原理，它是充分考虑地下开采与地面复垦措施的耦合，通过合理减轻土地损毁的开采措施和沉陷前或沉陷过程中的复垦时机与方案的优选，实现采矿与复垦同步进行的一种复垦思想，是对以往沉陷稳定后再复垦"末端治理"技术的重要革新。

2）土壤重构原理与方法：土壤重构即重构土壤，是采取适当的采矿和重构技术工艺，应用工程措施及物理、化学、生物、生态措施，重新构造一个适宜的土壤剖面和土壤肥力因素，在较短的时间内恢复和提高重构土壤的生产力，并改善重构土壤的环境质量。国内外学者一直把矿区复垦土壤重构作为土地复垦与生态修复的核心任务，也是检验复垦工程成败的关键。我国学者提出了"分层剥离、交错回填"的土壤重构原理并构建了数学模型，同时成功应用于露天矿采矿—复垦一体化土壤重构、采煤沉陷地复垦土壤剖面重构和煤矸石山生态修复的土壤剖面重构。

3）地貌重塑理念：国外的土地复垦发源于露天矿。露天矿开采对地表景观破坏最显著，因此，国外对地貌重塑很重视，要求恢复近似原地貌的生态景观。我国露天矿较少，对地貌重塑的重视程度较低，也因此对地貌重塑的研究起步较晚，主要针对采矿习惯性形成的台阶型地貌进行研究，并侧重于水土保持的研究，近年来主要是引入国外的仿自然地貌重塑的理念与方法，正处于吸收、消化阶段。

（二）矿区生态环境监测与评价领域的进展分析

在地表形变及沉降监测与评价方面，合成孔径雷达测量（Synthetic Aperture Radar，SAR）技术在矿区地表形变及沉降监测中的应用取得了突破，主要表现在：基于知识的矿区形变 SAR 信息提取技术、多尺度多平台时序 SAR 影像地表沉降信息获取方法、SAR 与全球导航卫星系统（Global Navigation Satellite System，GNSS）、激光雷达（Light Detection and Ranging，LiDAR）及无人机数据的融合方法、利用多平台 SAR 数据和三种基于水平形变假设的地表形变联合监测方法和矿区地表沉降、建筑物沉降及结构物形变监测的自动化监测系统。

在地下煤火及煤矸石山自燃监测与评价方面取得的突破主要包括：地下煤火地空一体化探测技术、煤火靶区及地裂缝 UAV 精准监测技术、煤矿区矸石山边界信息提取及温度异常信息监测诊断技术。

在矿区其他生态环境要素监测与评价方面，主要进展包括：多源遥感矿区环境及灾害动态监测技术与评价预警系统、基于 NPP 矿区生态环境监测与评价系统、矿区地表环境损伤立体融合监测及评价技术、采煤工作面沉陷裂缝损伤与生态因子监测技术、风沙区一体化的地表环境监测体系与土壤水分监测技术、基于物联网技术的矿区生态环境监测系统。

在矿区生态环境评价方面，主要进展包括：矿区环境综合评价与时空变化规律分析、地表环境损伤评价及整治时空优化技术、煤矿生态环境损害累积效应评价方法、煤炭开采对植被–土壤物质量与碳汇的扰动与评价方法。

（三）矿区开采控制技术领域的进展分析

为了减轻开采沉陷对土地资源和生态环境的影响，国内外专家学者通过了大量的试验与研究，总结出 5 种开采沉陷控制理论与控制方法，即煤矿覆岩离层注浆控制技术、采空区充填开采控制技术、条带和房柱式等部分开采控制技术、协调开采沉陷控制技术和保水开采控制技术。

目前，我国开采沉陷控制对土地资源和生态环境的影响技术处于国际领先水平，但理论与技术体系仍有待进一步深入研究，如：离层注浆控技术需要进一步研究离层形成的机理、发展演变规律，离层从形成到停止是一个动态过程，要提高注浆减沉效果和质量，就必须掌握离层形成随开采的时空关系，离层出现的时间、位置和最佳注浆时间等；充填开采技术需进一步研究减少对综采工作面产生不利影响的固体废弃物高产、高效充填工艺；部分开采煤柱应力应变特征与长期稳定性研究以及煤柱留设理论方法的研究，适于深部最佳采、留比优化设计方法研究；保水开采应进一步注重基础地质勘查与研究，综合勘察水资源、煤炭资源以及相关地质采矿条件，在三维空间中研究水、煤、岩等的组合形态、赋存特征等以及物理、力学、水理性质，提高导水裂缝带发育高度的预测精度。

（四）煤矿区土地复垦与生态修复规划设计领域的进展分析

煤矿区土地复垦与生态修复的规划设计研究的重点和成果表现在：探讨规划的内容与深度；提高矿区土地破坏调查和评价精度、实现可视化展示毁损和拟毁损土地；提高土地复垦适宜性评价精度；实现复垦土地空间结构优化与布局的自动化和可视化；与开采协调的动态复垦规划方法；复垦措施的优选与设计原则及方法；引入土地整治规划、土地毁损生态风险评价、生态农业规划、景观生态规划、矿城协同等理念应用于复垦规划设计；植物品种的筛选和群落设计；复垦成本估算方法等。

国内在煤矿塌陷地边采边复和景观生态修复规划方法上成果显著；土地复垦规划设计向系统化、整体化和高效化相结合的生态发展阶段迈进；并需加强山地小型矿山的生态修复规划设计研究。与国外相比，我国土地复垦规划设计存在的差距表现为：未能与采矿规划有机结合、采前环境本底调查薄弱、区域地貌的融合等。我国在3S技术、虚拟技术等信息技术支持下的土地复垦规划设计理论与方法紧随国际潮流，土地预复垦规划设计理论与方法上处于国际领先地位。

（五）土壤重构与地貌重塑领域的进展分析

矿区地貌重塑与土壤重构是矿区生态系统恢复重建的基础性和关键性工程。矿区土壤重构核心是进行成土条件的改造、土壤层次的改造、改善土性、调节土壤水肥气热等为核心的土壤剖面重构。土壤重构技术包括表土剥离、表土存放、土壤剖面重构、表土覆盖的工程过程，以及土壤物理改良等。井工煤矿沉陷地土壤重构研究主要方向有：高潜水位区沉陷地土壤重构技术，包括采用挖深垫浅法进行土壤重构，利用煤矸石、坑口和电厂的粉煤灰、露天矿排放的剥离物、垃圾、沙泥、湖泥、水库库泥和江河污泥等固体废弃物充填采煤沉陷地，以及引黄河泥沙充填复垦采煤沉陷地技术研究等。露天煤矿土壤重构研究的重点是"分层剥离、交错回填"的土壤剖面重构原理，以及针对我国大型露天矿地貌形态损毁极其严重、排土场岩土侵蚀剧烈、土壤肥力贫瘠等特殊问题，采用"采矿—剥离—排弃—造地—复垦"等一体化工艺，包括分层采剥技术、分类排弃技术、分区整地技术、黄土母质直接铺覆工艺、堆状地面排土工艺、生土快速熟化工艺等，再造一层永久性的良好土壤剖面构型等。煤矸石土壤重构研究重点是矸石山黄土薄层覆盖法、加速煤矸石山基质风化成土，以及采用物理改良、化学改良和生物改良来提高重构土壤肥力的研究等。

（六）植被恢复与生物修复领域的进展分析

矿区植被恢复是矿业企业对矿山建设、生产过程中造成的植被破坏进行修复，是矿业企业的社会责任和义务。生物修复可划分为植物修复、动物修复、微生物修复。

随着国内外植被—生态—微生物—土壤修复模式和方法的不断完善，筛选和鉴定出更多适合不同类型矿区废弃地种植的先锋植物和土壤种子库，探索微生物和生物菌群的优势功能，进一步推动菌根技术在矿区土地复垦与生态重建中的应用，以新技术新方法为手段，瞄准国际微生物学发展的前沿和热点，重点考察土壤微生物的遗传多样性和功能多样性的空间分布特征，探讨土壤微生物的地理分布格局及其形成机制，深入土壤微生物生态系统的研究，进一步研究植物—微生物根际化学交流过程，了解根际微生态圈等，是目前矿区废弃地生态恢复研究的重要方向。

（七）矿区建筑物保护与构建技术领域的进展分析

矿区受损建筑物防治与修复关系到矿区的和谐稳定、关系到矿区土地的高效集约利用、更关系到矿业城市的可持续发展，已经成为矿区土地复垦和生态修复学科发展研究的重要领域。随着采矿业的快速发展，矿区抗变形建设理论更加成熟，使得理论和实践应用成果更加丰富，矿区抗变形建筑设计理念更加深入人心，成为各个矿区建筑物下采煤和不搬迁采煤的普遍做法。标志是《建筑物、水体、铁路及主要井巷煤柱留设与压煤开采规程》的出现。

在矿区土地建筑利用技术发展方面，为解决采动过程中土地利用的难题，发展了矿区土地动态预复垦技术。随着城市化进程的加快，国家基础设施建设日新月异，使得城市建设用地日益紧张，矿区受损土地的建筑利用也面临着许多新的难题。为统筹好资源利用和城市开发又发展出了矿区受损土地城市建设系列技术。又开展了矿区受损土地建设城市景观湿地的技术研究，并在多个矿业型城市进行了实践，取得了良好效果。矿区受损土地向着地上、地下综合利用的方向发展，未来地下空间将被开发成地下城市，矿区建筑物的防治与修复必将面临新的挑战。

（八）煤矿沉陷区水资源及水生态修复领域的进展分析

沉陷水域是由于煤矿开采以后，造成地表下沉低凹积水而形成的类似湖泊的封闭水体，其主要的水源为地下水及大气降水，对沉陷区水资源环境特征开展了一些基础研究工作，包括：矿区沉陷区水域重金属污染研究、矿区沉陷区水域富营养化污染研究、矿区沉陷区水域其他污染研究、矿区沉陷区水域营养物质研究、矿区沉陷区水域生物研究、采煤沉陷区水体生产力、采煤沉陷区水环境评价。

（九）土地复垦与生态修复政策与管理领域的进展分析

1988年11月8日国务院颁布《土地复垦规定》，我国由此确立了土地复垦制度。《土地复垦规定》是《中华人民共和国土地管理法》的实施配套法规，首次规定了土地复垦的含义以及"谁破坏，谁复垦"的原则。1998年在国土资源部成立了耕地保护司和土地整理中心，负责全国的土地复垦工作，我国开始有了土地复垦工作的管理部门，土地复垦事业走上规范化道路。2011年3月5日《土地复垦条例》由国务院公布施行，标志着我国土地复垦事业步入了制度化、规范化和法制化的新阶段。为贯彻落实《土地复垦条例》，国土资源部还批准发布了《土地复垦方案编制规程》7 项行业标准。经过近 30 年的探索，我国煤矿区土地复垦与生态修复政策制定不断趋于成熟，管理手段与时俱进，相关工作也取得了较大进步，煤矿区土地复垦率由原来的 1% 提升到 2011 年的 25%。

三、本学科发展趋势和展望

（一）未来发展趋势及重点研究方向

"富煤、少气、贫油"是我国能源的现状，因此煤炭资源在今后很长一段时间仍占据主导地位。所以，煤矿区的土地复垦与生态修复不仅过去是我国土地复垦与生态修复的先锋，今后仍将是我国土地复垦与生态修复的重点。随着煤炭开发强度的加大和战略西进，煤矿区土地复垦将会出现一些新问题需要解决。展望未来的发展，煤矿区土地复垦将呈现蓬勃发展的态势，并将在以下几个方面得到广泛关注：

1）西部生态脆弱矿区土地生态环境的影响及其修复；
2）东部煤粮复合区耕地复垦与生态修复技术；
3）关闭矿山的土地复垦与生态重建。

（二）学科发展战略措施

绿色发展已经成为我国的发展理念。矿区的土地复垦与生态修复是保障矿区绿色发展、绿色矿山建设的关键，因此，本学科应得到更多的重视。建议学科发展战略措施是：①科学定位学科归属、完善学科内涵和体系；②加大科技投入支持力度；③打造国家级甚至国际级科研平台；④加大科研团队的建立及人才培养；⑤扩大试验示范基地建设；⑥扩大国际和地区科技合作与交流。

第十八节　矿物材料

一、引言

矿物材料是以天然矿物或岩石为主要原料，经加工、改造获得的功能材料。矿物材料学科从材料科学与应用角度研究矿物的结构、理化特性、功能及应用性能，通过加工改造优化矿物的结构、理化特性，提升其功能与应用价值。矿物材料学科是矿业工程学科的延伸和发展，与岩石与矿物学、材料科学与工程、化学与化工、生物医药、环境科学与工程、机械、冶金、电力与电子、能源科学与工程等学科交叉。

矿物材料不仅是现代高新技术产业不可或缺的功能材料，还是冶金、化工、轻工、建材等传统产业结构调整和产业升级必不可少的新材料以及生态修复与节能环保功能材料，广泛应用于航空航天、电子信息、机械、冶金、建筑、建材、生物、化工、轻工、食品、环境保护、生态修复、快速交通、通信等现代产业领域，对于高值和高效利用矿产资源、优化传统产业结构、节约能源和资源、保护环境、恢复生态以及促进高新技术产业发展等均具有重要作用。

本文综述了近年来矿物材料学科的主要研究进展和研发热点，并对国内外发展状况进行了分析比较，提出了促进矿物材料学科发展的建议。

二、本学科近年的最新研究进展

（一）碳质矿物材料

近年来，我国在碳质矿物材料，尤其是石墨烯的制备与应用技术领域，取得了比肩世界先进水平的重要研究成果。

天然石墨材料主要进展体现在电池阴极材料、石墨插层复合材料、柔性/膨胀石墨、新能源材料等功能材料的加工与应用方面。清华大学石墨材料科研团队开发的微膨改性鳞片石墨负极材料、微晶石墨负极材料、低温负压解理石墨烯及石墨烯基导电剂应用技术成果实现了在可快速充电和宽使用温度范围锂离子电池中的应用，推动了我国天然石墨资源深加工技术和锂离子电池材料技术的发展。在国家"十二五"科技支撑计划支持下，18家国内高校、科研院所及企业合作完成的鳞片石墨基础原料绿色制备技术及典型工程示范、高纯鳞片石墨制备技术与应用、低硫高抗氧化性可膨胀石墨及高导热柔性石墨板制备技术开发与示范、新型负极材料制备技术及产业化、先进金属 – 鳞片石墨复合材料开发及示范等科研成果大幅度提升了石墨行业深加工技术水平。

我国石墨烯领域的文献发表量和专利申请量已位居全球首位或前列。近几年石墨烯的主要进展是规模化制备及其在复合材料、微电子、光学、能源、生物医学等领域的应用研究。这些研究在实验室阶段均有突破，中试规模的制备与应用研究也取得一些重要进展。北京化工大学研发的液相剥离法创新性地采用石墨衍生物作为分散剂，采用超重力高速剪切法、以水为溶剂，在常压下制备层数多在七层以内的少层石墨烯，并建成了中试线；北京航空航天大学率先在世界上采用搅拌驱动流体动力学装置制备石墨烯，并申请了发明专利。在国家自然科学基金和重点基础研究发展计划支持下开展的以石墨烯为载体的复合光催化材料的制备、结构与性能研究取得了重要进展。这种复合光催化材料能有效抑制光生电子－空穴对的复合，增加参与反应的电子/空穴数目，显著提高了材料的光催化活性。

（二）黏土矿物材料

近几年，在蒙脱土改性与插层复合材料、纳米 TiO_2－蒙脱土复合材料、霉菌毒素吸附材料、膨润土环保与生态材料、二维纳米高岭土功能填料、煤系高岭土超细和煅烧增白技术、高岭土基分子筛和催化剂、凹凸棒土纳米棒晶解聚－分散－改性和应用、海泡石吸附环保材料、埃洛石生物功能材料等黏土矿物材料的科学研究和技术开发方面取得显著进展。

中南大学矿物材料科研团队揭示了矿物结构转型制备介孔材料的结构演变过程及调控原理，解析了矿物表面改性的电荷、离子属性、键性质的匹配与组合机制，提出了矿物基复合材料的设计思路与性能调控机理，建立了典型层状非金属矿物的优化结构模型，初步形成了具有矿物材料特色的理论计算与模拟体系；中国科学院兰州化学物理研究所凹土科研团队开发了对辊处理—制浆提纯—高压均质—乙醇交换一体化工艺，实现了棒晶束的高效解离和纳米棒晶的分散，从应用基础、关键技术到高值产品开发，形成了具有自主知识产权的技术创新链；合肥工业大学与中科院广州地球化学研究所发明了一系列凹凸棒石黏土－金属（氧化物）纳米复合催化材料制备及其在内燃加热式生物质气化、液化、气化炉内原位催化裂解焦油、生物质气化－催化裂解焦油一体化装置中的应用方法和技术；北京大学和北京科技大学开发了的劣质煤系高岭土生产高性能耐火材料的高效转型技术；中国地质大学（北京）以提纯膨润土为原料制备阴－非离子型、阴－阳离子型有机纳米膨润土和有机硅嫁接膨润土填料，并建成了中试生产线。

（三）多孔矿物材料

近年来多孔矿物材料在高性能分子筛、硅藻土负载纳米 TiO_2 复合光催化材料、沸石复合抗菌材料、硅藻土健康环保材料、膨胀珍珠岩防火保温材料等方面取得显著进展。

复旦大学科研团队通过优化合成条件和控制晶种溶解与诱导过程制备的多级孔 ZSM-5 沸石催化剂在甘油脱水制丙烯醛的反应中表现出良好的选择性和使用寿命，进一步在纳米微晶堆积结构中引入硼，得到了具有较多弱酸位、在甲醇制烯烃反应中具有超强抗积碳能力的多级孔 B-Al-ZSM-5 催化剂。北京大学、清华大学研制了一种新型的序批式生物沸石反应器，采用装填沸石的生物反应器系统强化处理焦化废水，取得了良好的应用效果。中国矿业大学（北京）非金属矿物材料团队攻克了中低品位硅藻土物理选矿提纯产业化关键技术以及用硅藻精土为原料的医药化工高浓度含盐废水和油田采油污水处理的污水处理剂以及纳米 TiO_2/硅藻土复合光催化材料产业化制备与应用关键技术，并建成了年产 12000 吨硅藻精土示范生产线、6000 吨/年污水处理剂示范生产线和 1000 吨纳米 TiO_2/硅藻土复合光催化材料示范生产线。

(四)钙质矿物材料

近几年在石膏装饰材料、重质碳酸钙表面改性与功能复合、轻质碳酸钙晶型控制、纳米碳酸钙粒径控制、天然大理石异型材、人造大理石固化技术等方面取得显著进展。

华中师范大学与企业合作通过调控脱硫石膏晶体生长工艺条件得到了高强度六角形柱状晶体 α-半水石膏,经表面修饰后和PVC进行复合,获得高强度力学性质的仿生微晶材料。中国地质大学(北京)、清华大学相关团队开发了微米级重质碳酸钙颗粒表面包覆纳米碳酸钙以及碳酸钙-TiO_2复合颜料的制备与应用技术。武汉理工大学科研团队揭示了方解石与不同金属离子在机械力化学反应中的差异性,并应用于Cu和Ni、Co和Cd的分离和分选以及水溶液中Fe、Cu、Zn、Ni、Cd在中性pH条件下的共沉淀净化。中国科学院电子研究所研发的微波固化技术显著缩短了人造石的固化时间,显著提高了生产效率和节约了能源。

(五)镁质矿物材料

镁质矿物材料近几年的主要研究进展是:电熔镁砂高效制备和清洁生产以及镁质阻燃材料制备与应用、滑石增强填料表面改性、超细活性水镁石阻燃填料、西藏水菱镁石资源的结构性能与应用等。

辽宁丰华实业有限公司及辽宁科技大学等单位科研攻关团队研制并建成了300立方米大型全密闭哑铃形自动化清洁生产重烧氧化镁竖炉,实现完全高效利用菱镁矿资源生产优质重烧镁砂和清洁生产;在国家重大基础研究项目支持下,东北大学联合企业科研人员优化了菱镁矿熔炼制备电熔镁砂的工艺。中国矿业大学(北京)与企业合作研发了无机复合超细活性水镁石阻燃填料以及氢氧化镁/重质碳酸钙复合阻燃填料的制备与应用技术,并实现了产业化生产和推广应用。

(六)纤维矿物材料

近年来纤维矿物材料的主要进展是硅灰石矿物纤维的制备与表面改性以及玄武岩纤维的规模化稳定制备关键技术。

中国矿业大学(北京)非金属矿物材料团队与企业合作攻克了高长径比超细硅灰石纤维的规模化制备工艺与装备、表面无机包覆与有机改性及其在复合材料中的应用技术,实现了硅灰石矿物纤维的规模化生产和在造纸(代替部分纸浆)以及PP、PE、尼龙等中的推广应用。浙江石金玄武岩纤维有限公司与东南大学相关科研团队研究解决了制约我国连续玄武岩纤维产业发展的技术瓶颈,开发成功了大漏板、过程智能化监控、专用浸润剂等连续玄武岩纤维高性能、低成本、低能耗的电熔炉生产技术及装备。

（七）硅质矿物材料

近几年硅质矿物材料的主要进展是高纯石英砂、球形硅微粉和高纯金属硅制备技术以及多晶硅生产工艺与装备优化、高纯熔融石英材料及光学石英材料制备与应用技术。

中材高新材料股份有限公司攻克了高纯熔融石英材料制备技术及应用技术，解决了用石英原料制备高纯熔融石英材料的规模化生产技术以及大尺寸石英陶瓷坩埚的产业化技术，实现了太阳能多晶硅铸锭用大尺寸石英陶瓷坩埚的完全国产化。中国电子科技集团公司第二研究所攻克了多晶硅片制造关键设备技术；哈尔滨工业大学攻克了无氯烷氧硅烷法制备多晶硅工艺技术；昆明理工大学攻克了真空冶金法制备低成本多晶硅工艺技术；天士力控股集团有限公司攻克了冶金法制备低成本多晶硅工艺技术；锦州新世纪石英（集团）有限公司攻克了高纯石英砂制备低成本多晶硅技术。武汉理工大学研发了利用脉石英制备超高纯石英的混合酸热压浸出与真空焙烧纯化技术。

（八）云母矿物材料

近几年云母矿物材料的主要进展是云母绝缘材料以及功能填料和颜料的表面改性。

武汉理工大学和湖北平安电工材料科研人员针对云母纸生产线前处理工艺、制浆和纸浆分级以及生产线智能控制关键技术进行了联合攻关，实现了厚度、纸浆浓度、水分以及烘干温度等参数的在线监测和智能控制，优化了云母纸的配方，同时制定了云母粉径厚比的测定标准及检验方法。该技术成果的应用和产业化示范显著改善了我国云母纸的质量及稳定性。

三、本学科国内外研究进展比较

国内外矿物材料学科的基本现状是发达国家拥有先发技术基础和较明显的技术优势。美欧日等发达国家从20世纪70年代中期开始开展功能矿物材料的制备、材料结构、性能与应用研究，并在80年代逐渐形成了规模化生产和应用；90年代开始广泛推广应用于工业、农业、通信、航空航天、航海、军工、环保、生态、健康领域，不仅促进了传统建材、冶金、轻工、化工、机械、建筑业的结构调整和产业升级，而且关联高技术与新材料产业、新能源、节能环保、生态建设、生物医药、电子信息等方面，产业规模不断扩大；到21世纪初，部分产业（如膨润土矿物材料）形成了年销售额超过百亿美元的产业集群。我国虽然在20世纪80年代也开始研发矿物材料，但基础研究薄弱、技术和装备落后。进入21世纪以来，特别是近几年我国矿物材料学科无论是基础研究还是技术开发，与发达国家的差距持续缩小，在石墨烯、超细煅烧高岭土、硅藻土健康与环保材料、凹凸棒石矿物材料、人造大理石和石英石、镁质矿物材料、功能矿物填料、矿物/聚合物复合材料等领域已达到国际先进水平，部分甚至达到国际领先。在检索统计的近十年国内外石墨矿物材料、石墨烯、黏土矿物材料、多孔矿物材料、钙质与镁质矿物材料、纤维矿物材料、硅质材料，云母材料等论文发表数量方面，国内研发机构均名列全球前10位，其中，石墨烯、石墨矿物材料、膨润土矿物材料、硅藻土矿物材料、沸石矿物材料、凹凸棒石矿物材料、石英材料等研究领域，国内研究机构名列全球第一。但是，我国矿物材料科技发展的总体水平与世界先进水平仍有一定差距，尤其是产业化技术、材料应用性能与需求的适应性以及应用技术方面，差距还较大。这些差距正是我们今后需要努力突破的方向。

在矿物材料产业发展方面，因资源种类、储量、品质及国情的不同，各个国家在矿物材料技术研发与产业发展的重点又有所不同，大多数国家是以本国优势资源为依托发展矿物材料与制品产业。欧洲、美国、日本等发达经济体在矿物材料产业方面仍处于前端或高端。近年来，我国矿物材料产业呈现快速发展态势，产能上已成为全球生产大国，但产业层次仍处于全球中端或中低端，在产品档次和应用性能方面与发达国家尚有一定差距。

四、本学科主要研究热点

近几年，石墨烯、石墨矿物材料、黏土矿物材料、多孔矿物材料、钙质矿物材料、镁质矿物材料、纤维矿物纤维、硅质矿物材料、云母矿物绝缘材料等均有研发的热点或重点。这些研发热点均与新能源与节能材料、环境治理材料、生态与健康材料、高硬度与高强度材料、无机阻燃材料、电工与电子材料以及复合功能材料等相关。

石墨矿物材料领域，材料品种与应用是最重要的研究方向，近几年电池阴极材料、柔性石墨、超高纯石墨、等静压石墨等是研发热点。石墨烯已成为全球材料学科最重要创新研究领域，主要研发热点是大规模、低成本制备与应用技术以及石墨烯复合材料。

黏土矿物材料的研究热点集中于膨润土、高岭土和凹凸棒石矿物材料。近几年，膨润土侧重于品种与应用性能的研究，特别是膨润土环境与生态修复材料、防水材料、霉菌毒素吸附材料、复合材料等；凹凸棒石侧重于结构和性能以及纳米棒晶解离、分散和高值化利用；高岭土矿物材料侧重于功能填料、石油化工催化材料以及煤系煅烧高岭土与应用等。

多孔矿物材料的研究热点是沸石和硅藻土矿物材料。沸石矿物材料主要以合成沸石及其在石油化工催化、吸附分离、环保中的应用为主要研究重点。硅藻土矿物材料的主要研发重点是环保健康功能材料、调湿材料、节能材料和高性能过滤材料。

石膏矿物材料、碳酸钙制备与应用是钙质矿物材料的研究热点。石膏矿物材料的主要研究方向是品种与应用；碳酸钙的主要研发重点是纳米碳酸钙制备与应用、粒度与晶型调控以及生产工艺优化。

镁质矿物材料的研发热点主要是滑石复合材料的制备、改性及其应用；电熔镁砂节能与清洁制备技术；水镁石阻燃材料的制备与应用。

纤维矿物材料近几年的研发热点是高长径比硅灰石矿纤的制备、复合与表面改性及其应用以及玄武岩纤维矿物的制备与应用。

高纯和超高纯石英砂、球形硅微粉、单晶硅、多晶硅、金属硅等硅质材料是近年来的研发热点。石英原料的主要研究方向是高纯加工技术；单晶硅、多晶硅、金属硅等硅材料的主要研究方向是制备工艺与设备、材料结构与性能以及应用。

现代电工、电器、电子用云母纸及制品等高性能云母绝缘材料是近几年云母矿物材料的主要研发热点。

五、对本学科发展的建议

（一）健全矿物材料学科体系

中国矿物材料学科是2000年前后由地质与矿业院校自主创建起来的，目前只有硕士点和博士点，只能培养少量矿物材料或矿物材料工程专业的博士和硕士研究生，矿物材料行业的企业严重缺乏本专业的本科生及相应的专业人才。建议在高等学校相应的矿业工程、材料科学与工程等一级学科下建立矿物材料工程或矿物材料与应用本科专业；在矿业工程、地质和材料科学与工程类科研院所，特别是非金属矿类研究院所以及大专院校设立相应的矿物材料与应用研究中心或重点实验室；同时鼓励矿物材料行业的大型企业和企业集团建立矿物材料与应用技术研究或工程技术中心。

（二）强化矿物材料科学研究

虽然近年来我国矿物材料科学研究取得了显著进步，但是，与世界发达国家相比，在矿物材料科学与应用基础研究方面仍存在较大差距，需要继续强化科学研究。建议国家自然科学基金委和科技部增强矿物材料与应用资助力度；地方政府根据区域资源特点立项支持矿物材料与应用科研项目，并配套支持国家自然科学基金委和科技部立项的矿物材料与应用项目；大专院校和国家科研院所利用教育部、地方教育部门专项科研经费和科技部院所科研专项经费持续立项支持矿物材料与应用基础研究与应用基础研究。

（三）加大矿物材料技术开发力度

目前，在国际矿物材料产业中我国矿物材料产业仍处于中下游位置。生产与应用技术落后是主要原因。在未来相当长的时间内，产业化关键技术，特别是应用技术仍是我国矿物材料行业的主要研发方向。建议科技部在国家层面的科技研发计划安排中相应提高矿物功能材料应用项目的比例；国家有关部门和地方政府积极扶持矿物材料行业企业建立技术创新中心，并立项支持成果产业转化。

第十九节 建筑学

一、引言

建筑学是研究建筑物及其内外空间与环境的学科,兼具社会性、技术性、艺术性等多重属性。通常情况下,建筑学是指与设计和建造相关的艺术与技术的总和,并区别于一般的建设技艺。在不同历史时期,根据不同的社会需要,建筑学一直与时俱进,持续发展,并在提升和改善人居环境中发挥了基础性的重要作用。传统的建筑学研究建筑物、建筑群以及室内家具的设计,风景园林和城市乡村的规划设计。随着学科的发展,城乡规划学和风景园林学从建筑学中分化出来,成为与之并列的一级学科。

建筑学主要包括建筑设计、室内设计、建筑构造、建筑历史、建筑理论、城市设计、建筑物理和建筑设备等方面的内容。建筑学的内涵和外延目前仍然在不断发展中,但其围绕人居环境提升和改善的定位基本未变。

基于我国建筑学学科近年来的最新发展,结合国外建筑学研究前沿动态,本课题按照六个方向展开研究,分别论述每个方向的国内外研究现状和动向,并对其未来的发展趋势进行分析和展望。六个方向分别是:①新型城镇化背景下的城市设计;②全球化进程中的建筑文化;③应对全球气候变化的建筑环境可持续发展;④城乡统筹发展背景下乡村建设;⑤信息化时代建筑数字技术;⑥产业现代化背景下的建筑工业化。

二、本学科近年的最新研究进展

(一)新型城镇化背景下的城市设计

城市设计主要研究城市空间形态的建构机理和场所营造,是对包括人、自然、社会、文化、空间形态等因素在内的城市人居环境所进行的研究、实践和实施管理活动。

在中国史无前例的快速城市化进程中,很多城市都不同程度地经历了城市规模的急剧扩张,城市的功能结构、空间环境和街廊肌理发生了显著的变化,出现了规划管理乏力、城市建设无序、文化传承缺失等问题,而其中城市规划与建筑设计、景观设计和基础设施建设之间的系统协调又是关键问题之一，亟待加强城市设计研究。进入21世纪以来，城市设计突破了以往主要关注物质空间视觉秩序的局限，进入关注人文、社会和城市活力的新阶段。当前，中国城市设计发展呈现出学术探索空前活跃，理论方法探索与西方并驾齐驱、工程实践面广量大、技术水平后来居上的发展趋势。

首先，中西方学者在城市设计理论和方法研究上开始出现齐头并进的发展态势，西方学者继续在城市形态分析理论、城市设计方法论等相关领域展开探索。中国城市设计学科最初总体顺应以美、日为代表的国际城市设计发展潮流起步，20世纪90年代开始，逐步建立起中国城市设计理论与方法架构；新千年伊始，随着快速城市化的进程和城市建设发展社会转型，城市设计项目实践得到了长足的发展，中国城市设计出现了一些体系性的新发展。理论探索主要反映在城市设计对可持续发展和低碳社会的关注、数字技术发展对城市设计形体构思和技术方法的推动，以及当代艺术思潮流变对城市设计的影响等方面。

其次，与西方相比，中国城市设计实践呈现出后发的活跃性、普遍性和探索性。欧美城市化进程趋于稳定，其城市设计实践多为局部性项目。而中国特色的城市设计实践探索则呈现出明显的丰富性。中国与西方发达国家的城市发展时段相位的不同导致中西方城市设计实践的不均衡性。20世纪90年代中期以降，中国城市设计项目实践呈现"面广量大"的现象，反映出四大核心内容：概念性城市设计、针对未来城市结构调整和目标完善的城市设计、城镇历史遗产保护和社区活力营造、基于生态优先理念的绿色城市设计。

（二）全球化进程中的建筑文化

"建筑文化"的讨论可有三层含义，一是社会文化特质内化于建筑后，所表现出的形貌和气质；二是建筑如何影响人们的空间行为及人际关系；三是对建筑传统与时尚的系统思考、传播及其理论化。

近年来，在求新求变的当代建筑思潮占据话语权的时代背景下，建筑界讨论比较集中的话题主要在两个命题：一是"如何批判性地反思建筑现代性和现代主义的激进建筑文化"；二是"如何批判性地传承和复兴'在地性'的传统建筑文化"；这两大命题可大致包括四方面内容。

其一是对现代主义建筑批判性继承的论述。对现代主义批判的批判也是螺旋式反思的一种尝试。总的来说，这种对"现代主义批判"的批判，反思了现代之后建筑出现的复制、仿古、标新立异等形式主义倾向。其二是对现代之前传统建筑文化的再认识和再提炼，表现为对传统、地域性、身份识别性的新探究。其三是对建筑创新问题的理论探讨。对于如何创新，今天的建筑理论家已经不再从哲学、艺术等层面讨论为新而新，而是更多从传统和原型中汲取营养与古为新。同时，建筑创新问题的讨论还聚焦于多样性，多元化的追求，这也早已是现代建筑文化发展的一个时代主题。其四是为对全球在地性、地域性和人本主义的讨论。

（三）应对全球气候变化的建筑环境可持续发展

气候变化加剧了洪涝、干旱及其他气象灾害，并导致高温、暴雨等异常极端天气增多，影响人类生存环境及可持续发展。因此，应对气候变化已成为建筑学学科的核心课题之一。在城市空间规划方面，注重土地集约发展模式，提高能源利用效率，倡导绿色循环经济和绿色交通，采取生态城市设计方法，实现延缓和适应全球气候变化的城市发展目标；在绿色生态环境方面注重节能减排、雨洪管理、生态保护与修复和景观营造；在建筑物理环境方面注重不同气候条件的物理环境需求，营造舒适的建筑室内外环境，减少建筑对自然资源的依赖；在综合防灾减灾方面注重构建系统的防灾减灾体系，从城市、街区及建筑层面加强建筑环境的韧性。

在城市空间规划方面，从最早的生态城市理念雏形到以有机疏散理论、城市人文生态学为代表的城市规划思想中都蕴含着生态城市的思想火花。我国学者结合国内外生态思想，逐步形成了具有中国特色的生态城市研究体系。在绿色生态环境方面，随着我国城市化进程的明显加快，基于我国国情与景观生态学理论，景观生态安全格局、绿色基础设施、绿色景观营造技术等研究成为热点。在建筑物理环境方面，国内外研究热点主要包括建筑节能、绿色建筑、建筑室内外声光热环境等内容。在综合防灾减灾方面，国外关于城市防灾减灾的理论研究和实践成果较多，主要集中于灾害法律法规制定、城市防灾规划研究、城市防灾空间设计及城市灾害应急管理救援等几个方面。我国在该领域也取得了一定的研究成果，形成了自己的灾害管理与救援体制，对于单灾种的研究已经比较深入，近年来也逐渐开展了一些关于综合防灾的研究课题。

（四）城乡统筹发展背景下的乡村建设

农业、农民与农村"三农问题"是中国国民经济与社会发展的难点。自2006年以来，中国乡村先后历经了社会主义新农村建设、新型农村社区建设、美丽乡村建设三大发展阶段，同时穿插着传统村落保护、特色小镇建设、扶贫安居工程等一系列重大专题工作，取得了显著的成绩也带动了建筑学界乡建研究热潮。

国外与乡村建设相关的研究主要包括乡村土地利用、乡村管理与政策、乡村保护、城乡关系、乡村生态、乡村社会和乡村环境等内容，同时还关注乡村遗产、乡村可持续性、乡土建筑、乡村景观和保护等。

国内的乡村建设研究基于我国国情，以问题为导向，结合城乡统筹和新型城镇化战略，取得了积极的成果。主要研究热点包括传统村落、新型农村社区、乡村治理、美丽乡村、乡村旅游、城乡关系、乡村复兴、村镇规划管理等。在建筑层面，研究主体多围绕传统民居展开，一方面向更大范围的乡村聚落拓展，一方面向更深层次的绿色、节能技术延伸。

（五）信息化时代的建筑数字技术

建筑设计与创造和革新紧密相关，随着信息数字技术的介入，其方法也正处于不断变迁的状态，新的理论、系统、方法不仅使其成果新颖独特，同时诠释着全新的设计理念，信息化时代的建筑数字技术将对建筑学产生深远影响。相关研究成果突出体现了先进数字技术引领下建筑学的无限可能性，创造出众多兼具科学性和想象力的实践成果。

在数字建筑的理论和方法探索上，近些年来的研究主要围绕几个方向展开：非线性系统理论与数字建筑设计的结合，信息论与控制论对数字建筑设计的影响，德勒兹哲学思想对数字建筑设计的影响，数字建构的理论和方法，参数化建筑设计与算法生形，物联网和建筑智能环境等。

在我国大规模城市化建设的背景下，数字建筑的实践探索也在理论与方法研究的推动下蓬勃发展。这些实践项目充分利用了近年来我国数字建筑理论与方法层面的研究成果，践行了参数化设计、数字建构、工匠技艺、地域特色、环境融合等理论方法，推动了整个建筑产业链的进步与升级。许多实践项目也成了当地著名的地标建筑。

（六）产业现代化背景下的建筑工业化

建筑业是我国国民经济的支柱产业。经过30多年的改革发展，建筑业对经济社会发展、城乡建设和民生改善做出了重要贡献。但同时，我国建筑行业的现状仍然不容乐观，劳动生产率总体偏低，资源与能源消耗严重，污染问题突出，建筑施工人员素质不高，建筑寿命短，建筑工程的质量与安全存在诸多问题。这些问题产生的根源就是传统落后的建筑生产方式，而建筑工业化正是建筑行业转型升级的有效路径之一。

建筑工业化是世界发达国家产业发展现代化的标志之一。工业化住宅的建设方式最早由欧美等发达国家提出并得到发展。美国、日本、德国、瑞典等国家是世界范围内建筑工业化发展水平领先的国家，在建筑工业化的道路上有着不同的发展方向，例如，美国高度成熟的木结构工业化住宅体系、欧洲多国的预制装配式混凝土体系、德国的工业化模板现浇混凝土体系等。

国内的建筑工业化研究近年来方兴未艾，从设计理论到施工技术，从结构体系到内装体系，高校、科研院所、设计企业、施工企业等都开展了一系列研究，取得了积极的成果。

三、本学科发展趋势和展望

新时代的建筑学一方面在世界范围内要积极回应绿色生态可持续的发展理念及全球化背景下地域文化兴起提出的要求，另一方面在国内要成为引领新型城镇化的重要核心学科。在此背景下，建筑学学科将呈现若干重要的发展趋势，主要包括：学科成果成为提升

城镇建设水平的重要技术手段和管理支撑；以进化和发展的观点和视角看待建筑文化与遗产保护；绿色建筑理论与技术成为贯穿建设全过程的重要方针；乡村建设向科学和可持续方向发展；从数字建筑设计走向工业 4.0；建筑工业化成为我国未来建筑业发展的关键。

新型城镇化背景下的城市设计的主要研究任务包括：构建创新型城市设计理论体系，基于可持续原则的绿色城市设计，结合遗产保护的城市特色保护与有机更新，基于大数据支撑的城市设计发展等。

全球化进程中的建筑文化的主要研究任务包括：场所精神与地域主义，传统与现代，原型与原创等。

应对全球气候变化的建筑环境可持续发展的主要研究任务包括：可持续城市空间规划与环境，绿色低碳的建筑物理环境，城市与建筑综合防灾减灾等。

城乡统筹发展背景下的乡村建设括的主要研究任务包括：转型期乡村规划建设策略，乡村既有建筑环境品质提升技术，乡村空废环境绿色消解与再利用模式，现代农业新型镇村体系规划方法，传统建筑保护与传承模式等。

信息化时代的建筑数字技术的主要研究任务包括：算法驱动的建筑设计，数控建造，基于建筑性能和物联网的智能建筑等。

产业现代化背景下的建筑工业化主要研究任务包括：建筑工业化的技术标准体系，建筑主体结构的工业化技术，建筑部品部件的工业化技术，建筑工业化的配套装备与工艺，支撑建筑工业化的信息及数字技术。

第二十节　纺织科学技术

一、引言

中国是世界上最大的纺织品生产国、出口国和消费国，纺织工业是我国传统支柱产业、重要的民生产业和创造国际化新优势的产业，是科技和时尚融合、衣着消费与产业用并举的产业，在美化人民生活、带动相关产业、拉动内需增长、建设生态文明、增强文化自信、促进社会和谐等方面发挥着重要作用。2016 年纺织行业规模以上企业实现主营业务收入 73302.26 亿元，主营收入占规模以上工业企业的 6.37%。中国纤维加工总量超过

5400万t，纺织品服装出口占全国货物出口比重12.88%，中国纺织品服装出口额约占全球纺织品服装交易额的38%。

近年来，纺织领域科技进步显著，我国常规纤维的差别化、多功能化水平显著提升，生物基纤维原料的产业化、绿色化生产技术取得突破，碳纤维、间位芳纶、超高分子量聚乙烯纤维、聚苯硫醚纤维等高性能纤维稳步发展；国产纺织装备高速化、连续化和自动化水平进一步提升，纺织产品开发力度加大，产品品种增加、品质提升；产业用纺织品制备技术取得显著进步，产业化步伐加快，应用范围扩大，增长较为迅速；印染加工领域积极开展绿色生产技术创新，少水及无水印染加工、短流程工艺、生态化学品应用等清洁生产技术得到推广应用；依托工业物联网技术，纺织智能制造正在给行业带来深刻的变化。2013—2016年纺织科学技术学科荣获国家技术发明奖二等奖2项、科技进步奖一等奖1项、二等奖11项。目前我国建有与纺织密切相关的国家级研发平台9个，其中国家重点实验室2个、国家工程技术研究中心6个、国家工程实验室1个。

在纺织人才培养方面，全国有40多所高校设有纺织工程本科专业，有60多所高校开设了服装设计与工程本科专业，有20多所开设了轻化工程（染整工程方向）本科专业。全国有7所高校具有纺织科学与工程一级学科博士学位授予权，有19所高校具有纺织科学与工程一级学科硕士学位授予权。

二、近年来本学科国内发展情况

（一）纤维材料工程

在生物质纤维方面，我国加大引进吸收和自主开发Lyocell生产技术力度，产能已达数万吨级，"万吨级新溶剂法纤维素纤维关键技术及产业化"项目获得2016年中国纺织工业联合会科学技术奖一等奖；"海藻纤维制备产业化成套技术及装备"项目获得2016年中国纺织工业联合会科学技术奖一等奖。在合成纤维方面，"新型共聚酯MCDP连续聚合、纺丝及染整技术"项目获得2014年国家技术发明奖二等奖；"管外降膜式液相增黏反应器创制及熔体直纺涤纶工业丝新技术"项目获得2016年国家技术发明奖二等奖。在高性能纤维方面，2016年5月，中国第一条千吨级T800原丝投产，"千吨级干喷湿纺高性能碳纤维产业化关键技术及自主装备"项目获得2016年中国纺织工业联合会科学技术奖一等奖；"干法纺聚酰亚胺纤维制备关键技术及产业化"项目获得2016年国家科技进步奖二等奖。我国开发的"连续油水分离功能材料及装备"于2017年获得第45届日内瓦国际发明展金奖。

（二）纺纱工程

在高效能精梳技术、特种纤维纺纱、特高支纺纱技术以及纺纱流程的连续化和智能化方面取得进展。"高效能棉纺精梳关键技术及其产业化应用"项目获得2014年国家科学技术进步奖二等奖，"苎麻生态高效纺织加工关键技术及产业化"项目获得2016年国家科技进步奖二等奖，"汉麻高效可控清洁化纺织加工关键技术与设备及其产业化"项目获得2016年中国纺织工业联合会科学技术奖一等奖，"特高支精梳纯棉单纺紧密纺纱线研发及产业化关键技术"项目获得2014年中国纺织工业联合会科学技术奖一等奖，"粗细联合智能全自动粗纱机系统"项目获得2015年中国纺织工业联合会科学技术奖一等奖，"环锭纺纱智能化关键技术开发和集成"项目获得2016年中国纺织工业联合会科学技术奖一等奖。

（三）机织工程

在高效环保浆纱技术、纯棉高档面料和特殊结构机织物开发方面取得进展。"经纱泡沫上浆关键技术研发及产业化应用"项目获得2015年中国纺织工业联合会科学技术奖一等奖；"纯棉超细高密弹力色织面料关键技术研发及产业化"项目获得2015年中国纺织工业联合会科学技术奖一等奖；"大褶裥大提花机织面料喷气整体织造关键技术研究及产业化应用"项目获得2016年中国纺织工业联合会科学技术奖一等奖。

（四）针织工程

在互联网技术应用、智能化提花技术、无缝成型技术、针织装备高速化方面取得一系列突破。"支持工业互联网的全自动电脑针织横机装备关键技术及产业化"项目获得2016年国家科学技术进步奖二等奖，自主研发的互联网针织MES系统于2017年4月通过中国纺织工业联合会组织的科技成果鉴定；"织物智能提花工艺技术的创新与产业化应用"项目获得2015年中国纺织工业联合会科学技术奖二等奖；"HYQ系列数控多功能圆纬无缝成型机"项目获得2015年中国纺织工业联合会科学技术奖一等奖。

（五）非织造材料与工程

在新型熔喷非织造技术、医卫防护材料、环保除尘材料、土工材料、过滤材料等方面进步显著。"新型熔喷非织造材料的关键制备技术及其产业化"项目获得2014年国家科技进步奖二等奖；"医卫防护材料关键加工技术及产业化"项目获得2016年中国纺织工业联合会科学技术奖一等奖；"垃圾焚烧烟气处理过滤袋和高模量含氟纤维制备关键技术"项目获得2016年中国纺织工业联合会科学技术奖一等奖；"宽幅高强非织造土工合成材料关键制备技术及装备产业化"项目获得2015年中国纺织工业联合会科学技术奖一等奖。

（六）纺织化学与染整工程

在纺织化学品方面，"基团功能强化的新型反应性染料创制与应用"项目解决了长期困扰国际染料行业的难题，获得2016年国家技术发明奖二等奖。在印染技术方面，"筒子纱数字化自动染色成套技术与装备"项目获得2014年国家科技进步奖一等奖；"高速数码喷印设备关键技术研发及应用"项目获得2016年浙江省科技进步奖一等奖。在纺织废弃物回收再利用方面进展显著，"丝胶回收与综合利用关键技术及产业化"项目获2013年国家科技进步奖二等奖。

（七）服装设计与工程

在应用高新技术方面取得进展，如人体测量技术、服装虚拟设计、智能穿戴技术等。"服装三维测体试穿系统的开发及应用"项目获得2014年中国纺织工业联合会科学技术进步奖三等奖。具有防火隔热、防水透气、防毒抗菌等多种功能防护服的研发方面取得一系列成果，"防护服的多功能设计研发及性能评价"项目获得2013年中国纺织工业联合会科学技术进步奖二等奖。在服装制造流程自动化、信息化、智能化也取得了一些集成创新成果。

（八）产业用纺织品

在医疗卫生用纺织品、大气和水处理及污染治理用过滤纺织品、安全防护纺织品等制备技术取得显著进步，"医卫防护材料关键加工技术及产业化""垃圾焚烧烟气处理过滤袋和高模量含氟纤维制备关键技术"和"多功能飞行服面料和系列降落伞材料关键技术及产业化"3个项目均获得2016年中国纺织工业联合会科学技术奖一等奖。

三、本学科国内外研究进展比较

（一）纤维材料工程

近年来纳米纤维材料、生物基纤维、高性能纤维等领域发展受到国内外的普遍关注，我国与世界先进水平的差距正在缩小。奥地利兰精公司拥有全球最大的6.7万吨/年的生物质再生纤维Lyocell生产线，产能已达22万吨/年，我国于2014年建成1.5万吨/年的Lyocell纤维生产线打破了国外公司多年垄断，但生产能力存在较大差距；我国差别化纤维种类跟国际先进水平相当，但是我国在差别化纤维研发技术和纺丝设备上与国外仍有差距；我国与日本在碳纤维研制方面的差距在缩小，但仍然存在明显的差距。2016年我国间位芳纶产能已达8500吨，产能全球排名第二，仅次于美国杜邦。

（二）纺纱工程

在纺纱生产高速化方面，与国外相比，我国在制造精度、技术先进性及高速运转的稳定性等方面仍有一定差距。纺纱关键专件水平方面，我国纺纱专件器材制造企业产品质量总体上取得了长足进步，逐步缩小并部分超越国外先进企业，但大多数纺纱专件器材与进口同类产品相比在质量、寿命上尚有差距。纺纱在线检测技术方面，我国也有较快发展，但相比国际领先水平，我国在在线检测和控制技术的精密性、智能化等方面还有待提高。

（三）机织工程

在机织工程领域，我国与国际先进水平在自动化控制、高速化、低能耗方面还存在一定差距。国外在织机的控制技术精度、节能等方面处于领先水平，喷气织机的织造速度超过了 2000 转 / 分，而国内最高为 1200 转 / 分。从国外机织 CAD 软件研发现状来看，开发水平高、界面美观，功能强大，并且许多软件包已具备了纺织品设计与服装设计一体的功能，这些方面国内差距明显。

（四）针织工程

高效、高速生产仍然是针织生产的最大特色。国内外针织装备不断提速，但国内装备在核心技术上存在一定差距。在针织装备数字化技术方面，德国斯托尔和日本岛精等国外品牌的针织横机生产技术处于国际领先水平。与国外先进的针织装备相比，我国的经编机、圆纬机和横编机在生产高速化及性能稳定性等方面与国际先进水平仍存在相当大的差距。

（五）非织造材料与工程学科

近年来国内外都关注于双组分纺粘非织造布的研究，美国 Hills（希尔斯）公司研发的双组分纺粘技术是目前国际上较先进的双组分纺粘非织造技术，而我国开发了纺粘与其他工艺复合非织造材料，但在自主研发纺粘新技术方面与国外存在一定差距。在废弃非织造材料特别是聚酯产品回收方面，我国已成为再生聚酯纤维的第一生产大国，但是我国在废弃聚酯的回收利用率、技术水平和产品档次方面尚待突破。在非织造专用纤维方面，国外在技术和产品上占据明显的优势，而在梳理的高握持力针布开发方面，我国针布产品的某些指标已经达到世界先进技术水平，针布齿条质量已与最为先进的英国 ECC 公司相当。

（六）纺织化学与染整工程学科

近年来我国与国外纺织化学与染整工程学科的研究开发差距主要在整体解决方案推广和使用等方面。在纺织化学品开发技术方面，国外优势主要体现在新型生态染料、环保助剂以及高效持久的功能性后整理剂的开发和使用。在染整加工技术方面，国外在智能化技术、自动化技术以及成套技术开发等方面研究水平相对较高。

（七）服装设计与工程学科

我国是服装生产和出口大国，但是在可持续发展服装设计与工程技术、智能服装和功能服装方面，我国与国外先进水平存在较大差距。服装 CAD 系统方面，国际上欧美、日本以及韩国等国家和地区服装 CAD 发展相当快，而我国的服装 CAD 普及率相对较低，且核心技术的开发和研究不足。

（八）产业用纺织品学科

我国各类产业用纺织品种类齐全，基本与国际水平相当，但在产品性能和产品创新上仍存在欠缺。在工艺技术和装备方面，捷克的全球首条 Elmarco 静电纺纤维生产线已经投放市场，美国克莱姆森大学与土耳其伊斯坦布尔大学合作开发了高性能汽车专用噪声防护材料生产线，而我国产业用纺织品加工装备及工艺研发投入不足，在自动化、信息化、智能化等方面有待提高。另外，我国的高端产业用纺织品，特别是高端医疗用纺织品的纤维原料工程化、产业化能力还比较弱。

四、本学科发展趋势及展望

近年来，我国纺织工业加快转型升级和结构调整，科技创新能力不断提升。高性能纤维、生物质纤维和差别化、功能化纤维成为纤维研发的主要方向，新型自动化、数字化和智能化纺纱、织造、针织、染整和服装设备将得到快速发展，信息网络技术不断深入纺织工业，大大提高了纺织生产力和纺织服装制造水平。

在纤维材料工程学科领域，纤维材料高新技术的重点包括化纤高效柔性、多功能加工关键技术及装备，高性能纤维及复合材料的制备技术，生物基原料高效合成制备技术和尼龙 56、PLA、PTT 等生物基合成纤维产业化技术，以及纳米纤维材料及应用产业化技术等。此外，加强探索废旧纺织品等再生资源的综合利用产业发展机制的新模式，建立废旧纺织品的循环利用体系，提高废旧纺织品再生利用率也是今后的发展重点。

在纺纱工程学科领域，高效纺纱技术、智能化纺纱设备和复合纱线品种的研发是未来发展方向。纺纱业将从纺纱技术通用化、纺纱生产连续化和自动化、纱线制备技术和装备的高端化等方面不断创新发展，使企业实现节约能源、节约资源、节省用工、提高生产稳定性和保证产品高品质，提升产品附加值。

在机织工程学科领域，需进一步创新高效环保浆纱技术、自动穿经技术和织造关键技术，实现环保、低耗，提高劳动生产率。研究新型织造工艺技术、新产品和设备，探索织造工序连续化、自动化、高效化；进一步研究机织 CAD 技术，在建模方法上向动态建模和网络化发展。

在针织工程学科领域，将重点发展针织全成形技术、智能化针织技术、针织物联网技术，实现装备智能化、生产模式智能化以及智能针织产品；实现全成形四针床电脑横机的国产化，以及全成形时装、全成形服饰、3D 异形结构的产品设计；实现针织物在线真实感仿真以及真实场景效果的虚拟试衣。

在非织造材料与工程学科领域，重点研究方向包括高速、高产量、低成本生产技术，实现资源节约；以棉和麻纤维为原料的天然纤维非织造产品的加工方式，实现产品的多样化；进一步优化非织造产品环保、节能生产工艺；研发环境友好型的非织造布用生态型黏合剂以及环保型后整理剂。

在纺织化学与染整工程学科领域，生态环保仍然是该领域面临的压力和挑战，亟待形成适合生物酶前处理加工的成套加工技术和整体解决方案；开发和应用智能化印染连续生产和数字化间歇式染色整体技术；继续发展非水介质染色技术、无染料染色技术；持续研究印染废水脱色和降解技术；进一步研究生态纺织化学品开发和零排放技术。

在服装设计与工程学科领域，将研究基于VR的服装体验设计与产品开发，为服装行业从制造经济转向升级为服务经济和体验经济提供途径；研究基于物联网的服装柔性定制生产体系关键技术，实现顾客定制需求的面料、版型与工艺生产架构的融合和衔接；研发防护服装、功能服装和智能服装，满足人们的特殊需求和高品质生活的需要。

在产业用纺织品学科领域，将继续呈现产品应用快速拓展并向中高端升级的发展趋势。今后增强用纤维基复合材料应用范围不断扩大，尤其是碳纤维和高强高模聚乙烯增强复合材料的研发应用；环保用高效滤料研发和废旧滤料回收技术的市场更为广阔；高端生物医用纺织品的研发力度将会加大；土工用纺织材料的应用普及程度也会继续提高。

第二十一节　制浆造纸科学技术

一、引言

制浆造纸工业是与国民经济和社会事业发展关系密切的重要基础原材料工业，纸及纸板的消费水平在一定程度上反映着一个国家的现代化水平和文明程度。特别是在全球环境危机所催生的低碳经济和可持续发展理念背景下，基于制浆造纸工业的森林生物质产业链及其生物质燃料、生物质材料、纸基功能材料产品，已经成为化石能源和材料的重要替代资源，方兴未艾的制浆造纸工业转型升级有望成为绿色循环经济的重要推动力量。

作为当代造纸技术的奠基者，我国东汉蔡伦所发明的造纸术及其在古丝绸之路沿线国家的传播发展，为推动人类文明进步以及世界文化、科学和信息的传承发挥了极为重要的作用。我国现代制浆造纸工业则在经历工业时代前期的缓慢发展之后，于近30年间步入快速发展期，重新成为世界造纸大国。我国已经成为众多跨国造纸企业及关联产业的生产研发基地和重要目标市场。

本报告主要论述制浆造纸科学技术学科近年来各领域的发展概况、最新进展、与国外先进技术水平的差距、未来的发展趋势等，提出了我国制浆造纸科学技术学科的研究方向建议。

二、本学科近年的最新研究进展

根据国务院学位委员会颁布的学科目录，制浆造纸工程学科隶属于工学门类一级学科，轻工技术与工程中的二级学科，具有相对独立、自成体系的理论和知识基础构成。目前，制浆造纸工程学科已形成了一个以制浆工艺、造纸工艺、制浆造纸机械装备、制浆造纸化学品、环境保护、纸基功能材料为核心领域的现代化应用科学技术体系。

截至2015年年末，制浆造纸行业内具有较高影响力的主要科研机构约37家。其中行业性专业科研院/所和工程设计院/公司21家，设有制浆造纸专业的高等院校24家，国家认定的制浆造纸企业技术中心9家。我国制浆造纸工程领域已形成了大专－本科－硕士－博士－博士后流动站的完整人才培养体系，是我国制浆造纸学科高层次人才的最重要来源。

（一）制浆科学技术

化学法制浆技术近几年在降低蒸煮温度、提前蒸煮终点、强化氧脱木素等方面的技术取得了一些进展。紧凑连续蒸煮具有连续、低温蒸煮、液比高、脱木素选择性好，系统启动迅速、生产消耗低、产品质量稳定等优点，成为大型制浆厂采用的主流技术。置换蒸煮对制浆厂来说具有较好的适应性。DDS置换蒸煮是低能耗置换间歇蒸煮技术，蒸煮均匀，浆料质量稳定。该技术扩展了初级蒸煮的作用，提高了纸浆得率，能够得到卡伯值低、强度高的纸浆。秸秆经过大液比立锅置换蒸煮（立锅中央置换循环管），大幅降低用碱量，蒸煮终点提前，滤水性改善，可大幅提高黑液提取浓度，提取的草浆进行氧脱木素，形成秸秆制浆新技术。

化学机械法制浆技术，近年来，主要研究和使用的方法有：BCTMP和APMP或者P-RC APMP方法。BCTMP作为世界上先进的化学机械浆生产工艺被广泛应用于我国各大企业中。目前，主要的研究在于纸浆质量优化和污染物控制，APMP/P-RC APMP作为主要的化学机械法制浆技术，其技术改进和优化适用于各种造纸纤维原料，是我国化学机械浆生产所采用的最主要工艺。在化学机械浆生产中，添加渗透剂可以提高制浆效率。以$Mg(OH)_2$或MgO为碱源的过氧化氢漂白，改良PM过氧化氢漂白，提高白度的RT预处理，植入OBA增白的过氧化氢漂白技术及可回用H_2O_2的中高浓组合漂白技术成为技术改进和发展的方向。

提高废纸回收利用率的关键是脱墨技术，其中脱墨剂的研究和应用成为关键。随着新型油墨品种不断增多，新的纸张表面化学助剂不断应用。在脱墨剂中加入表面活性剂，会提高废纸的脱墨效果。生物酶常用于废纸制浆的预处理，以增强后续脱墨效果。

无元素氯漂白（ECF）和全无氯漂白（TCF）技术是化学浆漂白的主流方向，添加臭氧的漂白已渐普通。

（二）造纸科学技术

近年来，制浆造纸工业的技术进步主要体现在压缩成本和提高产品品质上。以数字化、网络化、智能化为主要特征的"工业4.0"时代对造纸科学技术的发展产生了积极的推动作用。

木材中浓打浆技术日趋成熟。对于厚壁纤维针叶木浆打浆低浓预处理和中浓打浆相结合的打浆方式成为造纸企业的主流打浆工艺。打浆技术装备由传统的圆盘磨浆机向着锥形磨浆机和圆柱形磨浆机方向过渡。纸机的稀释水流浆箱一直在不断完善和优化中。高速文化用纸机多采用长网+顶网成形器、混合型夹网成形技术、立式夹网成形技术，可以改善纸张的两面差、Z向结构和填料的分布。

与普通压榨技术相比，新型压榨具有较长的压区滞留时间、较高的脉冲能力和脱水能力。在大幅提高纸张干度的同时，能够保持纸张松厚度特性。新型压榨主要包括大辊压榨、靴式压榨以及组合压榨。

现代高速纸机的干燥部出现了许多新技术：采用单排烘缸组代替传统的双排烘缸组，烘缸之间增设纸幅稳定器、无绳引纸系统、真空引纸传送带，采用热吹风管与袋式通风装置等。

现代涂布技术主要包括计量式膜转移施胶式涂布、喷雾涂布和帘式涂布等。近几年，国外出现了一系列先进的涂布方式与设备，维美德公司相继开发的OptiCoater涂布机、OptiCoater Jet喷射式涂布、OptiBlade新概念刮刀涂布等。

（三）制浆造纸化学品研究现状和进展

高效和功能性化学品是制浆造纸过程节能降耗减排的重要技术手段，针对制浆和清洁漂白需求，改性蒽醌（AQ）类蒸煮助剂、四乙酰乙二胺（TAED）类氧漂白活化剂和木素过氧化物酶类生物酶漂白剂的研究取得阶段性成果；研究开发了废纸脱墨用固体脱墨剂和高效的中性脱墨技术；二氧化氯制备技术与装置国产化研究开发已取得突破，拥有自主知识产权的技术成果成功应用于工业生产。

制浆造纸废水处理化学品的研究是目前的热点，研究开发了聚铝和聚铁类废水深度处理用高效混凝剂及其应用技术。

蒸煮助剂仍以AQ及其改性物为主。在深度脱木素技术中，蒸煮助剂正在由使用单一的蒸煮助剂向使用性能优良的复配型蒸煮助剂的方向发展。离子液体独特的性能及其良好的溶解和分离能力决定了其在纤维素工业和制浆造纸领域必将发挥越来越重要的作用。目前，臭氧漂白在发达国家已经实现了工业化。用氧漂活化剂进行漂白是对传统过氧化氢漂白的一个重大革新。

纸机抄造过程的高性能新型助留助滤剂、合成表面施胶剂、杀菌防腐剂等方面的研究较为深入，取得了一批新技术成果。

制浆造纸行业环保技术升级迫切。水处理絮凝剂在我国发展十分迅速，从低分子质量到高分子质量，从无机物到有机物，从单一到复合，形成了系列化和多样化产品。

（四）污染治理研究现状与进展

制浆造纸行业对自然环境所造成的污染仍比较严重，尤其是对水环境的污染，一直是工业污染防治的重点。分为源头控制和末端治理两方面。源头控制方面，目前国内相关的制浆造纸清洁生产技术主要包括高效黑液提取、深度脱木素、氧脱木素、ECF、TCF、低白度漂白等。末端治理方面，目前国内制浆造纸废水的处理技术一般分为一级处理技术、二级处理技术、三级处理技术。

制浆造纸行业中的持久性有机污染物主要包括 AOX 和二噁英。目前制浆造纸行业对持久性有机污染物的削减技术以源头控制为主，末端治理为辅助手段。

（五）制浆造纸装备科学技术进展

近年来，我国通过成套引进或引进关键部件的方法，促进了我国制浆造纸装备制造能力和科技水平的提高，缩小了与国际先进水平的差距。我国目前拥有世界上领先的制浆造纸技术和装备，如连续蒸煮、氧脱木素、二氧化氯制备、现代化高速纸机，深度废水处理系统，QCS、DCS、MCC、PLC 等运行自动控制和机械故障自诊系统。成套装备的规模在扩大，稳定性、可靠性在提升。我国制浆造纸装备企业开始国际化布局，"走出去"稳步推进。"中国技术""中国制造"越来越受到国外造纸企业的青睐。

（六）纸基功能材料科学技术进展

近年来，纸基功能材料领域的技术创新和产品开发取得突破性进展，一些技术含量高的产品填补了国内空白，如芳纶纸、空气换热器纸、热固性汽车滤纸等。

随着纳米纤维素及纳米技术的发展，其在制浆造纸工业展现出了广阔的应用前景。以纳米纤维素为原材料，通过化学、生物、物理、复合等方式可以制备附加值更高、性能更加优异、生物可降解性与生物相容性极佳的纳米纤维素基功能材料，是纳米纤维素未来发展的一大热点。

三、本学科国内外研究进展比较

我国现已发展成为世界上纸及纸板的生产大国，但仍不是世界上的造纸强国，我们与世界发达国家的造纸技术水平，相差仍然较大。

（一）制浆技术方面

由于装备研发条件的优势，欧美等国家在高得率制浆技术和低能耗化学制浆新工艺方面的研发成效仍然领先于国内，溶剂法制浆、生物制浆、废纸生物脱墨等新技术也处于发展中。受新能源开发潮流的推动，国外研究的热点是将传统单一的制浆造纸过程转化为能够同时生产纸浆和纸、高分子材料和化学品、生物能源的复合型生物质提炼过程。国内在吸收、跟踪、借鉴先进技术的同时，根据自身制浆纤维原料的特点进行了大量适应性和改良性研发应用，取得良好成效。特别是在麦草、竹材等非木材纤维原料清洁制浆和无元素氯漂白技术方面进展显著，自主创新性成果丰富，技术水平居国际领先。

（二）造纸技术方面

重要的造纸技术研究主要是由国外造纸设备公司推动的，并通过新型设备予以承载和体现新技术特点及优势。为改善纸及纸板产品品质和生产过程节能降耗，在节能磨浆工艺、新型稀释水流浆箱、靴式压榨、强化干燥等方面不断优化。同时，开发的帘式涂布和喷雾涂布等新一代涂布加工技术步入商业化应用阶段。国内在速生材高得率化学机械浆和低品质废纸浆扩展应用技术方面成果明显，起到了缓解国内优质纤维原料匮乏的成效。

（三）制浆造纸装备制造技术方面

缺乏从基础理论研究以及从工艺到装备和控制等的系统性、成套新研究，而尽管差距在逐步缩小，但总体上我国明显落后于国外。国内制浆造纸装备的技术研究和制造主要采取吸收、消化和再创新等措施，研究开发和升级了一批制浆造纸装备。但由于核心技术缺失，研发和集成能力相对薄弱，目前大型成套制浆造纸设备市场仍被国外跨国公司垄断。

（四）制浆造纸化学品技术方面

我国制浆造纸化学品发展速度加快，但高效专用型产品的开发和应用技术的研究仍十分缺乏，表面施胶、中性施胶、高效脱墨、新型填料等方面，与国际水平差距仍然较大。国外的产品在类型、环保等方面上均处于领先水平，我国制浆造纸化学品目前仍处于低端产品生产，中高档产品少，特别是绿色化学品合成及应用方面落后较多的状态。

（五）污染治理技术方面

废水 $CODCr$、$BOD5$ 减排和固体废弃物资源化利用是国内外共同的研究重点。由于《制浆造纸工业水污染物排放标准》（GB 3544-2008）的颁布实施，废水深度处理和回用技术成为研究热点。废纸制浆造纸企业采取强制封闭循环和内部净化处理相结合方式，在废水超低排放或零排放技术方面取得进展；非木材原料化学制浆黑液提取及碱回收技术趋于成熟；制浆和水处理污泥等固体废弃物的浓缩技术与生物质能源利用技术成果丰富。

（六）纸基功能材料技术方面

我国特种纸生产企业主要以中小型企业为主，其技术研发水平和综合实力与国外还有一定差距，产品技术创新主要停留在模仿阶段，原创技术与产品偏少，导致国产自主品牌缺乏，削弱了国际市场竞争力。

四、本学科发展趋势和展望

（一）本学科发展目标和前景

面对资源和环境制约，本学科技术的发展将以"资源低消耗、过程低排放、产品可再生、废弃物资源化"为目标，以发展循环经济、创新发展模式、建设资源节约型制浆造纸工业为核心，重点开展资源的高效和循环利用、污染治理、节能减排技术与装备研究，促进制浆造纸工业实现产业发展与环境、社会效益的完美统一。

（二）学科未来发展趋势

针对我国制浆造纸工业纤维资源短缺，能源和水资源匮乏，而环境保护要求日益严格的现状以及产业政策要求，制浆造纸学科技术必将依据我国制浆造纸工业原料结构特点和企业技术水平的现状，进一步加大科学研究与技术开发力度，通过跟踪研究国际前沿技术，发展自主创新的先进技术，加快制浆造纸科技进步的步伐，促进我国制浆造纸工业的持续发展。学科研究成果的推广应用将进一步降低纤维资源和能源的消耗，减少污染负荷。

未来的制浆厂将是一个复合型植物纤维生物提炼厂，可得到纸浆、能源和多种化工产品。制浆造纸工程学科的发展将为我国制浆造纸工业发展成资源节约型、环境友好型的绿色产业提供有力的技术支撑。

（三）本学科研究方向建议

为加快制浆造纸科技进步的步伐，从我国制浆造纸工业现状和发展需求以及行业政策导向出发，建议今后重点开展以下几个方面的制浆造纸科学研究和技术开发。

1）纤维资源高效与循环利用技术。以废纸、速生材和非木材为主要对象，研究开发新型纤维分离技术和低品质纤维制浆造纸利用技术，提高纤维原料利用率，减少纸张回收和再生过程中的损耗和技术障碍，充分节约纤维资源。

2）环境友好型制浆造纸关键技术。以制浆造纸过程清洁化技术为重点，研究开发生产过程污染减排先进适用技术、节水节能技术、造纸废水高效低成本处理和再生回用技术、固体废弃物资源化利用技术等，实现制浆造纸工业的降耗、减排、节能目标。

3）高性能纸基功能材料生产技术。开发高性能纸基功能材料重点产品，满足高技术领域对纸基功能材料的需求；拓展纸基功能材料技术含量和功能特征；丰富纸基功能材料品种和应用范围。

4）高度集成和性能明显提升的大型、先进、专用技术装备研制。以市场需求为导向，研制开发节能型清洁制浆成套设备；低能耗速生材高得率制浆成套设备；宽幅、高速先进纸机，高度集成的自动化制浆造纸生产线。

5）高效、专一、功能性造纸化学品的开发应用，包括绿色化学品、生物质化学品、纳米材料的开发应用。

6）制浆造纸行业循环和低碳经济关键技术的开发。围绕纤维资源高效利用和废弃物资源化目标，开展生物处理技术、膜材料处理技术、生物质能源技术的应用研究与产业化实施，建立制浆造纸工业生产过程循环可持续和低碳化发展模式技术体系。

第二十二节　光学工程

一、引言

传统光学工程基于几何光学和波动光学，拓宽了人的视觉能力，形成了以望远镜、显微镜、照相机、光谱仪和干涉仪等为代表的光学仪器，至今仍在生产和生活中发挥着重要作用。

光学工程当前已发展成为以光学和光电子技术为主，与信息科学、能源科学、空间科学、材料科学、生命科学、精密机械与制造、计算机科学及微电子科学等紧密交叉和相互渗透的学科。现代光学新原理、新技术和新材料的提出加速推动着科学、工程和应用等多方面的发展。

在科技基础方面，光学工程在空间尺度、时间尺度、光谱尺度上向两极拓展，不断接近甚至突破理论极限，新器件层出不穷，新材料不断涌现，新技术不断发展，已形成并渗透到大量基础交叉学科领域。在工程和应用方面，应用的深度和广度在继续拓展。光学工程在通信、遥感、工业制造和国家安全等方面正在积极影响和改变各国科技、工业和军事力量。

二、本学科近年的最新研究进展

（一）固体激光技术向多极化发展

固体激光技术的发展趋势体现在提升峰值功率、压窄脉冲宽度、延伸激光波长范围、增加能量密度等方面。2016年具体发展态势为：①激光峰值功率提升到拍瓦级：强激光功率密度迈向1025瓦/平方厘米，激发的电子特征能量达到1太电子伏特（TeV）；②激光脉冲宽度达到阿秒：激光器超短光脉冲达到80阿秒，其脉宽比光穿过一个原子的时间还短，可用于探测和操控极端超快的电子动力学过程；③激光波长向两级延伸：一方面向短波长拓展，产生深紫外、X射线甚至γ射线波段的激光器，另一方面向长波发展，拓展到太赫兹波段；④激光能量提升到武器级：美海军已将30千瓦级的激光武器系统样机安装在前沿浮动基地上，开展了实战测试，并将开展新型高能激光武器系统舰上测试，输出功率可达150千瓦。

（二）新型光学系统和光学仪器不断涌现

光学仪器和光学系统是光学工程领域技术发挥功能最主要的手段，随着新材料、新技术、新理念的出现和应用，其发展速度越来越快。

在衍射极限方面：2016年美国科罗拉多大学演示了空间分辨率达2η的多光子—空间频率调制成像技术，产生了纳米级图像，进一步突破光学显微成像分辨率极限；在超快时域范围方面，美国华盛顿大学成功研发出快门速度达到1000亿分之一秒的单镜头压缩光速摄影机；在波段覆盖范围方面，科罗拉多大学新型电润湿变焦距光学元件已从可见光跨进中红外波段。

（三）光波精密调控将获得突破

新材料和新工艺的应用也加速了现代光学系统的发展和光波调控的突破。2016年美国哈佛大学使用高纵横比的二氧化钛纳米阵列构成"超表面"以控制其中光波相互作用的方式，得到了数值孔径高达0.8的透镜，可在可见光谱范围内高效率工作，实现亚波长分辨率成像；美国国防高级研究计划局（DARPA）和美国国家航空航天局（NASA）联合投资研制一个轻型光学系统，在硅材料上通过极精确的激光烧录上千个望远镜阵列，成功研制新型轻质光学望远镜，即分块式平面光电侦察成像探测器系统；另外，为研制新型光学材料，各科技大国均已启动利用二维超表面光学、三维容积光学以及全息等方式来调控光的重点项目。

（四）量子计算技术已初见端倪

近些年量子计算技术研究已经开始出现具有实用价值的成果，2016年更是取得一系列突破：在量子光源方面，中国科技大学科研团队实现了十光子纠缠；在原型机芯片集成方面，美国麻省理工学院的科研人员研制出一款量子计算原型芯片，中国科技大学研发了半导体量子芯片；量子计算原型机方面，美国马里兰大学制造了一台由五比特量子信息组成的新型量子计算机，能执行一系列不同的量子算法；在光学谐振腔制作方面，国家标准与技术研究所和马里兰大学的科研人员已经开发出一种纳米光子频率转换器，可在单光子量级实现光波转换。可以预测，量子计算技术有望在不久的将来实现应用。

（五）激光增材制造推动工业制造水平到全新的高度

目前激光增材制造技术所使用材料已涵盖各类金属合金、非晶合金、陶瓷以及梯度材料等，所制造的产品已包括锻件类金属零件、微米尺寸光学器件、人体组织、建筑物、电动汽车等复杂构件，其应用范围大幅度拓宽，尤其在高端领域优势突出，技术更新迅速，通过开发4D打印结构体、新型陶瓷、零重力打印、五轴增材制造等技术实现适应更宽应用范围、满足各种个性化需求的精密光学加工制造，推动工业制造水平到全新的高度。

（六）空间激光通信技术成为实现天地一体化的重要手段

随着空间遥感、空间开发和天地一体化进程的加快，空天互联网信息时代即将到来，因此需要建立传输速率快、信息量大、覆盖空间广的通信网络系统，采用波长短的激光进行天地间和空间卫星之间的通信是实现这一通信网络系统的重要手段。欧洲已经启动了"数据中继系统"项目，将提供一个快速、可靠、无缝的电信网络，可根据需要在适当的地点和适当的时候从卫星实时获取数据，并于2016年成功发射EDRS-A中继卫星，日本也计划于2019年发射激光数据中继卫星。

（七）光学观测在空间监测中越来越受到重视

近年来，太空碎片不断增大的威胁促使全世界关注空间监视，而光学观测手段以其精度的优势，一直是空间遥感监视的重要手段。2016年，国际上启动了一系列地面对空间目标光学观测工程重大计划，包括NASA启动的"宽视场红外巡视望远镜"计划，欧洲南方天文台签署的在智利建造39米全球最大地基光学望远镜的协议。可以预计，未来对空间目标的高分辨率光学观测在空间监测中将越来越受到重视。

三、本学科国内外研究进展比较

在全球化的背景下，我国光学工程学科也得到了长足的发展。近年来在国家科技创新战略引领下，我国光学工程取得了一系列具有重大国际影响的科技成果，部分技术还实现跨越式突破。但在应用基础、中高端光电子器件和光学仪器设备的产业发展方面，与国际发达国家仍存在一定差距，很多高精度仪器仍需要从国外进口。

我国的光学工程领域的科研以高校、中科院团队和部分央企科研院所为主体，近年来我国加强了基础研究积累，与国际研究团队紧密合作，部分领域课题实现了重大跨越式突破，取得了国际领先的成果。诸如超快超强激光物理领域、光谱及应用、介观光学和纳米光子学、特异材料与应用、光的操控和精密光谱等一些具体问题的研究中我国科研人员已经做出了具有较大影响的成果；在激光与物质的相对论性和高度非线性相互作用、激光粒子加速、阿秒脉冲激光器、激光聚变基础物理、新型激光等离子体光源等方向我国科研人员已走在国际前沿，国内多个研究机构已建成了一些从太瓦级到几百太瓦级、甚至到拍瓦级的超快超强激光装置。

2016年，我国在国际上首次实现多自由度量子隐形传态、量子通信安全传输，首次在一根普通单模光纤中利用超大容量超密集波分复用实现信息传输80千米，首次实现亚纳米分辨的单分子光学拉曼成像，首次获得太阳大气可见至近红外7波段同时高分辨力层析图像，在硅基光波导芯片上首次集成"旋涡光束"发射器件阵列，并成功研制世界唯一实用化深紫外全固态激光器。

近几年国家对光学光电子产业的发展给予了大力的支持，促进了制造业的升级。借助巨大的市场需求和科研生产能力，国内生产的一些光学光电子器件也有了较高的技术水平，并在国际上具有了一定的市场份额，国内光电子技术发展较好的地区也已经形成了光学器件的产业链，拥有了较为完善的产业集群。

在激光器的整体研究水平上，我国相对于先进国家有一定的差距，但在某些方面，国产激光器也有其优势，如国产紫外激光器，其在稳定性和可靠性方面并不输于进口激光器。同时，在针对客户特别需求的定制方面，我国激光器具有更大的灵活性。

我国现代光电子行业整体的发展带动了国内光学镜头行业的发展。2000年以来，国内安保、互联网等行业发展迅速，一些世界一流厂商快速崛起，我国发展成为全球光学镜头需求最多的国家，激增的市场需求使国内涌现出一批优秀的镜头制造厂商，这些厂商在产品研发、设计和生产工艺等方面的持续投入，不断推出分辨率更高、成像质量更稳定的光学镜头，很快成为全球光学镜头产业的生力军，国内镜头的发展也带动了国内光学冷加工产业的迅猛发展，我国已成为全世界最大的光学冷加工产能承接地和聚集地。

虽然我国已成为世界上最大的光学和光电子市场，但我国高端光电子产业发展仍处于中低水平，很多高性能器件还依赖进口。高端光学仪器产业、激光制造业和其他光学辐射产业中核心光学产品很多被国外掌控，同时一些光学工程产业严重缺乏配套的关键设备和产业群支撑，对于庞大光学工程领域的工程应用和产品产业发展还缺乏掌控和引导能力。

四、本学科发展趋势和展望

（一）我国将在基础科研领域形成一批重要成果

在"神光"等强激光装置计划长期支持下，我国在激光聚变驱动源和高亮度光源设施建设方面居于国际先进行列，2016年上海超强超短激光试验装置项目成功实现5拍瓦激光脉冲输出，达到国际领先水平，预计2017年将率先实现10拍瓦激光输出目标，2018年建成面向用户的实验装置，有望保持世界领先水平。

（二）我国激光增材制造产业将进入快速发展阶段

我国的金属材料激光增材制造处于世界先进水平行列，率先在国际上突破钛合金、超高强度钢等大型整体承力关键结构件制造工艺、质量控制等关键技术。2016年我国实现了人体组织、钛金属部件、微光学器件和建筑等的激光增材加工生成。未来，我国将进一步开展制造机理、质量保证、质量检测等方面的研究，建立新的加工标准，优化软件，扩充数据库，加速激光增材制造技术和传统加工技术的有机结合，推广其向各领域应用。激光增材制造产业进入快速发展阶段。

（三）我国空间激光通信将步入实用化

2016年随着"墨子"卫星的升空，我国在激光通信关键技术得到了验证，空间激光通信向应用又迈进了一步。未来我国空间激光通信将实现天基、空基、海基、地基等多平台的信息获取与分发，并通过星间、星地激光通信链路与地面光纤骨干网连接，形成天地一体化激光宽带空间信息网络，实现天、空、地各类信息平台的高速、广域传输分发和即时接入。

（四）光纤通信将在技术和应用层面实现多项系统级创新

随着互联网、物联网和泛信息化的进一步发展，在高性能光电子器件和硅基光电集成器件等的支撑下，未来我国光纤通信将在以下几个方面得到发展：①通信速率将得到提高（1Tbps或者更高）；②通过软件定义光网络、光网络虚拟化等先进技术形成具有可编程性和资源灵活调控能力的全光网络；③在各项先进技术的协同与光融合方面提升整体性能；④通过多种技术手段和工程方法将使用成本降低。

（五）光学技术将进一步提升我国军事斗争和反恐维稳能力

光学技术在军事和反恐领域具有广泛的应用需求，我国在此方向取得了长足的发展。近年来，随着我国光学工程领域新技术、新材料、新工艺、新系统的不断涌现，我军的态势感知、信息传输和准确攻击等能力得到了极大的提高。未来，光学技术将在军事领域发挥更大作用。

第二十三节 农学（基础农学）

一、本学科近年的最新研究进展

经过国家科技计划持续支持和科技人员的长期努力，围绕农业环境保护、农产品加工、农业耕作制度等学科的重点领域，国内农业高校和科研单位团结协作和联合攻关，在产地环境质量控制与修复、农业应对气候变化、农业气象与减灾、设施栽培农业、食品加工、农产品加工质量安全、食物营养与功能、农产品贮藏与保鲜、作物栽培与生理以及作物生态与耕作等学科方向取得了一系列重要研究进展。

（一）前沿技术研究

面向国际科技前沿，在基础和前沿技术研究领域突破了农业固碳减排机理、农业气候资源区划与利用、主要粮油产品储藏过程中真菌毒素形成机理及防控基础、农产品加工过程安全控制理论等重大理论与方法，抢占现代农业科技制高点，显著提升了相关领域的学术水平与国际影响力。

（1）农业固碳减排机理

在建成中国农田土壤有机碳及温室气体排放数据库基础上，定量评价了农田土壤有机质对作物生产力的控制作用，并从土壤学、生物学和生态学等多学科研究阐明了其多效应生态系统功能，提出了土壤有机质功能演替假说；发展和推进了农田有机碳动态演变及固碳潜力定量评价的方法学，定量评估了农田土壤有机碳固定及其自然与农业技术潜力，并确证有机无机配施的合理施肥是农业固碳减排的最有效途径，延伸发展了农业固碳减排计量方法。

（2）农业气候资源区划与利用

在国家科技基础性专项的支持下，开展了气候变化背景下农业气候资源分析与利用的研究，利用信息技术手段，按照农业气候资源统一的规范标准，整合农业气候资源数据，规范农业气候资源指标生成方法，完成了中国农业气候资源数字化图集的编制工作。在此基础上，精选1000余图幅，编辑出版了《中国农业气候资源图集》四卷（包括综合卷、光温资源卷、作物水分资源卷和农业气象灾害卷），以及《中国主要农作物生育期图集》

一卷。该系列图集系统反映了气候变化引起的农作物种植区域、作物生育期和各作物生育期气候资源变化的特征,并首次编制了作物水分资源卷和农业气象灾害卷,为合理利用农业气候资源、优化农业布局、加强农业风险管理提供了科学依据,在我国农业气象研究领域具有里程碑意义。

(3)主要粮油产品储藏过程中真菌毒素形成机理及防控基础

研究了我国主要粮油农产品储藏过程中真菌菌群变化规律、真菌毒素形成的分子机理、真菌毒素的防控策略研究。具体是明确了花生和小麦储藏过程中真菌菌群的演变规律;揭示了黄曲霉不产毒的分子机理,阐明了水分通过HOG(高渗透甘油)信号途径、温度通过TOR(雷帕霉素靶标)信号途径调控真菌生长和毒素产生的分子机制;揭示了花生白藜芦醇、乙烯抑制黄曲霉毒素合成的分子机理,解析了真菌毒素的微生物降解和酶解机理。建立了真菌生长活动早期监测预警体系和黄曲霉毒素产生预警模型。

(4)农产品加工过程安全控制理论

阐明了典型加工过程中危害物形成的分子基础和调控机制,发现了新危害物和潜在危害物的生成途径和转化规律,明确了食品加工过程中对目标危害物干预、阻断与控制的关键分子机制,解析了食品加工过程中食品组分–加工条件–危害程度的相互作用规律。

(二)攻克瓶颈问题

面向国家重大需求,针对制约我国农业转型升级的全局性重大瓶颈问题,攻克了养殖废弃物微生物发酵床处理技术、智能植物工厂能效提升与营养品质调控关键技术、营养功能成分分析检验技术、高效果蔬品质劣变控制技术、作物肥水高效生理与技术等一批对我国农业发展和农业高技术产业开发带动性强、覆盖面广、关联度高的核心技术、共性技术及配套技术,取得了一系列具有重大产业化前景的核心自主知识产权。

(1)养殖废弃物微生物发酵床处理技术

利用发酵床内的微生物将养殖粪尿进行原位或异位分解,从而降低污染物的外排;一体化发酵技术针对养殖场的固体和液体废弃物污染问题采用一体化发酵装置,实现对废液的大规模连续发酵;种养一体化模式主要是将沼液或者养殖废水收集后堆置于发酵池内,并添加生物腐植酸菌剂进行微生物发酵;保氮除臭免通气槽式堆肥发酵是在畜禽粪便等有机固体废弃物原料中,添加筛选的高效腐熟菌剂和调理剂等最终将其转化为无臭无味、富含有机质和植物营养元素的优质有机肥。

(2)智能植物工厂能效提升与营养品质调控关键技术

率先在国际上提出了植物"光配方"思想,探明了基于LED的植物光配方优化参数,发明了以AlGaInP红光660纳米芯片与InGaN蓝光450纳米芯片为核心的多光质组合(R/G/B/FR)植物LED节能光源;开发出基于植株发育特征的水平与垂直方向可移动式LED动态光环境调控技术,显著降低了光源能耗。提出充分利用室外自然冷源进行植物工厂环境调节的"光–温耦合节能调控"方法,发明了基于室外冷源与空调协同调温的植物工厂节能环境控制技术及配套装备。率先探明了植物工厂营养液自毒物质的特征机理，发明了 UV-纳米 TiO_2 协同处理营养液自毒物质的技术方法；发明了基于采收前短期连续光照与营养液氮素水平控制的蔬菜品质调控技术，降低叶菜硝酸盐含量达 30% 以上，并显著提高了 Vc 和可溶性糖含量。

（3）食品营养功能成分分析检验技术

应用代谢组学技术使得食品组分分析变得细致，可以检测到成百上千种有机化合物单体。这种"化整为零"的策略是人们了解食品具有独特口味、质地、芳香或色泽的分子基础。基于上述技术和理论，建立了基质辅助激光解吸电离-飞行时间质谱快速分析三文鱼肌肉组织磷脂质组学的方法，快速分析三文鱼肌肉组织中磷脂酰胆碱和磷脂酰乙醇胺的分子种类，并根据磷脂脂肪酸链长度与不饱和度对三文鱼进行营养评价。

（4）高效果蔬品质劣变控制技术

从生物大分子氧化与修复、能量代谢、次生物质代谢与调控、信号分子合成与作用、不同贮藏条件下品质变化规律和生理应答机制等角度阐明了果蔬采后损失的机理及控制机制。在上述理论研究基础上，研发出利用信号分子（1-MCP、NO 和 AiBA）抑制衰老激素合成和诱导耐冷性的技术，使柑橘、叶菜类、果菜类等果蔬保质期延长了 60% 以上，显著提高了果蔬保鲜效果。

（5）作物肥水高效生理与技术

研发集成的水稻精确定量施肥技术、小麦精确定量施肥，大幅度了减少氮肥施用量，肥料利用率提高 10% 以上；研发的大田作物缓/控释肥、生物肥、有机复合肥、功能性肥等新型肥料，减少了肥料损失、提高了肥料利用率。建立了冬小麦节水高产高效技术体系，形成 3 种节水高效栽培模式，与传统高产技术相比，每公顷节约灌溉水 $1500m^3$，水分利用效率提高 20%。

（三）基础农业学科研究

面向"三农"建设主战场，针对解决农业和农村经济发展中全局性、方向性关键性的重大科技问题，集成了农业气象灾害监测预警、风险评估与防控、设施蔬菜连作障碍防控、作物高产高效栽培等一批科学意义重大、行业影响深远、具有明显时代特征的技术体系，引领农业科技的跨越式发展，大幅度提升我国基础农业学科科技的创新水平，为我国农业和农村可持续发展提供了有效的科技支撑、雄厚的技术储备和强力的技术保障。

（1）农业气象灾害监测预警、风险评估与防控

农业气象灾害监测预警研究已经从单一指标和单一方法逐步提升到多指标、多方法的立体、实时监测预警体系，构成了天—空—地三维监测网；在信息耦合上，向集成地面气象、农业气象、田间小气候观测以及农情、灾情和地理信息系统等多元信息方向发展；在技术研发上向模型化、动态化和精细化方向发展。农业灾害保险与巨灾保险机制与产品研发取得突破性进展，形成了迄今有关农业气象灾害风险领域最全面和系统的研究成果；在

灾害防控方面，主要以北方和西南地区干旱、低温灾害防控技术研究为突破口，建立了分区域、分作物、分灾种、分季节的适合当地农业特点的干旱与低温防控技术体系，为各地区重大农业气象灾害应急和防控提供科技支撑。

（2）设施蔬菜连作障碍防控

针对设施蔬菜连作障碍、蔬菜农残高以及制约健康可持续生产的瓶颈问题，围绕连作障碍成因不明、土壤连作障碍因子消除困难、蔬菜对连作障碍因子抗性弱等三个核心问题，揭示了连作障碍高发成因与规律，发现了连作障碍防控的突破口。明确土壤初生障因消除和蔬菜根系抗性增强是防控核心；攻克了土壤连作障碍因子消除技术难点，实现从化学农药消毒向环境友好型消除的重大技术变革。实现了化学农药零投入的土壤连作障碍因子系统消除；发明了蔬菜根系抗性诱导技术，突破了优质蔬菜连作难的技术瓶颈。解决了蔬菜优质品种因线虫等高发而难于推广的产业瓶颈；创建"除障因、增抗性、减盐渍"三位一体连作障碍防控系统解决方案，为设施蔬菜安全可持续生产提供了技术保障。

（3）作物高产高效栽培

利用稻瘟病敏感品种和抗病品种间作、玉米和马铃薯间作控制病害；通过禾本科/豆科作物间作体系种间根系氮素竞争使得禾本科作物获得更多的土壤矿质氮，同时使豆科作物固定更多的大气氮素，利用种间相互作用，提高了养分利用效率。发展了间套作全程机械化管理，实现间套作生产的全程机械化。

二、本学科国内外研究进展比较

结合农业学科有关国际重大研究计划和重大研究项目，研判国际学科发展新趋势与新特点，比较分析我国农业环境保护、农产品加工、耕作与栽培等学科在国际上的总体水平和学科发展状态，明确与国际水平相比的优势与差距，可为制定学科发展战略、确定未来发展重点方向提供必要的基础支撑。

随着国际现代农业和生物学、医学、工程学、信息学等相关学科的发展，农业环境保护、农产品加工、耕作与栽培等学科发展总体呈现国际化、全球化趋势，多学科交叉、融合与渗透日益加强，学科发展不断向微观结构深入、向宏观综合联系扩展，理论创新不断加速，技术创新不断向机械化、信息化、规范化、定量化、规模化、集约化方向演进，科技创新驱动产业升级特征明显。

总体来看，近年我国在农业环境保护、农产品加工、耕作与栽培等学科实施了一系列国家科技计划，建立了一批分支较为齐全的研究机构和基地，形成了一支颇具规模的研究队伍，取得了显著的研究进展。在农业固碳减排、节能日光温室、食品功能成分检测分析、间套作控制作物病害研究和应用等方面的技术处于世界先进水平。然而，与发达国家相比，我国在农业环境保护等方面的立法不健全，对法律责任的规定较为模糊；农学基础学科发展的科技研发整体投入强度不足，在研究设备和手段、基础资料积累、基础理论研究等方面差距明显，一些新兴前沿的研究方向起步较晚，处于追赶阶段；产业核心技术供给能力仍显不足，许多高端和成套装备对外依赖度高，缺乏配套技术和技术的标准化；成果转化效率偏低，企业自主研发能力明显不足。

三、本学科发展趋势和展望

在分析农业环境保护、农产品加工、耕作与栽培等农业基础学科未来发展趋势的基础上，针对新的战略需求，分析未来 5 年学科发展的重点领域及优先方向，明确了有待解决的重大科学问题和核心关键技术难题。

农业环境保护学科方面，重点发展农业环境风险管理理论，研发环境友好肥药、生物降解地膜、残膜回收机具、新型生态饲料等，集成创新污染物资源化循环利用技术、土壤污染控制与修复技术、畜禽废弃物无害化处理与资源化利用新技术及产品、农业固碳减排技术、农业气象灾害损失评估与灾害风险防范综合技术、设施栽培农业技术等。

农产品加工学科方面，重点发展食品加工制造理论，探索新型加工技术，开发智能化、数字化核心装备，发展加工过程中食品营养品质评价、控制与监测技术与装备，突破中式主餐工业化关键技术，解析食品营养组分对健康影响的作用机制，大力加强营养功能食品品质改良与制造技术研究，积极推进农产品贮藏保鲜相关科学问题与关键技术装备研发，并开展全链条技术集成应用。

农业耕作制度学科方面，加强作物栽培生态生理生化研究，重点创新作物高效、优质、绿色、机械化、轻简化、抗逆减灾理论与技术，研究构建新型轮作休耕模式、间套作模式，研发作物信息化智能化栽培技术，研发区域土壤耕作制以及相配套机具，建立农田固碳减排技术体系。

此外，还从加强法律制度保障、加强学科规划布局、健全相关技术标准、实施区域差异化管理、重视基础性长期性科技工作、持续提升科技创新效率与引领支撑水平、推动关键技术集成配套与推广应用、加强人才团队培养和国际合作交流等方面，提出促进农业环境保护、农产品加工、耕作与栽培等农业基础学科发展的战略思路与对策措施。

第二十四节 林学科学

一、引言

党的十九大提出了习近平新时代中国特色社会主义思想,强调生态文明建设是"中华民族永续发展的千年大计","要坚持人与自然和谐共生,要像对待生命一样对待生态环境"。林业是生态文明建设的重要力量,肩负着维护生存安全、淡水安全、国土安全、物种安全、气候安全和建设美丽中国的历史重任。林业科学学科主要是以森林和木本植物为研究对象,揭示其生物学现象本质和规律,进行森林资源培育、保护、经营管理和利用等的学科。

二、本学科近年的最新研究进展

(一)森林培育

我国人工林培育面积6933万公顷,位居世界第一。基于适地适树、良种良法理论基础上,提出了遗传、立地、密度、植被、地力等五大控制的育林体系,使人工林培育质量得到了提升。森林营造上抗逆栽培生理生化机理研究进入功能基因分析与调控阶段,创新形成了混交林树种间养分关系及作用链理论。用材林水肥精准调控理论与技术得到发展,困难立地森林营造进入微立地评价及充分利用自然力促进植被恢复时期。森林抚育创新了生态公益林精准抚育理论与技术体系,构建了以目标树经营为特征的近自然森林经营技术。森林多功能培育更加关注碳汇林、城市森林、污染土壤生物修复等领域。

(二)林木遗传育种

林木遗传育种学科在现代生物学理论技术支撑下进步迅速。树木全基因组测序,材性和抗逆等重要性状基因调控网络的构建、关键基因功能与调控因子是学科前沿领域。充分利用各世代优良种质的改良策略，分子辅助育种、优良种质体细胞胚发生与无性系化体系为技术研究热点。我国倍性育种技术具有优势，传统育种理论与技术处于跟跑地位。育种目标多样化，在生物组学基础上对树木重要性状遗传调控机制开展研究，开发密切关联的分子标记，实现良种培育效率与效果的新突破。

（三）木材科学与技术

木材解剖学通过 DNA 条形码技术识别濒危树种取得新进展；木材细胞壁微尺度表征、多壁层结构及微力学性能研究获得新突破；木材与水分关系、木材传热传质和干燥机理、木材弹性力学和黏弹性力学方面获得可喜成果，形成了木材纳米尺度表征、木材功能化理论、木材仿生学等交叉学科研究领域。在木材改性、人造板、木材产品和胶黏剂制造等关键技术等方面取得了重要突破，"人造板及其制品环境指标的检测技术体系"等六项成果取得明显突破。

（四）林产化工

林产化学加工工程学科已从传统的松脂、活性炭等研究拓展为生物质利用，即利用作物秸秆、林业剩余物、种植的能源植物及非木质资源（淀粉、天然树脂、油脂）等为原料，通过绿色化学方式生产生物基产品的一门新兴产业和交叉学科。研究领域拓展到生物质能源、生物基材料、生物基化学品、生物质提取物、制浆造纸等几大方面。研究内容在国际生物质利用和多学科技术创新上取得新突破，推动了我国生物质产业可持续发展。

（五）森林经理

森林经理学科紧跟国际前沿，应用新理论和新方法，在森林经营理论与技术模式、森林资源调查监测、林分生长和收获预测、林业资源信息管理、森林经营方案编制及天然林结构化经营等方面取得了一系列重要进展，取得了"东北天然林生态采伐更新技术""与森林资源调查相结合的森林生物量测算技术""森林资源综合监测技术体系"等重大科技成果。构建了森林生长精准预测及经营响应、森林质量天空地一体化监测评价技术等森林可持续经营理论与技术体系。

（六）经济林

重要经济林树种的基因组学、重要产品成分的生物学代谢途径、自交不亲和性等基础研究取得重大进展。经济林种质创新与良种化工程技术取得了重要进步。银杏、杜仲、核桃等的良种选育和优质高效栽培等成果分别获得国家科技进步奖二等奖，特色经济林在不同地区的造林技术和生态治理方面取得较大进展。特色经济林产品精深加工与资源高值化利用、机械装备研制等取得新的进展。

（七）森林生态

森林生态学在应用基础研究和技术研发等方面取得了显著进展。中性理论、高通量测序技术、空间技术等新理论、新技术和新方法在森林生态学研究中的应用极大地促进了森林生物多样性形成、共存、维持和保育等研究的发展。推动了不同尺度森林对环境变化以及森林增汇减排等理论和技术研究。形成了空间森林生态水文学崭新研究领域。森林生态学与生物学、地学和社会科学等学科的交叉和有机融合使退化森林适应性生态恢复与重建、人工林生态经营以及森林生态系统服务科学评估等领域的研究越来越深入。

（八）森林保护

森林昆虫学科关注焦点从单个害虫种类向整个森林生态系统水平转移。对害虫的防治策略从防治已成灾的害虫，向调控害虫种群的发展、预防有害生物入侵的转变，着力恢复和提高森林生态系统对害虫的抵抗能力和缓冲能力。防治思路由以化学防治为主向以无公害防治为主转变。利用遥感卫星和无人机监测虫害发生、植物抗性育种、转基因抗虫等手段促进了森林昆虫治理技术的发展。对森林昆虫微观研究主要是对害虫特殊基因片段进行功能鉴定、基因编辑和验证；宏观研究是由个体到群落到生态系统的延伸。病原物与寄主、环境的互作研究不断深入，森林病害研究对象从病原个体不断延伸到森林生态系统，森林病理学与其他学科的交叉融合不断加强，病害防控的新理念和新方法不断发展。森林病害的生态调控机制及其应用森林病害的绿色防控技术、气候变化条件下森林有害生物的发生发展规律以及相应的控制技术等研究进展迅速。

（九）林业气象

森林水碳通量、树木稳定同位素水文气象学、树木年轮气候学等国际热点主题的研究水平与国外的差距逐渐缩小，像元尺度森林植被水热通量观测研究已接近国际先进水平。研究目标更加紧密结合林业生态工程建设、林业应对气候行动；研究手段注重长期定位观测与模拟模型相结合，研究方式注重与树木生理生态学、森林水文、森林土壤等交叉型学科的融合，观测技术逐步自动化和精细化。

（十）森林土壤

我国森林土壤学科在森林土壤资源分布、土壤生物、土壤碳氮循环、森林土壤地力退化及防治、森林土壤利用对全球气候变化影响及响应等方面取得了不同程度的进展。通过发展混交林、施用微生物肥料、测土配方施肥及集成有机物覆盖、微生物和配方平衡施肥深入，研究维护和恢复森林土壤功能的技术途径。在高效功能菌筛选、鉴定及施肥应用方面成效显著。

（十一）林业史

2014年，由尹伟伦院士任主编的《中华大典·林业典》全部出版。科技部科技基础性工作专项"中国森林典籍志书资料整编"启动。北京林业大学至今已建立起硕士、博士、博士后等较为完整的人才培养体系，同时建设了年龄结构合理、研究特色鲜明的教师团队。开设了《林业历史与文化概论》《林业史与环境史专题》等课程。2015年《中国林业史》教材列入国家林业局普通高等教育"十三五"规划。

三、我国林业科学学科发展趋势

（一）森林生态

森林生态学多过程、多尺度的深入研究，特别是生态学过程和机理、生态系统服务功能的形成机制等迫切需要与生物学、地学、经济学和社会科学等学科的有机融合。高通量测序和空间技术在森林生态研究中的应用明显加快，已成为现代森林生态研究的重要方法。建立长期的规范化、标准化监测森林生态系统的平台已经成为必然趋势。合理利用包括不良立地森林土壤资源、提高森林土壤生产力成为森林土壤研究的热点。系统研究森林土壤结构、变化过程和功能演变的规律和机制，维护和恢复森林土壤生态功能。林业气象时间尺度注重长期性和动态性。微观上要加强气候要素对林木生理生态、生理生化过程等与生长发育相关过程影响的研究，宏观上要加强林业工程建设区域性气候效应的监测与评估。

（二）森林培育与林木遗传育种

森林质量精准提升和生产力大幅提高是森林培育的根本途径及当前主要任务。低价低效的天然次生林具有很大的产量和质量提升潜力，完全可以通过各种抚育、改造、更新等培育措施加以挖掘。维护生物多样性对培育健康稳定的森林至关重要，生物多样性研究与利用将逐步深入。天然林与人工林同为森林培育对象，培育理论与技术应协同均衡发展。常规育种技术仍将是林木良种选育的主要手段。分子育种技术为常规育种提供更强的技术补充。树木性状形成的分子基础研究，将获得更多有育种价值的调控基因。安全高效的转基因技术的发展为基因工程育种及应用提供新的途径。

（三）森林经营与保护

探索发挥森林的多种功能的森林经营理论将成为学科研究的重要任务。森林资源调查研究重点是提高遥感森林分类精度和调查因子获取。森林生长模拟主要基于近代统计方法及计算机模拟技术。探索发挥森林的多种功能的森林经营理论将成为森林经理理论研究的重要任务。利用3S技术进行森林虫害和病害监测与预报，实现对有害生物进行的多种类型控制。现代生物技术对森林昆虫学将起到巨大推动作用。外来有害生物、原物与寄主互作的分子机制解析、优良抗病树种的选育与快繁等研究继续成为热点。

（四）木材科学与林产化工

木材科学与技术将从木质材料的解剖、物理、化学和力学等特性入手，进一步系统研究木材细胞壁构造与化学组成及其对物理力学性能的影响，构建木材细胞壁物理力学模型，阐明外部环境激励条件下木材结构的变化规律及其对物理力学性能的响应机制。木材改性、人造板加工、胶黏剂制备等关键技术的研究与推广应用，是木材加工研究的重点。木竹材的天然多尺度微结构与宏观功能的协同机制、木竹材仿生智能纳米界面的形成方法与原理等将成为木竹材研究的主要内容。林产化工将在木质资源加工关键技术、非木质资源的高效利用关键技术、新型生物基材料与化学品的创制与技术集成、特色林源提取物活性成分的筛选与高效利用技术和低质材及混合材高效清洁制浆技术等方面取得突破。

四、我国林业科学学科发展重点领域

1）森林生态：①森林生态系统关键生态过程与效应；②森林生物多样性形成、维持与保育；③变化环境下森林生态系统的响应、适应与恢复；④森林健康及其生态调控；⑤森林生态系统多目标管理与重大林业生态工程；⑥干旱地区森林与植被结构；⑦森林土壤主要生物元素的格局与循环过程；⑧根系驱动的森林土壤生态过程与机理；⑨森林土壤对全球环境变化的响应与适应；⑩森林土壤健康与生态服务功能；⑪构建森林土壤学数字化、信息化以及模式化动态管理系统；⑫森林生态系统水碳氮耦合关系；⑬森林生长对气候变化的反馈作用；⑭重要经济林产量及品质形成的微气象机理；⑮城市森林边界层物理特征与缓解热岛效应机制；⑯林业生态工程区域性气候效应及其影响机制。

2）森林培育与林木遗传育种：①高碳储量森林培育；②森林质量提升；③困难立地造林；④人工林长期生产力保持；⑤构建绿色高效经济林体系；⑥营造高效能源林；⑦实现森林功能多元化；⑧加强森林类型的布局；⑨林木种质生长、品质、抗逆性和适应性等重要性状的精准评价；⑩培育多目标林木新品种；⑪揭示控制林木生长、品质、抗逆等重要性状杂种优势和倍性优势形成机理；⑫应用 GWAS 分析、QTL 定位等解析复杂经济性状；⑬创新林木基因工程、细胞工程技术；⑭构建林木育种群体及高阶种子园；⑮构建良种规模化、标准化及无性繁殖关键技术体系。

3）森林经营与保护：①立地质量评价与潜力分析技术；②不同尺度森林多目标协同优化技术；③森林生长精准预测及经营响应；④森林质量天空地一体化监测评价技术；⑤高精度森林资源信息系统构建；⑥重大森林有害生物监测、预警技术；⑦森林有害生物成灾的生理生化与分子机制；⑧森林有害生物成灾的生物生态学机制及控制技术；⑨气候变化条件下森林生物灾害的发生与危害特性及控制技术；⑩外来有害生物风险评估、预警与控制技术。

4）木材科学与林产化工：①木材形成及材质改良的生物学与化学基础；②木材工业节材降耗、安全生产、污染检控等生产管控关键技术；③表面绿色装饰、环保胶黏剂、木材绿色防护与改性、低质原料清洁制浆等绿色生产关键技术；④木竹、木塑复合材料的研究；⑤木质家居材料健康安全性能检测与评价技术；⑥木质资源能源化化学转化关键技术；⑦非木质资源的高效利用关键技术开发与示范；⑧新型生物基材料与化学品的创制与技术集成；⑨特色林源提取物活性成分的筛选与高效利用技术；⑩低质材及混合材高效清洁制浆技术。

第二十五节　植物保护学

一、引言

植物保护学是保护国家农业生产安全，保障粮食安全、农产品质量安全、生态安全、人民健康，促进农业可持续发展的重要学科。我国是一个农业生物灾害频发、农业生态环境脆弱的农业大国。在全球气候变化和结构调整等背景下，我国农作物病虫害呈多发、频发、重发态势。据初步统计，2016—2017 年，我国主要农作物病虫草鼠害总体为偏重发生，发生面积约 4.2 亿～4.4 亿公顷次，防治面积 5.1 亿～5.4 亿公顷次。

面对上述严峻形势和国家重大需求，以及人们对生态环境与食物质量和自然资源需求的不断提高，2014 年以来，植物保护学科发展迅速，在基础及应用基础、关键技术创新及应用等方面取得了一批突破性重大研究成果和进展，在国家经济社会发展中的地位和作用不断得到提高。

二、本学科近年的最新研究进展

（一）基础研究水平进一步提高

1. 植物病理学科部分基础研究已步入世界先进行列

在部分真菌和卵菌病害病原菌的效应因子鉴定与功能、致病因子和植物抗病机理，植物病毒基因功能、症状形成及病毒运动机制、RNA沉默介导的病毒抗性和介体昆虫传播病毒分子机制，细菌群体感应机制和三型分泌系统，作物重要抗病基因标记定位、克隆和功能等方面的研究均与世界同步，甚至引领国际植病研究领域的前沿。

2. 农业昆虫功能基因组研究取得新突破

成功完成小菜蛾基因组测序与分析，是第一个世界性鳞翅目害虫的基因组；研究构建出约6.3Gb大小的东亚飞蝗基因组图谱，是迄今已测序最大的动物基因组；完成稻飞虱基因组测序与分析，获得了首个褐飞虱全基因组序列图谱；鉴定获得棉铃虫蜕皮激素途径中的多个受体、调控因子和效应因子，进一步阐明了昆虫蜕皮变态的分子调控机制，明确了棉铃虫的滞育调控机理。

3. 生物防治技术提出新策略、新思路

以植物对天敌的支持作用作为落脚点，提出了害虫天敌的"植物支持系统"概念；围绕栖息地生境，深入研究栖息环境对天敌昆虫保育的影响及其作用机理；从系统、群落、种群3个结构层次，探索研究天敌定殖的时空规律；研究植物免疫诱导蛋白的应用，提出"植物疫苗"的田间应用概念。

4. 入侵生物学在互作机制、大数据信息库创建等方面进展显著

明确烟粉虱发生分布格局与内共生菌侵染率的相关性，揭示昆虫寄主内共生菌可以借助于寄生蜂进行水平传播的新途径；集成创新了豚草持续治理技术体系；开展了苹果蠹蛾、紫茎泽兰、薇甘菊等的转录组、蛋白质组、小RNA组、代谢组、表观基因组等研究，为创建我国独立运行的入侵生物组学大数据平台，提供技术支撑和平台保障。

5. 植物化感作用机理研究取得新进展

明确侵入黄顶菊抑制小麦和棉花生长发育的作用机理；探明水稻化感品种与稗草、小麦和麦田杂草种间的化学识别机理；揭示作物在害虫胁迫下产生防御化合物以及害虫感知作物信号化合物的机理。

6. 农药学科基础研究进展显著

明确毒氟磷、宁南霉素和氰烯菊酯等药剂的潜在作用靶标；探明氨基寡糖、嘧肽霉素和氟唑活化酯诱导抗病分子机理；发现原卟啉原氧化酶的产物反馈抑制机制；获得AtHPPD与其天然底物HPPA复合物的三维结构；揭示了生长素诱导的TIR1泛素连接酶底物识别的详细分子机理；发现了赤霉素进出受体的新通道，提出了赤霉素受体GID1识别底物的全新机制。

7. 农药施用技术理论研究取得新的突破

率先建立了"表面张力和接触角"双因子药液对靶润湿识别技术；实现了农药雾滴雾化过程的数据化分析计算，建立了雾滴沉积能量守恒方程，实现在作物叶片上农药雾滴沉积分布的全程数据化成像。

8. 杂草生物学、相关基因和杂草群落演替研究成果显著

明确了我国4个杂草稻群体的起源及其独立去驯化起源方式；揭示了稗与水稻竞争生

态适应的分子机制；发现栽培稻基因渗入对杂草稻种群的遗传分化、适应性进化起着重要作用；研究发现休眠相关基因通过 microRNA 途径调控种子休眠和开花的新机制。

9. 鼠害防治基础理论研究，为制定鼠害生态调控策略提供了科学依据

探明了害鼠繁殖调控及种群动态规律、布氏田鼠等害鼠自然种群的近交回避机制、褐家鼠和布氏田鼠的生理、遗传调控机制，以及 VKORC1 基因是抗凝血杀鼠剂的靶标基因。

（二）关键技术创新及发展取得重大进展

1）植物病理学科。发掘植物抗病新基因，获得第一个抗玉米丝黑穗病基因 ZmWAK 以及抗稻瘟菌基因；开发了来源于真菌的植物免疫诱抗制剂阿泰灵（ALI）；建立了水稻条纹叶枯病和黑条矮缩病综合防治技术体系。

2）农业昆虫学科。明确了棉铃虫等靶标害虫对 Bt 作物的抗性机理，建立了天然庇护所利用的抗性治理策略；创建了西藏地区青稞与牧草害虫绿色防控技术体系；建立了长江中下游稻飞虱治理新对策和技术体系，以及环境友好的非化学方法防治水稻害虫的技术模式；组建了针对不同种植区域特点的小菜蛾无害化综合防控技术。

3）生物防治学科。建立了多功能、多指标合一的诱导植物免疫蛋白综合筛选评价体系；组建了重要天敌昆虫发育调控技术；革新了优质生物防治作用物大规模扩繁的核心技术；优化真菌制剂工厂化生产发酵工艺和后处理工艺。

4）入侵生物学科。研究组建了"主要农业入侵生物的预警与监控技术"；集成创新了入侵生物阻截防控技术。

5）植物化感作用学科。选育成功作物化感新品种，并利用化感作用物质开发了一批对病虫草具有控制作用的新型防控剂；组建了基于生态功能的水稻病虫草害防控技术体系和茶树害虫防控技术体系。

6）农药学科。建立了基于 PEG 介导的南方水稻黑条矮缩病毒筛选模型和基于 TMV-GFP 的抗植物病毒药物筛选模型；构建了较为完善的面向绿色农药分子设计的计算化学生物学技术平台；创制了具有自主知识产权的新型抗植物病毒剂——毒氟磷，构建了全程免疫防控技术；首次从植物代谢产物中创制了用于防治植物病害的天然蒽醌类农用杀菌剂——大黄素甲醚。创制出哌虫啶、环氧虫啶等新型杀虫剂，创制出喹草酮和甲基喹草酮等除草剂新品种。

7）农药应用工艺学科。研究成功干粉种衣剂生产新工艺及其系列新产品；制备出氟虫腈·毒死蜱微囊悬浮种衣剂，突破了固体活性成分难以微囊化的难题；研发了土壤熏蒸剂胶囊施药技术；创建了农药高效低风险技术体系；研发了农药喷雾量分布分段采集分析关键技术。

8）杂草科学。明确我国部分稻区稻田稗的抗药性水平及其产生抗药性的重要原因，组建了稻田杂草防控技术体系；发现麦田杂草对多种除草剂产生不同程度单抗、交互抗或多抗的抗药性；探明大豆田、棉田不同杂草种群产生抗药性的原因以及靶标基因单突变、双突变及其过表达是其产生靶标抗药性的重要机制。

9）鼠害防治学科。证明 ENSO 等气候条件是布氏田鼠等重要害鼠种类暴发的重要启动因子；开发出适用于不同环境及要求的毒饵站系统，提高了毒饵引诱性及毒饵使用效率，有效降低了杀鼠剂对周边环境及非靶标动物的影响。

三、本学科国内外主要研究进展比较

1）植物病理学科。我国在主要真菌卵菌病害病原菌逃避寄主抗性反应策略、病原物毒性变异机制、植物抗病性调控机理、重要抗病基因标记定位、克隆和功能研究等领域以及在病毒基因功能、病毒致病性、症状形成及运动机制、RNA 沉默介导的病毒抗性和持久增长型病毒的介体昆虫传播分子机制、病毒病害防控等方面的研究与世界同步，部分研究达到了国际领先水平。目前，我国克隆主要作物"持抗"或"一因多抗"关键抗病基因极少，利用基因沉默技术防治赤霉病等病害的研究与国外同行还存在较大差距，病害流行与监测预警研究投入较为薄弱。

2）农业昆虫学科。我国在昆虫基因组测序方面取得较好的成绩，但基础研究整体水平与国际先进有一定差距。国外转基因抗虫作物研发与产业化应用占显著优势，利用 RNA 干扰技术定向沉默昆虫关键基因的转基因技术已开始应用于新一代的抗虫作物研发，多种双翅目、鳞翅目、鞘翅目等昆虫被成功进行基因转化，但我国转基因昆虫研究尚处于起步阶段。

3）生物防治学科。我国在天敌昆虫个别产品和技术上处于国际领先地位，在微生物制剂应用方面，走在了世界前列。但我国在生物防治产业化程度方面与欧美发达国家还有显著差距，天敌昆虫和微生物农药产品数量少、生产规模小。

4）入侵生物学科。国际上，融合新兴科学技术的入侵生物防控技术创新和集成发展迅速，已集成区域防控新模式并进行应用，逐渐形成智能联动防控技术体系，成为入侵生物控制的主要策略。

5）植物化感作用学科。我国在水稻化感新品种选育以及作物化感品种抑草机制等方面已经达到国际前沿水平，但在作物 – 病原微生物和作物 – 害虫化学互作关系研究仍存在较大差距。

6）农药学学科。在药物分子合成技术，农药相关靶标的结构生物学、靶标与小分子作用机制研究等领域与国际先进水平处于"并跑"阶段。但与欧美日发达国家相比，我国具有自主产权的原创先导及具有全新作用机制的骨架仍然匮乏，缺少具有引领甚至颠覆性的农药新品种。

7）农药应用工艺学科。我国在土壤熏蒸药剂应用方面走在世界前列，但在基础研究、配套机具、检测技术等方面与发达国家相比，均有不少差距。国内种衣剂产品研发和种子包衣技术仍以生产仿制品种为主；国内在喷雾技术基础研究和喷雾技术研究相对不足。

8）杂草科学。我国在杂草稻起源，演化机制，栽培稻基因渗入对杂草稻种群的遗传分化、适应性进化的影响、休眠相关基因的调控作用等方面的研究都处于国际先进水平。我国急需深入研究杂草抗药性机制，提高代谢抗药性分子机理解析等研究水平。

9）鼠害防治学研究。我国鼠害控制在鼠害不育控制理念、TBS 的大规模改进与应用和毒饵站技术领域的研究已处于世界前列。但在鼠害基础理论研究方面还存在较大的发展空间，与欧美等发达国家仍有一定差距。

四、本学科发展趋势和展望

（一）植物病理学科

1）建立流行性病害病原物致病性变异与品种抗性变化的基础性长期性监测网络；
2）加强病害流行监测网络建设和病害流行预警预报研究；
3）发掘持久抗性基因或"一因多效"基因，提高抗病品种培育效率，注重"多抗"和"持抗"品种的培育；
4）开发如寄主诱导基因沉默技术（HIGS）等病害防控新策略；
5）研发和应用如基因编辑等新技术在作物抗病育种中的利用，建立重要农作物相关突变体资源库，挖掘具有重要价值的基因。

（二）农业昆虫学科

1）加强基因组学和基因编辑技术等在昆虫学科的研究与开发应用；
2）加强转基因抗虫植物、转基因昆虫的研发与应用；
3）加强迁飞昆虫异地监测预警与区域阻截技术等研发；
4）基于景观生态学、农田生态系统服务功能等理论和方法，促进害虫区域性治理的理念创新与技术发展；
5）创新农作物害虫综合防治技术模式以及基于"互联网+"的信息服务平台，提升农业害虫绿色防治技术水平。

（三）生物防治学科

1）天敌昆虫、生防微生物在农田生态系统中保育及控害特征、过程与机理研究；
2）天敌昆虫与生防微生物定殖性的时－空波动规律研究；
3）滞育对天敌昆虫的延寿保育机制研究；
4）生防真菌、细菌、病毒、线虫等与寄主识别信号及传递机制研究；
5）主要生防功能基因及表达调控研究；
6）筛选高效微生物菌株及工程菌株，丰富天敌种类；

7)优化生产工艺,革新助剂结构,延长生防产品货架期。

(四)入侵生物学科

1)新入侵生物快速分子识别与扩散阻截扑灭技术;
2)完善已入侵生物的生防抑制与持久生态修复调控技术;
3)完善主要入侵生物动态分布与本底信息,创新智能化监测新技术;
4)编制入侵生物分布危害的最新研究报告及重大疫情报告;
5)入侵生物数据资源采集规范与质量控制规范,编制入侵生物野外数据调查规范;
6)绘制入侵物种的专题电子图集,建立一体化的入侵生物数据库及信息共享平台。

(五)植物化感作用学科

1)全面解析重要化感物质的生物合成途径及其分子调控机理;
2)揭示化感物质的作用机理及其在整个生态系统中所发挥的生态学功能;
3)以作物－有害生物为研究对象,系统剖析生物识别相关化学信号并作出相应化感响应的化学与分子机理;
4)深入研究各种生物与非生物因子对作物化感物质生物合成及其化感作用强弱的影响;
5)研发基于植物化感作用的有害生物治理技术。

(六)农药学科

1)完善绿色农药全创新链,以开展基于我国农业生产重大需求的产品创新,并发展、推广应用先进的综合使用技术;
2)应用基因技术、分子生物学、结构生物学等生物学技术,促进农药筛选平台、新先导化合物发现和新型药物靶标验证等的快速发展;
3)加强以基因编辑为代表的基因工程技术与农药创制的紧密结合;
4)加强农药相关的多尺度环境与生态安全研究,开展多领域,多学科交叉的农药靶标化学生物学研究。

(七)农药应用工艺学科

1)深入开展农药应用工艺学理论研究;
2)融合信息化和智能化技术,研发高效施药技术,逐步提高施药的精准和防效。

(八)杂草科学

1)开展基于基因组及表观遗传组学的农田恶性杂草演化与致灾机制研究;
2)杂草抗药性和生态适应性分子机制研究;
3)重要杂草基因组测序、相关基因功能解析、基因修饰(沉默)等理论基础研究;

4）构建以生态控草为中心，可持续控草技术体系。

（九）鼠害防治学科

1）加强害鼠种群的生态功能研究；
2）深入解析害鼠种群发生及波动的调控机制；
3）加强环境友好型鼠害控制技术和控制策略的研究。

第二十六节　水土保持与荒漠化防治

一、引言

我国是世界上水土流失最为严重的国家之一，也是世界上荒漠化面积最大的国家之一，实现水土资源的可持续利用和生态环境的可持续维护，是经济社会可持续发展的客观要求，也是当前我国急需破解的两大问题。水土保持与荒漠化防治（Soil and Water Conservation and Desertification Combating）学科在我国始建于1958年，学科奠基人关君蔚院士编写了新中国第一部《水土保持学》，确定了水土保持的知识体系框架、名词术语。1980年成立我国第一个水土保持系，1984年由国家教委批准了水土保持学科，1989年被评为国家重点学科。1992年成立我国第一个水土保持学院，全国发展至10所水土保持专业院校。至此，随新中国的进展，基于人类可持续发展的需要在我国"水土保持"已经从一门可有可无的选修课，逐步发展成为重点专业课、水土保持专业、水土保持系、水土保持重点学科、水土保持重点开放实验室、直到现在的水土保持学院，取得了长足的发展。1994年国家高等院校专业调整后将"水土保持与荒漠化防治"确定为国家重点学科，包含水土保持与荒漠化防治两个方面的内容，是一个覆盖全部陆地国土的完整的应用基础学科。

二、学科建设与人才培养

水土保持与荒漠化防治学科也是多学科结合的交叉性学科,属农学门类中的环境生态类专业,是理、工、农、水、林等专业交叉性学科,也是我国目前仅有的三个环境生态类专业之一,与国家环境建设、生态安全和国土资源保护密切相关。2013年,由"教育部高等学校自然保护与环境生态类专业教学指导委员会"制定《高等学校水土保持与荒漠化防治本科专业教学质量国家标准》,进一步规范专业设置,形成了鲜明的专业特色。

在人才培养方面,水土保持高等教育质量的提高促进了专业人才培养规模的扩大,同时培养了大批水土保持领域的高层人才并投入到我国经济建设中,也带动了全国其他高校与科研单位水土保持高层次人才培养的蓬勃发展。此外,我国高等教育对该学科的重视增强,水土保持学科地位得以不断巩固。据统计,设有水土保持专业的本科院校从20世纪50年代北京林业大学1所高校,发展至包括西北农林科技大学、内蒙古农业大学、南京林业大学、东北林业大学等24所高校。另外,我国台湾省的中兴大学、屏东科技大学均设有水土保持系。全国研究生教育也从80年代初仅北京林业大学、西北农林科技大学与内蒙古农业大学设立硕士点,发展至包括北京师范大学、中国农业大学、中国科学院水利部水土保持研究所、中国水利水电科学研究院、中国林业科学研究院等46所高等院校招收水土保持与荒漠化防治硕士研究生;北京林业大学、中国农业大学、西北农林科技大学等14所高等院校及研究院所设有水土保持与荒漠化防治博士点。但是,由于水土保持专业是一个交叉学科,各部门和不同学者对此有不同的理解,此外不同省份的水土保持任务和要求以及对学生培养的规格要求不同。因此,虽然同为水土保持与荒漠化防治专业,各院校的人才培养标准并不统一,其根据各自的教学科研工作基础和现有的办学条件等,保持了各自的办学特色,从而使水土保持的培养目标和课程体系呈现出多样化的局面。目前,大体存在三种类型,分别是以复合应用型和拔尖创新型人才培养为主,以工程应用型人才的培养为主和以培养本地区急需的水土保持与荒漠化防治专业高级人才为主。

在师资队伍建设方面,由于水土保持是一项综合的系统工程,涉及农、林、牧、水、土、经济学、社会学等众多领域,需要自然地理学、土壤学、生态学、生物学、地图学与地理信息系统、土木工程、环境科学、经济学等众多学科的支撑。因此,本领域广开门路,吸纳和引进不同学科的人才,形成了一支高水平综合学术队伍。据统计,我国现有设立水土保持与荒漠化防治专业的本科和研究生高等院校和科研院所中,水土保持高等教育专任教师547人,教授占29%,副教授占36%,讲师占22%,其他占13%。其中中国工程院院士1名,博士生导师85人,硕士生导师265人。

在课程设计方面,大部分院校水土保持与荒漠化防治专业形成了以土壤侵蚀学、水土保持工程学、荒漠化防治工程学、林业生态工程学、水土保持规划学和开发建设项目水土保持6门课程为核心,以生态学、水文学、水力学、土力学、工程力学、气象学、土壤学、地质地貌学、水土保持植物学、计算机技术应用和 GIS 技术与应用 11 门课程为支撑的课程体系。但是，不同院校相同课程的学时分配差异悬殊，这主要与各高等院校的师资结构、办学理念和人才培养方向有关。

近几年，水土保持与荒漠化防治专业就业形势总体发展良好。《中华人民共和国水土保持法》的正式施行，以及"一带一路""京津冀"一体化和长江经济带等战略部署，学生工作就业迎来了新的发展机遇，就业面不断拓宽，从以往的本行业就业局面发展至多行业就业新模式，主要服务于国土资源、水利、农业、林业、电力、矿业、石油、航空、环境保护等行业部门以及相关的企事业单位。同时，本科毕业生的就业选择呈多元化趋势，除了直接就业外，读研深造的本科生占到毕业生总人数的一半以上，还有 10% 左右毕业生选择出国深造。这与高校其他专业有共性之处，也有其专业的独特性。随着水土保持事业的发展，其行业科技含量大幅度提高，用人单位对毕业生的学历要求也呈明显的上升趋势。从近几年北京林业大学水保研究生就业情况分析看，80% 以上的毕业研究生进入国企、事业、国家机关工作；20% 左右的研究生进入私营企业或选择基层就业服务项目；研究生的留京比率也远远超出本科毕业生，约为 30%。

水土保持与荒漠化防治专业具有综合性、实践性、应用性强的特点。其中，野外实践教学是水土保持教学的一个重要组成部分，是拔尖创新型人才培养的关键环节。目前，涉及水土保持方向的野外台站有 29 个，这些野外台站的成立对推动我国水土保持学科的发展起到了重要且积极的作用。水土保持专业野外实践教学更应充分利用国家级高水平科研平台的优质资源，以野外台站为依托，积极开展野外实践教学基地教学活动，实现科教融合，为我国"双一流"建设提供借鉴作用。

除此之外，作为联系广大水土保持科学技术工作者的桥梁和纽带，中国水土保持学会是发展我国水土保持科学技术事业的重要社会力量。学会成立以来，先后接待过来自美国、英国、法国、德国、尼泊尔、日本、俄罗斯、朝鲜、泰国、韩国等多个国家的学术团体及专家，组织召开专业性的国际学术交流会议，共同开展水土保持及相关研究领域的学术交流。普及和推广水土保持科学技术，为广大水土保持工作者服务。学会还致力于加强海峡两岸水土保持领域的交流与合作，与中国台湾地区相关的民间组织及专家学者有着密切的协作关系，共同组织一系列的海峡两岸水土保持学术交流活动，促进了海峡两岸水土保持方面的互动交流及学科的发展与创新。

三、本学科近年的最新研究进展

伴随着 60 多年的学科发展，其研究领域也不断拓宽，形成了具有我国特色的水土保持学科领域，其内涵不仅包括解决水土流失、荒漠化、湿地、开发建设造成的脆弱和退化生态系统等传统生态安全问题，也涉及亟待解决的新的生态环境与社会问题。在我国，关于水土保持与荒漠化防治学科的研究主要集中在 2 个方面，分别是基础理论研究和科学技术研究。其中，基础理论方面主要体现在水力侵蚀、重力侵蚀、风力侵蚀、流域管理和城市水保等研究；技术研究进展主要体现在水土流失和荒漠化防治的长期实践过程中，不断提升了水土流失和荒漠化的防治水平的科学基础支撑技术研究，如黄土高原坡面降雨径流调控与高效利用技术、黄土高原林草植被可持续恢复与营造技术、长江上游坡耕地水土保持微地形改造技术、治坡+降坡+稳坡三位一体崩岗治理模式等。同时，在"中国水土流失与生态安全综合科学考察"基础上，紧紧围绕黄河中游、长江上游、东北黑土地保护、石漠化治理、南方崩岗治理等国家重点工程和大型生产建设项目水土流失治理中急需解决的关键技术问题，相继开展了多项国家"973"计划项目、国家科技支撑计划项目、水利部公益性专项、中科院西部行动计划，国家自然科学基金创新研究群体项目以及教育部科研创新团队项目等的一系列国家级重大研究项目。项目研究区涉及我国水土保持工作的七大片区，水土保持科学研究工作呈现领域不断扩大、项目不断增加的良好局面，水土保持技术也随之取得重大突破。

在学科队伍建设方面，北京林业大学水土保持与荒漠化防治学科的队伍发展情况，首先，确保专业教师数量和结构合理，向多元化方向发展；其次，重视师资队伍建设，树立"队伍建设服务学科建设、学科建设促进队伍建设"的思想，形成按学科建设主线有序配置师资队伍资源的共识，建立起分层次分重点、共同目标共同投入的人才引进和培养机制，创新师资管理模式，建立有利于优秀人才成长和施展才华的运行机制，改善和优化师资队伍结构，努力开创师资队伍建设新局面。

在社会服务方面，水土保持涉及农业、林业、水利、国土、环保等部门以及生产建设各行业，任务艰巨、关系复杂。针对水土保持面临的新形势和新问题，审视水土保持的区域、部门、行业特点，以新修订的《中华人民共和国水土保持法》为依托，以水土保持方案报告、水土保持监测报告、水土保持技术转让、水土保持知识科普为媒介，充分发挥水土保持社会服务作用，推动水土保持事业的发展。此外，也向基层和生产单位进行水土保持科学普及与培训教育，并开展青少年科普教育宣讲，全方位推进科学普及传播工作。

四、本学科发展趋势和展望

如今，党的十九大报告把生态文明建设提高到前所未有的战略高度，竖起了生态文明建设的里程碑。"双一流"建设、"一带一路"倡议、"京津冀"一体化和长江经济带等战略正逐步实施，国家重大水利工程建设如南水北调、三峡工程后续问题等一系列在建设生态文明和美丽中国的进程中，国家在生态环境建设领域存在着的重大现实需求，水土保持与荒漠化防治学科大有可为。同时，大面积的水土流失亟待治理、人为水土流失尚未有效遏制，民众对生态环境要求的普遍提高但参与意识不够，科技的支撑作用仍然薄弱，水土保持信息化能力有待提高以及原创性标志性成果偏少，人才培养质量亟待提升等一系列问题，又对水土保持提出了更为紧迫和更高的要求，水土保持需要在新的历史时期做出新的回应。

新时期的历史使命给予了水土保持与荒漠化学科新的机遇与挑战，如何继承和发扬学科优势，继续为国家生态环境建设事业服务是一个涉及学科前途命运而又亟待解决的重大问题。唯有进一步深入探索学科发展，加入世界一流学科行列，注重行业需求。大体可以从加强水土保持相关法律法规的宣传力度，提高人民法制观念；引进和开发新技术，提高水土保持现代化水平；组建多学科交叉的科研和教学团队，促进多学科交叉融合；开展基础研究工作，实现可持续发展理念以及注重青年人才培养，打造高层次学科带头人等方面应对，实现学科的可持续发展。同时，水土保持必须紧跟形势进一步推动改革发展，从传统粗放式管理向现代精细化管理转变，推动信息技术与水土保持深度融合，及时、准确地掌握所需水土保持信息数据并实现各级共享，提高管理能力与水平。利用"天地一体、上下协同、信息共享和人才融合"等信息化来实现现代化，从而更好地为政府决策、经济社会发展和社会公众进行服务。

第二十七节 茶 学

一、引言

自2010年以来，政府对茶业科研经费投入明显增加，茶业科技创新平台提升，茶业科技创新实力不断增强，总体上取得较大进展，尤其是在茶树分子生物学、茶树种质资源与育种、茶树植保、茶叶加工等学科领域研究进展显著。

二、本学科近年的最新研究进展

（一）茶树分子生物学研究

我国科学家在世界上率先完成茶树基因组的测序，公布了茶树全基因组序列；应用转录组学、蛋白组学和代谢组学等方法获得了丰富的组学数据，在茶树次生代谢、生长发育与抗性、生殖生物学以及叶色变异机理等研究领域取得重要进展。明确了茶树中儿茶素没食子酰基化途径及其相关基因、对苯丙烷代谢途径和类黄酮合成途径的多个结构基因和调控基因、茶树咖啡碱生物合成的关键酶 N- 甲基转移酶基因、芳樟醇等重要萜类化合物代谢及其相关基因等；对茶树抗寒、抗旱的分子机理以及 miRNA 的调控已获明显进展；在转录组水平上探索了茶树对病虫危害的显著反应，对一些与非生物胁迫相关的茶树抗性基因（如 E3 泛素连接酶基因、脱水素基因等）也开展了研究；对白化、黄化等品种或基因型在叶色转化过程中转录组、蛋白组以及代谢组水平上进行了深入的探索，着重研究了叶色转变过程中鲜叶中次生代谢的变化规律。

（二）茶树种质资源研究

截至 2016 年年底，国家种质杭州茶树圃和国家种质勐海茶树分圃已收集保存国内 20 个省市、9 个国家的茶组植物资源 3400 多份，其中野生资源约占 10%、地方品种约占 60%、选育品种和育种材料占 30%，是目前世界上保存茶树资源类型最多、遗传多样性水平最丰富的茶树种质资源平台。近年来，为了提高种质鉴定评价的效果和效率，在鉴定评价方法和技术的标准化、遗传多样性分析、核心种质构建和优异优质资源的发掘等方面都取得了较大的进展。同时，遗传多样性的检测手段日益成熟和多样化，可以从形态多样性、生化成分多样性和 DNA 水平多样性等角度和层次揭示茶树种质资源的遗传变异。特异品种资源的开发和利用已展现出巨大的市场潜力，如紫娟、黄金茶、中黄系列的新品种等。

（三）茶树育种研究

在探明不同茶树品种开花、结实率特点基础上，我国茶树育种专家提出了双无性系茶树人工杂交体系；^{60}Co-γ 射线诱变育种和太空育种技术得到更好地利用；农杆菌、基因枪等转基因技术在定向创造抗虫、抗病、抗寒、抗盐、低咖啡因等性状改良上得到有效应用。

基因组学、代谢组学、蛋白质组学技术以及色谱新技术的应用，使品质鉴定更精准；茶树 – 昆虫互作的化学生态学技术、抗性酶及其相关基因标记技术在茶树品种抗性鉴定中得到应用；光合系统活力鉴定技术和计算机模型技术在茶叶产量鉴定中的应用，提高了育种效率。茶树育种程序的进一步优化，使国家茶树品种的育种周期缩短了 6 年以上。2010—2016 年间，全国累计有 37 个无性系茶树品种通过国家级鉴定，56 个通过省级审（认）定或登记，31 个获得植物新品种权。

茶树良种的繁育技术多样化，茶树胚培养技术、组培苗直接生根技术、覆膜扦插技术、地膜配套保温棚技术、无纺布育苗袋等的综合应用，缩短了育苗周期、加速了珍稀良种种苗繁育，促进了无性系茶树良种化进程。

（四）茶树栽培研究

茶园土壤研究方面，除了关注长期植茶后茶园土壤有机质变化特点外，对茶园土壤在全球碳循环中所起的作用更加重视；分子技术手段也开始应用于茶叶土壤方面的研究。在茶树养分功能和调控技术方面，对氮、磷、钾等主要营养元素的营养功能及其在茶叶品质成分代谢中的作用和茶树吸收特性等有了更深入的认识；在茶树养分转运子基因克隆、氮营养分子生理机制、抗环境胁迫的分子基础等方面的研究逐渐深入。

茶园有机肥施用技术开始大面积推广；茶园土壤和茶叶重金属含量调查从区域性扩展到全国范围进行；茶园机械化程度越来越有利于名优茶的生产。提出了优质绿茶机采茶园的树冠培养模式、采摘适期指标，机械化采摘及分级处理技术，研制出了新型便携式名优茶采摘机、鲜叶筛分机等关键设备。

（五）茶树植保研究

随着对病虫害形态特征的再鉴定，以及分子生物学技术在茶树有害生物分类研究中的应用，对假眼小绿叶蝉、茶尺蠖等我国茶树主要害虫的种名以及炭疽病等茶园主要病害的病原菌有了重新的认识。茶小绿叶蝉监测预警平台已经建立。

水溶性农药在茶叶生产中的风险性引起关注，筛选出溴虫腈、茚虫威、唑虫酰胺、高效氯氟氰菊酯·噻虫嗪等低水溶性农药。化学农药在茶树上的减量化和合理化应用成为共识。2016年科技部启动了国家重点研发计划项目"茶园化肥农药减施增效技术集成研究与示范"，其目标之一就是到2020年茶园化学农药减量施用25%。

神经电生理学和昆虫生理学的结合加深了对刺吸式口器害虫如茶小绿叶蝉为害性的理解；分析技术的发展加快了茶尺蠖、灰茶尺蠖性信息素的研究，并成功鉴定出其成分，研制出了灰茶尺蠖高效性诱芯。

茶树病虫害绿色防控技术有了较大发展。研制出的数字化粘虫色板、天敌友好型LED杀虫灯等设备，实现了茶园害虫诱杀的精准化、高效化，最大限度地降低了对天敌昆虫的误杀，保护了茶园生态环境。

（六）茶叶加工研究

在茶叶加工基础研究领域，明确了茶叶在萎凋和揉捻过程中的物性变化规律，杀青机结构参数对杀青叶水分和叶温的影响；探索出儿茶素氧化的新途径，发现了黑茶渥堆后出现了普洱茶素、茯砖茶素等新的儿茶素类衍生物，获得了普洱熟茶和安化黑茶的香气特征物质。

在加工技术领域，绿茶加工主要针对杀青技术展开了系统研究，红茶加工的研究重点在于发酵技术，黑茶加工的研究重点在于渥堆技术，研究内容集中于能源改进、精准调控、品质优化、连续化作业等。关键工序的工艺特性及在制品理化变化规律日益明确，食品加工高新技术不断被融合。茶叶精制主要集中在色选、匀堆和包装技术的开发。现有色选机已发展到全彩色 CCD，能真实还原毛茶的颜色信息，可达到较好的色选效果；自动拼配匀堆机得到广泛应用，茶叶包装设备得到快速发展，实现了自动称量、自动充填、抽真空（充惰性气体）及制袋包装的全过程自动化。

（七）茶叶深加工研究

在茶叶功能成分提制技术创新方面，酶工程、超声波提取、微波提取、超高压提取等新技术的组合应用，有效提高了茶多酚等有效成分的提取效率，超临界 CO_2 和亚临界提取技术使茶叶提取物实现了有机溶剂零残留。酶法合成茶氨酸、甲基化儿茶素、甲基化茶黄素等取得了新进展。

速溶茶提制技术方面，提出了速溶茶萃取新装置、速溶茶用酶解新工艺、连续真空冷冻干燥方法等一些新技术、新方法和新装置，为速溶茶风味品质提升奠定了更好的技术基础。

茶饮料加工技术方面，酶工程的应用降低了茶汁苦涩味、提高了茶汁浓度和鲜味，糖类物质的添加提高了茶浓缩汁体系稳定性以防范沉淀物形成。

EGCG、茶氨酸、茶树花、茶叶籽油被我国列为新资源食品，为其在食品领域的大量应用突破了技术瓶颈和法规障碍。研制了一大批含茶成分的功能型休闲食品以及个人护理品等。

（八）茶叶质量安全与检测研究

农药残留检测技术在快速、简单、绿色和高通量样品前处理等方面取得长足进展。

茶叶标准体系初步建立。国内建立四级标准制度，构成了我国的茶叶标准体系的构架。中国参与并主导 ISO 标准中的绿茶和乌龙茶制订。茶树种苗、茶鲜叶要求等基础性标准逐步完善，全产业链标准不断延伸，从茶园到茶杯全程控制标准基本形成。

茶叶质量安全风险评估走上正轨。茶叶中污染物的评价实现了从危害评估向风险评估的转变，非致癌化学污染物暴露风险评估模型采用，依据评估结果取消了茶叶中稀土的限量。

茶叶质量安全与风险控制研究取得成效。茶树上母体农药及其降解产物在茶叶种植、加工与冲泡过程中的迁移转化规律已逐渐清晰。

（九）茶与人体健康研究

在茶叶保健功能有关的功能性成分方面，除儿茶素、茶氨酸、茶黄素等常规成分以外，花青素、甲基化儿茶素、聚酯型儿茶素等儿茶素衍生物成为研究热点，在普洱茶中发现了儿茶素 8-C 位 N- 乙基 -2- 吡咯烷酮取代的 8 个新的儿茶素衍生物（Puerins I～VIII），从茯砖茶中分离鉴定出了 7 种儿茶素的衍生物成分，为儿茶素的 B 环裂变的内酯化合物；在绿茶中发现 5 种聚酯型儿茶素主要为 Theasinensins A~E。

在饮茶与抗癌、预防心血管疾病、减肥降脂、糖尿病的预防和治疗等方面均取得一定的研究进展。关于茶叶抗癌的研究结果在实验动物和人体间存在较大差异的原因有了初步认识：一是活体动物和人体在血液中茶多酚类化合物的浓度上有很大差异；二是儿茶素类化合物在化学结构上的羟基数量导致其在人体上生物可利用性低。研究报道了绿茶能够有效地降低心血管疾病的风险；普洱茶、绿茶和红茶能够调节机体代谢内环境的平衡；普洱茶和茯砖茶具有显著降血脂功效；黄大茶对糖尿病有一定的预防效果。

（十）茶产业经济研究

近年来，我国茶产业规模不断扩大，茶产业经济的研究也逐步深入，呈现出"三个转变"的特点，即从定性分析到定量研究、以生产端为主向消费市场研究转移、由关注产业发展规模到关注产业发展品质转变。

在生产领域，茶叶生产成本增加、市场需求不足成为茶农增收的限制因素。在消费领域，我国茶叶消费市场规模不断扩大；年轻消费者成为我国茶叶消费增长的潜在动力，消费者偏好呈现多样化、个性化趋势；茶叶电子商务发展迅速，线上线下渠道相互融合，提升了茶叶消费量；品牌成为影响茶叶消费的重要因素。在国际贸易领域，卫生和植物检疫措施（SPS）措施已经成为我国茶叶出口的主要障碍，产业技术升级是突破贸易壁垒的根本途径。

在产业融合发展方面，2015年年底《关于推进农村一二三产业融合发展的指导意见》出台，作为茶业新型经营主体的茶庄园、茶香小镇、茶园综合体等迎来了新的发展机遇。"以茶促旅，以旅促茶"的茶旅融合发展模式对茶叶消费的带动效果逐步显现。

三、本学科的国内外研究比较

在国家科技计划的持续支持下，我国茶学学科和茶叶科技快速发展，在应用基础研究、共性关键技术研发、新产品开发等方面取得显著成效，部分领域如茶树分子生物学、茶树育种、茶树植保、茶叶加工与深加工已达到国际先进水平。与国外的差距主要体现在以下几方面。

1. 种质资源和基因资源挖掘广度和深度不够

近年来我国逐步开始重视特异种质资源发掘和功能基因的研究，但在抗逆性、抗病虫、功能性成分等种质资源发掘和利用上还不够深入。例如，日本重点鉴定和发掘抗炭疽病、抗桑盾蚧、高氮素利用率、高稀有儿茶素和花香型资源；印度重点开展抗茶饼病、抗螨、抗霜冻、耐水淹资源的鉴定和发掘；肯尼亚重点开展抗旱种质的筛选。我国则主要针对叶色变异、早生、耐寒、抗病虫、高香型、低咖啡碱、高茶氨酸和儿茶素等性状开展鉴定和发掘。与同时期国外相关研究相比，我国在茶树遗传转化方面的研究相对滞后，在全基因组和各组学研究领域具有明显优势，但在功能基因组研究领域尚处于跟踪阶段。

2. 原始创新明显不足

目前，我国茶学科研工作者仍是以跟踪研究为主，在茶树栽培和茶叶加工的基础性研究上缺少重大理论创新，在加工设备和新产品开发上仍缺少具有自主知识产权的关键技术。整体来看，我国围绕名优绿茶、乌龙茶、黑茶、白茶等茶类开展的研究处于领先位置；关于大宗绿茶的机械化生产技术与装备的研究以及深加工技术与新产品的研究仍落后于日本；C.T.C红茶的品种选育、栽培和加工技术研究、连续化自动化生产线的开发等与印度、肯尼亚等红茶生产国仍存在一定的差距。

3. 基础研究还有待于加强

近年来我国在茶树基因组测序、茶树次生代谢和茶树化学生态等领域取了重要进展，但在茶树分子育种、茶叶加工技术的基础理论、茶叶生物活性成分及其保健功能等方面研究相对薄弱。茶叶质量标准的研究缺乏系统性，也缺少完整的茶叶标准规划；茶叶质量安全风险评估方法尚不先进、数据缺乏共享、新型危害因子研究少。茶叶产业经济研究还刚起步，基础数据的采集和研究急需进一步加强。

四、本学科发展趋势和展望

（一）发展趋势

未来我国茶学学科发展和茶叶的科技创新必须紧紧围绕制约我国茶产业转型升级和可持续发展的三大问题：①劳动力短缺、生产成本不断攀升的问题；②质量安全和生态安全压力加大的问题；③产品结构不尽合理和资源利用率不高的问题。一是要加强应用基础研究。系统开展茶树全基因组功能挖掘、茶树逆境响应机制和调控、茶叶品质形成机理与调控等方面的研究，力争取得理论方法突破，为应用技术研究提供原始创新动力。二是要加强关键技术研发。重点开展茶叶全程机械化生产技术研发与应用、绿色高效生产技术研发与应用、茶叶多元化加工利用技术研发与应用，为技术更新换代和产业转型升级奠定基础。三是要加强学科深度交叉融合。主动吸收中医药学、食品学、经济学等其他相关学科的新理论、新方法、新技术，不断拓展茶与健康、茶叶深加工、茶产业经济等学科交叉新领域，力争取得新技术、新方法和新产品的突破。

（二）重点研究领域

（1）茶树分子生物学

开展茶树特有性状调控基因的发掘与精细功能鉴定研究；采用多组学研究策略解析茶树特异性状的形成机制；研究建立稳定可靠的茶树遗传转化技术体系，破解茶树基因功能同源鉴定的技术限制；研究建立与发展高效的茶树基因组编辑技术，实现茶树基因的定点编辑与定向分子设计育种。

（2）茶树种质资源

加大我国野生、地方品种种质资源收集力度，完善国外资源的引进和保存体系；开展种质资源的表型和基因型精准鉴定，深入剖析重要经济、品质性状的遗传机制，挖掘出各优异性状关键效应基因及其优异等位变异。

（3）茶树育种

重点开展多样化目标的茶树品种选育，如适合不同品质需求的茶树品种、具有不同抗性的茶树品种、适应田间作业机械化和采茶机械化的专用茶树品种。进一步加强茶树品种品质、抗性和产量精准早期鉴定技术研究；加快分子标记辅助、转基因等分子设计新技术研究，建立高效快速生态友好的繁育技术体系。

（4）茶树栽培

开展茶园土壤微生物遗传多样性研究、茶园土壤酸化过程及机制研究、茶园土壤质量评价指标的筛选与量化研究；根据不同茶叶区域的土壤养分含量、茶树需肥特性研制生产茶树专用配方肥。加强化肥减施、有机肥替代的应用基础理论和关键技术研究。

（5）茶树植保

重点开展减少化学农药的用量应用基础理论和关键技术研究；茶叶农药残留和新污染物的控制技术研究；茶树主要害虫的性信息素和防控技术研究；茶树害虫的色、光物理防治技术研究以及以嗅觉、味觉为基础的茶树害虫的化学防治技术研究；进一步完善茶园主要病虫害测报预警平台，建立和完善茶园绿色防控技术体系。

（6）茶叶加工

重点开展不同种类茶叶加工品质形成机理与调控技术研究，标准化精准化茶叶加工工艺参数研究，高效、节能、智能型茶叶加工设备与生产线研制，茶叶精制智能化技术装备研发等。

（7）茶叶深加工

重点开展茶饮料高保真制造与保鲜技术研究、茶叶功能性成分的绿色提制工艺研究、茶及功能成分在医药、保健品、护理品、纺织印染、空气净化、畜禽养殖等领域的应用技术研究和产品开发。

（8）茶叶质量安全与检测

建立数字化的品质评价方法和系统，开展国内外标准的比较研究，开发产地环境评价与控制技术；创制适用于茶叶农药残留检测的新材料，开发高通量、自动化样品前处理技术，搭建茶叶农药残留检测色谱质谱分析平台，构建茶叶农药残留大数据库。

（9）茶与人体健康

开展饮茶与衰老和代谢综合征相关的疾病预防与治疗研究，不同茶类在同一功能上的作用效果和差异的研究、不同茶类和不同活性成分对人体主要疾病的预防和治疗功效、作用通路或作用靶点差异等研究，为合理饮茶提供更精准的科学依据。

（10）茶叶产业经济

重点开展提升我国茶叶品牌竞争力的途径和模式研究、茶叶三产融合发展模式研究、"一带一路"沿线国家茶叶消费市场研究、中国茶产业国际竞争力提升战略研究等。

第二十八节 毒理学

一、引言

毒理学旨在研究化学、物理和生物等因素对生物系统和环境生态系统的有害作用及其机制，评估和预测这些因素对人体健康和环境系统的危害，为制定有关标准、管理方案及预防措施提供科学依据。毒理学既是一门基础学科，也是一门与人类健康、环境安全及社会经济发展密切相关的应用学科，综合应用实验观察、临床观察和流行病学调查等方法，通过研究分析外源性物质的暴露、代谢转化、毒性作用及其机制，以保障人类健康与生态环境安全，为促进社会和谐和经济可持续发展提供重要的科技支撑和管理决策依据。

近年来，随着工农业生产、新兴科技和社会经济的快速发展，人类生产和生活以及环境中所暴露的各种有害因素出现显著增加，给人类健康与生态系统带来巨大的威胁，也极大地促进了毒理学的发展。随着细胞生物学、生物化学与分子生物学、系统生物学和转化医学等学科与新兴技术的发展，以及毒理学与化学、物理学、数学、材料科学和计算机科学等学科交叉融合，深刻影响并促进了现代毒理学的发展，产生了毒理基因组学、毒理蛋白质组学、计算毒理学以及转化毒理学等新的毒理学学科发展方向。2007年，美国国家研究委员会（NRC）发表了一份在毒理学领域具有里程碑意义的报告，即《21世纪毒性测试：愿景与策略》（简称TT21C）。该报告阐述了当前主要基于动物实验的毒理学研究与风险评估方法所面临的挑战，指出21世纪毒理学发展的重点将由整体动物试验转向非动物测试方法的重大变革。

二、本学科近年的最新研究进展

（一）毒理学学科建设概况

我国现代毒理学的发展始于20世纪50年代，90年代得到了快速的发展。目前，我国毒理学的发展已经涵盖了人类健康、环境、工农业和军事等多个领域，形成了包括工业毒理学、食品毒理学、药物毒理学、环境与生态毒理学等多个分支学科。毒理学学科在整个国民经济与社会发展及国家安全中的重要作用更加突显，成为保障国民健康和环境安全、应对突发公共卫生事件和灾害救援、促进经济可持续发展以及维护国防安全和社会稳定的重要支撑力量。全国有100多所医药高等院校为大学生提供毒理学教学，有60多所院校提供毒理学研究生教育与培训，每年有数百名毒理学专业研究生毕业从事毒理学研究、教学与管理工作。目前，中国毒理学会是全球最大的毒理学学术团体组织，下属26个专业委员会和1万3千多名会员，学科发展建设已达到一个较高的水平。在科学研究方面，研究成果呈现良好势头。我国毒理学论文发表从2000年至2016年呈上升趋势，2016年论文发表接近3万篇，论文发表数约占全球的20%。尤其在意外大爆炸有害物中毒医学救援、群体中毒事件应急处置与救治、朝核危机的应急响应、大气环境污染防治机制等方面做出了重要贡献。

（二）毒理学基础研究进展

毒理学基础研究通过分析外源性物质的暴露、代谢转化、毒性作用及其机制，为风险评估和保障人类健康与生态环境安全提供支撑和依据。近年来，我国基础毒理学研究进展主要集中在细胞氧化应激、毒性通路与靶毒性效应、内分泌干扰作用、系统毒理学与生物标志、遗传损伤与修复等几个方面。

在氧化应激方面，发现细胞内承担精密调控的抗氧化系统是以Nrf2为核心的"抗氧化酶链"，即，Nrf2能够调控CAT、GPx、HO-1、PPARγ、GCL、GST等抗氧化酶，维持细胞抗氧化能力。在细胞的氧化体系中，转录因子SETD6具有调控氧化酶的核心作用。在氧化还原稳态失衡的调控与致病机制方面，发现了氧化损伤诱发炎症呼吸爆发是窒息性毒剂光气、氯气等致肺水肿的关键因素；氧化还原稳态失衡在镉中毒、肺纤维化、环境污染物致糖脂代谢紊乱等疾病发生发展中发挥关键作用。

毒性通路强调化学物对细胞内亚结构或大分子的干扰。OECD在毒性通路的基础上，提出了有害结局路径的概念，应用其阐述化学物诱导的靶器官毒性。目前已经有10个从受体激活到肝细胞脂肪蓄积的有害结局路径，AhR活化、AKT2活化、LXR活化、NFE2L2/FXR活化、PXR活化、CAR抑制、HNF4α抑制、脂肪酸β氧化抑制和NFE2/Nrf2抑制等为其分子起始事件。

在内分泌干扰物研究方面，发现了内分泌干扰物对精子发生障碍、男性不育及不良出生结局的病因和分子机制。发现烷基酚、植物雌激素等可显著增加男性不育患病风险；孕期木酚素暴露与子代出生体重等存在关联性。PFOS可通过靶向支持细胞，激活p38 MAPK信号通路，下调血睾屏障连接相关蛋白的表达及细胞定位，从而诱导血睾屏障结构和功能的破坏，使得PFOS能进入生精上皮，发挥直接的生殖毒作用。

利用组学技术、GC-MS和生物信息学等技术筛选中毒相关生物标志物方面取得进展，如发现若干基因表达量低的大鼠对乙酰氨基酚致肝损伤敏感，尿液丙二酸、L-苏氨酸、

3，4-二羟基丁酸、L-甲硫氨酸、甘氨酸、半乳糖、尿苷、p-D-吡喃葡萄糖苷可能为致肝损伤代谢标志物。在镉暴露组人群尿液中筛选了 9 个代谢产物，与镉暴露肾损伤早期生物标志相比，代谢组学分析能够更早地发现镉暴露对机体的效应。在多环芳烃和尿液中发现 1-OHPH 有望成为多环芳烃更灵敏的暴露标志。发现尿液 5-HT 可作为评估壬基酚和辛基酚暴露的生物标志；血清 PLG 水平可作为职业低水平苯接触的效应标志。发现血清吡咯加合物能有效地代表组织中的 2，5-HD 和吡咯加合物。

在毒物分析毒理学方面，我国学者在复杂介质化学品的非靶向分析方法探索上成果丰硕。相继开发基于电喷雾萃取电离技术、固相微萃取技术和高分辨质谱技术等的新型分析方法，实现了复杂环境介质中未知氟代、溴代污染物的非靶向识别。

在计算毒理学方面，分别采用支持向量机和毒性比率法，成功构建了 581 个化合物对梨形四膜虫生长抑制急性毒性以及 963 种麻醉毒性化合物和反应/特殊反应活性物质对鱼急性致死的预测模型；基于受体作用建立内分泌干扰效应分类模型可用于预测应用其他化学物质对这一受体介导效应的干扰作用。基于复杂网络技术，开展化学品与机体信号通路关联研究和基于组学数据的规律挖掘。

总之，基础毒理学的研究成果为风险评估、安全防护标准的制定和深入认识毒物的作用机制提供了重要支撑。

（三）应用毒理学研究进展

随着工农业生产、新兴科技和社会经济的快速发展，我国居民从生产生活环境接触到的有害因素越趋增多，通过环境、膳食和职业等途径暴露后可能对人类健康产生不良影响，加上毒理学与多个学科和新技术发展的交叉融合，促进了毒理学研究的迅速发展。近年来，针对社会发展需求，我国毒理学应用的发展主要体现在环境安全、职业安全、食品安全及药物安全等几个重点领域。同时，大力发展建立毒性评价体系的学科平台建设。

生态毒理学、环境健康暴露参数研究正在深入开展，并逐步应用于环境基准/标准。为制订符合我国国情的环境基准剂量，基本构建了我国流域水环境基准方法技术体系，并着手开展本土物种毒性数据及环境暴露参数的研究。广泛和深入的毒理学研究资料为多部环境、食品和职业安全领域法律法规的制定和完善提供了较充足的科学依据。2003 年发布、2010 年 1 月修订的《新化学物质环境管理办法》、2015 年国家卫计委《职业健康检查管理办法》及中国毒理学会《突发中毒事件应急医学救援中国专家共识 2015》等法规和专业指导文件的修订完善使我国环境、职业安全管理变得更加有序。

食品毒理学形成了具有自身特点和系统的概念、理论和方法体系，为新时期我国食品安全风险管理体系提供了重要的学科支撑。随着《食品安全法》2009 年颁布实施、2015 年修订，国务院食品安全委员会、国家食品安全风险评估中心、国家食品药品监督管理总局等机构的成立，我国的食品安全监管逐渐改变了"九龙治水"格局，有效地向集中化转变；通过实施一系列旨在保障食品安全的行动计划，初步建立了食品安全法律法规体系和食品安全标准体系，我国食品安全监测能力也得到了显著提升，食品毒理学在食品安全风险评估和风险交流中的作用日益彰显。

我国临床用药不良反应监测体系逐步完善，药物毒理学研究成果丰硕。药物毒理学研究在安全评价体系的规范化建设、实验动物管理、新技术新方法的应用、药物毒性作用机制研究等方面均取得可喜的进展。应急毒理学和中毒救治学科注重新理念和新技术的应用，建立了多种新型动物染毒方式及评价模型，并可通过毒性通路识别以及基于毒性通路的化学品毒性作用模式分析，对化学品毒性进行预测，为毒物的染毒损伤评价和风险评估提供了必要的技术支撑。

随着科技进步带来的现有检测手段的不断更新升级，毒理基因组、转录组、蛋白组和代谢组等"多组学"研究也随之进入了崭新的大数据时代，大量新型检测手段及数据的汇集，让应用毒理学学科的研究进展愈发快速。现代生物技术的飞速发展、《21世纪毒性测试：愿景与策略》(Toxicity Testing in the 21st Century: a Vision and Strategy，简称TT21C)及有害结局路径（Adverse Outcome Pathway，AOP）等理论体系的提出均大力推动了毒理学安全性评价迈入新的时代，促进了我国应用毒理学的快速发展。

（四）管理毒理学研究进展

近年来，我国在管理毒理和风险评估方面取得了明显的长足进步。国家先后对与健康和环境有关的法律法规进行了重要修订，如《中华人民共和国食品安全法》《中华人民共和国药品管理法实施条例》《危险化学品安全管理条例》《农药管理条例》等，有关监管部门也分别修订和颁布了一系列的配套法规、指南和标准。在这些配套文件中，对良好实验室规范（good laboratory practice，GLP）的修订，是在各监管部门各自多年实践的基础上做出了重大调整，采取了不同的管理体系和监管方式，使我们的GLP体系更加与国际接轨并适合我国国情，并呈现出一定程度的多样性管理模式。以《农药管理条例（修订草案）》为指针，修订的《农药登记管理办法》和《农药登记资料规定》，明确了风险评估的法律地位，提出各专业领域具体的风险评估技术资料要求，今后将在农药登记评审和再评价工作中，全面实施农药风险评估，开展风险管理。与上述法规配套的其他管理文件，如《农药神经毒性试验指南》《农药施用人员单位暴露量测试导则》和《农药透皮吸收评估指南》等一系列的指南、导则和标准，已经通过审定。所有这些法律法规及其配套文件，都较好地反映了国内外管理毒理学和风险评估的基本理念和最新进展，标志着我国与健康和环境相关产品的监管体系已经初步形成，为管理决策更加科学和合理创造了必要条件，将为我国预防和控制各类化学品对健康和生态环境的潜在危害打下了良好的基础。

三、本学科国内外研究进展比较

（一）描述毒理学

描述毒理学主要是通过动物毒性试验、体外试验、人体观察和人群流行病学调查方法，对外源性有害因素的毒性作用进行鉴定，为安全性评价和风险管理提供依据。近年来，与国外毒理学学科发展的比较分析，我国在描述毒理学研究方面呈现以下四个方面的特点，一是在毒性测试替代方法的发展、应用和管理取得了较快发展，成立了毒理学替代方法与转化毒理学专业委员会，多次组织召开专题研讨会；二是积极启动基于AOP的毒性测试与风险评估策略的研究，连续4年组织主办学术研讨会，宣传推广AOP工作；三是有毒有害化合物的研究对象，研究重点逐步转移到分布广、危害性高或毒性不明并且与社会经济发展和公众健康密切相关的优先化合物中，如持久性有机污染物、环境内分泌干扰物，特别是纳米材料和空气颗粒物；四是重点关注复合污染暴露毒性效应的研究。

（二）机制毒理学

机制毒理学研究化合物对生物系统产生损害作用的生化、分子和细胞机制。在研究手段上，以基因组学、转录组学和暴露组学为代表的高通量检测分析技术越来越多用于毒性作用和毒性机制研究中，实现了从整体和器官水平向细胞和分子水平的飞跃；积极开展基于毒性通路的毒性作用机制研究，形成了以人体生物学（人源细胞模型系统）为基础的、涵盖广泛剂量范围、高通量、低成本的体外测试方法；深入开展表观遗传调控机制研究，关注新的表观遗传调控因子、表观遗传调控网络以及遗传与表观遗传调控的cross-talk等国际前沿问题。

（三）管理毒理学

近年来，我国对外源性化学物风险管控的立法步伐不断加大，对化学品、医药产品、农药、化妆品等的管理工作，也正在向风险控制、风险管理的模式迈进。2015年4月国务院公布了《水污染防治行动计划》；2017年2月8日，国务院常务会议通过《农药管理条例（修订草案）》，与之相配套的《农药登记资料规定》也进行了修订；自2008年9月1日起，国家食品药品监督管理总局负责化妆品的监督管理职能；《化妆品卫生监督条例》在规范化妆品生产经营行为、加强化妆品监管方面发挥了积极作用。尽管我国管理毒理学和风险评估发展较快，国际合作与学术交流也日益频繁，但相比欧美等发达国家，无论在技术力量、基础研究、新方法的更新及数据的补充方面，还是在管理能力及法规标准建设方面，我国都仍存在一定的差距。

四、本学科发展趋势和展望

毒理学作为一门基础与应用并举的跨领域学科,已具备坚实的学科基础和系统的学科分支,特别是"大毒理"理念已经深入到环境与健康、医药卫生、军事与国防安全等很多大的领域以及药品食品化学品风险管理中,与人们日常生活更是气息相关,被赋予的社会责任重大。站在中国社会发展新的历史方位,毒理学学科发展也进入一个新的时代,需要更加完善毒理学科研、教育和管理体系,确立毒理学学科地位,加强毒理学基础及应用研究,对于维护环境友好、保护人类健康、支持国民经济的可持续性发展具有重要的意义。

毒理学研究是一个全链条模式,从毒性效应、毒作用机制,到风险评估、干预和防治技术的转化应用。毒理学研究要瞄准国家社会经济发展过程中出现的影响环境生态、健康和安全等的重大实际问题和科学支撑需求,从对环境污染物的关注逐步向重点人群健康效应的毒理学问题转变,提出毒理学整体计划方案、研究技术路线图。

1)创建或优化现代毒性评价新技术手段。加强对物质毒性毒理的"生物组学"大数据的发掘、分析、系统整合和转化应用研究;建立高通量毒性早期发现和筛选模型及技术系统;利用"In silicon"、定量结构活性模型(QSAR)计算分析手段,以及信息和智能技术,创立物质毒性的自动化和智能分析预测技术等;建立大数据与系统毒理学研究体系,深化其在毒理学通路和调控网络的描绘、从试验系统和组学数据导出通路信息中的技术应用。

2)毒性测试替代方法与转化毒理学已经成为 21 世纪毒理学的重要发展方向之一,需要继续加强相关的方法和评价技术手段的创新,促进以及替代方法的管理认可,从法规或管理层面上推动替代方法的发展和应用。

3)机制毒理学与靶效应是毒理学研究的永恒主题,加强对环境化学毒物低剂量长期暴露和混合暴露的毒效应及机制研究、应用系统毒理学理念拓展毒理学机制和毒效应研究的技术路径、毒性预测新的技术与体外模型与实验研究密切结合。

4)环境重点污染物的毒理学。在国家加大整治力度的同时,通过分子流行病学与实验毒理学的结合,深化对环境(大气、土壤和水)重点污染物的健康危害、毒理学机制和风险评估研究的力度。

5)风险评估策略与管理。管理毒理学已逐渐发展成为更加综合和多交叉现代学科——监管科学(Regulatory Science)。生命科学领域研究的新成果的不断融入和广泛应用,为各类化学物潜在危害的管理提供了更加可靠的科学依据和针对性更强的管理决策,更加推动对化学品管理工作向风险控制、风险管理的模式迈进。

鉴于目前我国毒理科学学科现状,亟待建立和完善我国的国家毒理学完整的学科体系,以及与之匹配的科技政策支撑和全新的协调机制,以适应我国社会经济发展的新形势,更好地发挥毒理学在国家发展中的科技支撑作用。借鉴国际经验,制定具有中国特色并能体现"一带一路"合作发展理念的《国家毒理学规划》,并纳入国家安全科学技术发展总体规划，在政府有关部门的主导下，逐步形成国家毒理学科学研究、国家毒理学教育、毒性灾害应急处置、国家中毒控制与信息咨询、毒物与中毒防治技术和法律援助和国家毒物管理咨询服务六大体系。在毒理学教育、人才培养等方面需要贯彻大毒理的理念，建立国家毒理学基础教育和学科人才培养体系，继续巩固毒理学作为预防医学基础学科的地位，同时也要兼顾和强化毒理学跨领域如环境科学、军事医学、农业等的基础和支撑学科地位，关注毒理学研究工作的实用性和社会性，以适应社会发展对毒理学学科和人才的需求。

第二十九节　公共卫生与预防医学

一、引言

截至"十二五"末，我国总体健康指标已优于中高收入国家平均水平。与此同时，快速城镇化、人口大量流动、老龄化、生态环境和生活方式变化以及区域卫生服务提供与卫生服务利用不平衡等，都给保障和促进健康带来一系列新的挑战。在推进健康中国建设、实施健康中国战略的过程中，公共卫生与预防医学学科比以往任何时候都发挥着更加突出、无以替代的关键作用。过去两年为"十三五"开局的关键时期，我国接连编制印发《"健康中国2030"规划纲要》《"十三五"卫生与健康规划》《中国防治慢性病中长期规划（2017—2025年）》等重要大政方针，在继续推进实施"艾滋病、病毒性肝炎等传染病防治"重大科技专项的基础上，启动实施精准医学专项、人群大队列建设和队列数据标准化建设、大气污染与健康、食品安全与生物安全关键技术、生殖发育与出生缺陷等一系列国家重大研发计划和国家重大专项，为公共卫生与预防医学学科发展注入了新的强劲动力。

二、本学科近年的最新研究进展

以人群为基础的大型队列研究、健康大数据和精准医学的发展，推动了流行病学学科的发展。借助日趋完善的公共卫生防治体系和信息化等现代手段，重点传染病得到了有效控制，但是，新发传染病仍然是潜在的严重威胁，如2017年分离出了高致病性的禽流感变异病毒株。慢性病是我国居民疾病负担之首，其流行病学特征不断发生变化，如脑卒中所导致的城市居民死亡率下降而患病率上升。我国大型前瞻性队列建立起步较晚，但这几年正在迅速赶上国际领先水平，如中国慢性病前瞻性研究（CKB 项目）于 2016—2017 年在顶级期刊发表了多篇研究文章，展示中国声音。我国地域辽阔，不同区域经济与社会发展状况、自然环境、民族与人文差异巨大，通过分别设置不同区域自然人群队列研究专项，确定重大疾病特别是严重危害人体健康的重大慢性非传染性疾病的危险因素和作用机制；未来通过建立生殖发育与出生缺陷队列、孕前和母婴队列、跨代遗传队列等，开展生命全程疾病与健康研究，为开展针对性的行为干预与生物医学干预提供关键科学依据。与此同时，大型自然人群队列与以临床为基础的单病种或专病研究队列互为补充，通过广泛应用生命科学领域、暴露评价、生物统计学以及人文社会科学的新技术新方法，将有助于对疾病发生发展的深入了解，有力推动疾病的综合防控。

环境与健康依然是社会热点问题。环境空气质量监测体系已覆盖了我国全部地级及以上城市及部分县级市，为我国开展大气污染健康研究提供了基础平台。与此同时，《中国人群暴露参数手册》的发布，为我国开展进一步的大气污染暴露评价工作提供了第一手的基础数据。土地利用回归模型和卫星遥感技术用于大气污染物监测，丰富了相应的研究手段。水污染是环境与健康的另一个重要领域，近年来，我国开展了苯乙腈、三溴 - 羟基 - 环戊烯 - 酮、含氮酚类碘类消毒副产物等饮用水新消毒副产物的识别工作，饮用水中抗生素污染受到广泛关注，环境混合暴露评估技术和方法得到较好发展，处于国际同步水平。

毒理学作为公共卫生与预防医学的基础学科，其发展与其他学科的融合非常明显。基于健康风险评估的化学物安全管理已经成为政府监管化学物的重要技术支撑，但是基础数据仍然缺乏。基因修饰、RNA 干扰和反义核酸技术、干细胞模型广泛应用于环境因素毒作用的评价和毒作用机制研究中。国内多个单位建立了相应的暴露组学研究平台，推动了学科发展，但要纳入安全性评价毒理学实验还任重道远。在纳米安全领域的研究几乎与国际同步，已经形成了一定规模和具有国际竞争力的纳米毒理和生物安全性研究的科研队伍。

经过近十年的快速发展，我国已经成为世界上最大的化学品生产和消费国，各类化学品急性中毒发病情况总体呈现下降趋势。2004—2016 年，在各类突发公共卫生事件中，中毒事件发生频次和发病人数仅次于传染病疫情居于第二位，但死亡人数则位居第一。随着高毒、剧毒农药在我国的全面禁限用，农药中毒的发病情况已明显呈现下降趋势。药物中毒中仍然以镇静催眠类药物的中毒发病人数最多，自杀是最主要的中毒原因。蘑菇中毒是目前危害最大的有毒生物中毒，其中含鹅膏肽类毒素的毒蘑菇中毒占蘑菇中毒死亡人数的 80% 以上。"毒物数据库与有毒生物标本库"以及"突发中毒事件卫生应急信息平台系统"建成促进了中毒预防与控制的水平，但是毒物鉴定技术仍有待于进一步提高。

近年来，在世界卫生组织全民健康覆盖及联合国可持续发展目标倡导下，国际上社会医学相关研究在初级卫生保健、健康社会决定因素和健康老龄化等方面取得了一些新认识和新进展。进一步强调了生命全周期和代际之间的动态健康、医疗向预防的转型、各层级卫生服务之间的协调与整合、全社会参与和治理，同时对社会决定因素的作用机制和路径、社会网络分析等新的研究方法进行了探索。中国社会医学工作者，响应联合国和世界卫生组织倡导，根据《"健康中国 2030"规划纲要》需要，积极研究医改中的关键问题和热点问题，关注重点人群、弱势人群的健康和卫生服务公平性；积极应对城市化、老龄化的挑战，加强慢性病和社会病防控及医养结合模式研究；围绕"强基层、保基本、建机制"，开展大量社区健康管理、家庭医生制度、分级诊疗、以人为本的卫生服务体系等研究；探索健康监测与测量新方法，"互联网+"在分级诊疗和健康管理中的应用；不断探索社会发展与健康的研究主题，同时把研究成果上升为政策建议和措施，通过专业学会的组织形式，以多种方式进行合作与交流，共同致力于中国社会医学学科建设与发展，为中国与全球健康做出贡献。

2016 年 7 月，全国爱卫会经国务院同意印发《关于开展健康城市健康村镇建设的指导意见》，把"营造健康环境、构建健康社会、优化健康服务、培育健康人群和发展健康文化"作为重点建设领域。2016 年 11 月，国家卫生计生委等 10 个国家部门联合下发了《关于加强健康促进与教育的指导意见》。同年，世界卫生组织与国家卫生计生委在上海召开了"第九届全球健康促进大会"，发表了《2030 可持续发展中的健康促进上海宣言》以及《健康城市上海共识》，进一步推进了健康教育工作的发展。中国居民健康素养水平逐年提高，平均每年增长 0.5 百分点。为了全面提高居民健康素养水平，国家卫生计生委制定了《全民健康素养促进行动规划（2014—2020 年）》。全民健康生活方式行动开展 8 年来，通过领导重视、社会合作、群众参与，经过媒体传播，互动推广，形成了"全民健康，全民参与"的氛围，对倡导改变慢性病相关不良生活方式起到了初步效果。互联网技术与新媒体相结合，运用短信、微信、APP 等智能通信手段开展健康信息推送、健康知识传播、自我疾病风险评估、健康管理等，可以有效促进个体健康生活方式和健康行为并提高全民健康素养；未来还应就如何有效利用互联网技术促进健康城市建设加强研究，通过互联网技术将健康社区、健康村镇、健康单位、健康学校、健康家庭等联为一体，真正形成建设健康城市、普及健康生活的美好局面。

妇幼保健是我国的基本卫生服务的重要内容，提供孕产期保健系统性和规范性服务是近年来孕产期保健管理的重点，女性心理健康也受到更多关注。近年在母胎医学、胎儿疾病预防、早期发现和干预技术方面的研究、发展产前筛查和产前诊断技术及强化规范管理以及开展危重孕产妇监测与管理等方面有所进步，其中早孕期超声软指标如鼻骨异常等和母血中胎儿游离 DNA 检测技术目前已经成为国内临床可供的筛查方法。随着国际上筛查技术的不断进步，目前我国宫颈癌筛查中使用的筛查方法也由以前的传统巴氏涂片细胞学检查和 VIA/VILI 检查，扩展到液基细胞检查、HPV 检测、DNA 倍体定量分析、实时筛查技术 TruScreen 以及人工智能细胞学筛查技术等。2016 年，HPV 预防性疫苗获准在我国上市用于预防子宫颈癌等相关疾病。

中国在降低儿童死亡率方面取得了令人瞩目的成绩，降低和消除可预防的新生儿死亡是未来妇幼健康工作的重点之一，在中国已经实施了超过10年的新生儿复苏项目取得初步成效。近年来我国早产儿发生率有逐年上升趋势，国家卫生计生委于2017年颁布实施了《早产儿保健工作规范》，对早产儿院前、住院期间和出院后管理流程进行了规范。我国儿童的营养状况日益改善，但是微量营养素的缺乏如铁、钙、维生素A、维生素B2等不足导致的营养性疾病仍然是影响婴幼儿健康的突出营养问题。通过开展爱婴医院复核管理干预和推广儿童营养改善项目，2013—2014年儿童贫血率下降了20.9%，生长迟缓率下降了16.8%。我国的儿童心理保健工作起步晚、发展快、不平衡，儿童心理行为测评工具的移植速度不断加快、形式不断更新。肥胖、脊柱侧弯、近视等依然是影响我国儿童青少年健康最突出的问题，身体素质水平仍需重视。

得益于人口健康信息化顶层设计的逐渐明晰和以云计算、大数据为代表的新技术应用，公共卫生信息化取得了一系列建设成果。国家级疾病预防控制信息系统进一步完善，省级疾控信息系统建设和应用水平大幅度提升，国家人口与健康科学数据共享平台公共卫生科学数据中心持续提供数据服务，部分省（市）的区县初步建成区域公共卫生大数据平台，实现公共卫生业务数据可视化。信息安全防护体系建设也向纵深发展，中国疾病预防控制信息系统已建成全国疾病预防控制业务数据中心和同城异地的数据容灾备份中心，全面实行国家级用户CA认证，系统运行在与互联网安全隔离的网络环境中。2012—2016年，国家卫生和计划生育委员会陆续发布多项卫生行业标准进一步规范和促进了公共卫生信息化建设。随着经济社会发展和政府的高度重视，未来必将有越来越多的区域实现公共卫生信息化，进一步推进预防医学与临床医学之间的交叉融合，催生更多的区域医疗联合体并将进一步发展为区域健康联合体，推进整合式医疗模式向整合式大卫生与大健康服务模式转变，为居民提供更为优质高效的医疗与健康服务。

精神疾患越来越受到重视。我国精神障碍患病率水平较高，但分类方案与诊断标准研究相对滞后。2004年开始的"中央补助地方卫生经费重性精神疾病管理治疗项目"以及2009年的"重性精神疾病患者管理"基本公共卫生服务项目加上部分地区的重性精神疾病患者免费的药物治疗大大促进了我国精神病性障碍患者的治疗覆盖率达到70%左右，但有效治疗覆盖率则仍然处于较低水平。进入21世纪以来，我国对精神卫生工作越来越重视，不仅出台了一系列的精神卫生政策，而且开展了很多全国性或地区性的精神卫生项目，但尚缺乏对精神卫生机构进行科学评估。整体上，精神卫生服务资源的短缺，治疗依从性低。

加强全球健康研究是我国经济发展走向全球化、海外利益拓展的迫切需要，是我国作为负责任大国形象、落实"一带一路"倡议、推动构建人类命运共同体的必然要求。近年来，我国公共卫生与预防医学学科发展越来越多地关注全球健康，国内的高等院校陆续设置了全球卫生的科研与教学机构开展跨学科培养与研究，而国家卫生计生委卫生发展研究中心全球卫生研究室重点面向决策部门开展全球卫生的研究与咨询服务。近年来，围绕卫生的议题，全球政府、学术界、产业界、非政府组织间交流愈加频繁，中国全球健康大学联盟、中国全球卫生网络等国内相关的交流与沟通平台也逐步形成。而2016年中华预防医学会成立全球卫生分会,成为我国全球卫生领域首个学术协会。随着全球卫生教学在国内高校逐步推开,2016年,人民卫生出版社出版了《全球健康学》系列教材,国内第一个全球卫生慕课——"全球卫生导论"在中国大学MOOC上线。近年来,中国学者一方面致力于将全球卫生的概念引入中国,另一方面从借鉴国外经验开始走出国门,除了在非洲、亚洲等地筹备建设教学科研基地外,还探索在一些国家建设人群观察性研究队列,总结凝练并在部分非洲和亚洲发展中国家推广介绍中国在妇幼健康、血吸虫病和疟疾等传染病防治领域的成功经验。同时,随着我国国际交往的不断扩大,援外技术和劳务人员往返于境内外,加之来华开展短期交流、长期工作乃至定居的外国人越来越多,显著加大了外来传染病的传入风险。如何快速评估和有效防范外来传染病的传入风险必将引起高度关注和深入研究。

第三十节 科技政策学

一、引言

科技政策学是研究科技政策的性质、产生和发展及相关问题的学科领域,它具有核心学术问题、学科基础、制度和组织保障。随着科学技术在国家社会经济发展中发挥日益重要的作用,科技政策学日益受到各国的重视。

随着科学技术与创新越来越紧密地联系在一起,科技政策也与创新政策联系和结合在一起。国际学术界把传统意义上的科技政策扩展为科学、技术和创新政策(science, technology and innovation policy),把与科技政策相交叉的那部分创新政策包含进来。相应地,科技政策学也包括与科技政策相交叉的创新研究。本文所谓的科技政策学完整的含义是指科学、技术和创新政策学,本研究包含技术创新、区域创新等相关内容。

科技政策学具有以下特点:①科技政策学研究的学科基础广泛;②重视研究方法、定量分析和数据;③应用性强,科技政策学问题大多从政策实践中来,也面向实践,为解决政策问题提供理论、实践和方法。

改革开放以来,我国科技政策学研究得到很大的发展。40年来,中国科技政策学研究不断发展，已经形成了相对系统的理论基础、学科体系和一支专业队伍，对中国科学技术及社会经济的发展起到了重要的作用。

中国科技政策的研究已从国内走向国外，产生日益重要的影响。

二、本学科近年的最新研究进展

本研究根据以下方法确定科技政策研究的最近进展：①科学基金机构资助的科技政策研究项目；②学术期刊发表的文章。

（一）国际研究前沿和热点

1. 美国国家科学基金会 SciSIP 资助的前沿方向

根据美国国家科学基金会 SciSIP 自 2007—2016 年资助的项目分析，美国对科技政策学的资助，重视交叉学科研究方法，聚焦科技政策的证据基础，如模式、框架、工具和数据集等。资助项目的研究内容涉及美国和其他国家有关投资、组织科学、工程和创新中长期存在的问题调研等。

2. 国际学术期刊发表论文反映的研究热点

根据本研究所选的国际 16 种有代表性的科技政策学术期刊为样本，得出的高频关键词有：创新、文献计量、引文分析、专利、R&D、合作、开放型创新、H 指数专利、科学计量等。

采取整数计数（whole counting）的方式，VosViewer 对大于等于 30 次的关键词进行共现分析（DE），共计 127 个关键词，6 个主题聚类分析。得到的结果是："创新"和"文献计量"是科技政策论文发表最重要的两个研究领域。

根据关键词频次，创新热点领域的内容包括创新、专利、R&D、合作、开放式创新、研究评价、治理、纳米技术、技术转移、生物技术、预见、技术、企业家精神。根据主题词聚类分析，创新热点领域可以分为 3 个大的部分：①创新、专利、开放创新和技术转移、创新政策、规制；②技术政策、技术预见；③治理，包括政治、权力、专业技能和伦理学。

根据关键词的频次，在"文献计量"热点领域的内容有引文分析、h- 指数、科研评估、合作等。根据主题聚类分析，文献计量学热点也可以分 3 个方面：①研究指向 / 问题：科研评估、研究生产力、合作；②研究指标：引文分析，影响因子、H 指数等；③研究工具与数据：如前汤森路透的 Web of Science 数据库、信息检索、机器学习、数据挖掘等。

3. 国际上科技政策前沿的最新进展

本文把国际上科技政策学研究的前沿和热点概括为以下几个方面：

1）科技政策问题：科技投资与回报，科学知识的生产和生产力的组织结构，大学和政府在技术转移和创新中的作用，创新激励机制和成果转化；

2）科技与创新问题：创新、专利、开放创新和技术转移、创新政策、规制；

3）对科学技术和创新的治理，包括政治、权力、专业技能和伦理学；

4）方法、工具和数据收集及处理。

（二）中国科技政策研究的最新进展

本研究从3个方面分析中国科技政策研究的前沿和热点：①国家自然科学基金资助的"科技管理与政策"方向的项目；②国际期刊发表的论文；③国内期刊发表的论文。

1. 科学基金资助的前沿方向

根据对2011—2016年国家自然科学基金委员会管理科学部"科技管理与政策"（G0307）资助的924项项目，对项目关键词进行了共词聚类分析。该领域的研究主要包括：科学计量学、科技管理、技术创新与管理、高技术发展与管理以及知识产权管理。研究热点方向科学计量学、自主创新、协同创新、技术创新、科技政策、创新政策、知识产权、创新网络、战略性新兴产业、创新生态系统。

2. 国际期刊发表论文反映的研究热点

在本文所选的16种期刊中，中国学者发表在2007年的只有21篇，到了最近4年，每年都超过100篇。尽管总量不多，但增加速度较快，10年的增加幅度超过8倍。表明中国科技政策研究成果在国际学术界不断增长。

通过对VosViewer对大于等于5次的关键词进行共现分析（DE），共计74个关键词，11个主题聚类分析。得到的结果是："创新"和"文献计量"是中国科技政策论文发表最重要的两个研究领域。

根据关键词的频率，创新并不是中国科技政策研究的重点（仅位列第五），这一点与全球范围内"创新"排名关键词第一的情形形成鲜明对比。主题词聚类进一步显示，我国为数不多的创新研究聚焦在"中国"，可以分为两个部分：①创新绩效、新兴市场；②专利、开放性创新。

根据关键词频次，文献计量学领域的热点有文献计量分析、引文分析、科学计量学、文本挖掘、文献计量、h-index、网络分析、合作、web of science 等。

根据主题聚类分析，中国在文献计量学领域的发表聚焦在传统的科学计量学指标测度和工具上。

3. 国内学术期刊论文反映的研究热点

根据对《科学学研究》《科研管理》《科学学与科学技术管理》《科技管理研究》《科技进步与对策》《科学管理研究》《研究与发展管理》《中国软科学》《中国科技论坛》这几种重要期刊2012—2016年这5年间的期刊数据，运用文献计量学、网络分析的方法挖掘近5年来科技政策领域相关的研究主题。根据对以上期刊的主题分析，词频排在前10的关键词，前10的关键词为技术创新、影响因素、创新绩效、协同创新、战略性新兴产业、科技创新，创新，产业集群，知识产权，自主创新。

除此之外，近年来，国内科技政策界对科技政策学理论与方法的探讨取得一定的进展，包括科技政策研究问题、领域、工具与方法等方面，提出了政策问题-政策过程中的关联性研究议程以及科技政策学的知识构成和体系。

（三）主要领域进展

科技政策各主要领域：科学社会学，科学计量学，科学评价，科技人才，技术创新，区域创新，创新创业和科学基础设施都取得重要进展。例如，在区域创新研究中，由中国科技发展战略研究小组和中国科学院大学中国创新创业管理研究中心编、1999年开始发布的《中国区域创新能力评价报告》，至今已经出版了近20年，2015年被纳入国家创新调查制度系列报告。自2015年开始，科技部每年出版《中国区域创新能力监测报告》。此外，北京和上海等具体区域的区域创新报告也出版。

三、本学科国内外研究进展比较

（一）基本状况比较

发文数量

本研究所选择的16本科技政策期刊2007—2016年所有发文12268篇中，发表最多的国家是美国，发文数量近3000篇，占比超过20%。紧随其后的是英国，中国位居第四位，占比超过7%。

（二）中国与国际研究热点的比较

通过对比本研究所选16本期刊的高频关键词，可以看出中国科技政策与国际研究之间的差异。从国际科技政策的热点看，"创新"排在首位，从中国的热点来看，"文献计量"排在首位。"创新"只排在第五名。从排在前20名的高频关键词来看，国际上研究热点与创新和科技有关的占一大半以上（超过15项），而中国正相反，占一少半（只有5项左右），大多数是文献计量学的热点。不同于创新研究，文献计量学不特别强调背景、制度以及文化等国别间存在很大差异的因素，所以更容易与国外交流，易于追赶研究前沿。

根据关键词共线的聚类，在创新的聚类中，国际创新热点领域可以分为3个大的部分：①创新、专利、开放创新和技术转移、创新政策、规制；②技术政策、技术预见；③治理，包括政治、权力、专业技能和伦理学。中国的创新聚类中，比较单一，没有涉及国际上的治理、技术预见、规制等内容，有些内容是国际上不是重点，如新兴市场、创新有效性，专利引证。

在文献计量的聚类中，国际研究的热点可以分3个方面：①研究指向/问题：科研评估、研究生产力、合作；②研究指标：引文分析，影响因子、H-index等；③研究工具与数据：如前汤森路透的 Web of Science 数据库、信息检索、机器学习、数据挖掘等。与国际的热点相比，中国的研究在文献计量研究还比较单一，虽然在议题上紧追国际前沿议题（例如，当前科学计量领域的新生指标 Altmetrics 在中国学者的研究中也呈现增长趋势），但中国在该领域的研究主要局限在对方法、工具和传统指标的分析上，尚未形成国际上"问题导向"的研究范式，计量用于科研评估的实践上目前还很缺乏。

（三）科技政策学科在世界上的地位

与国外相比，中国科技政策学科发展较晚，但已经取得突出的成就，在本研究所选的16种期刊中，中国学者发表的文章数量在 2007 年只有 21 篇，到最近 4 年都超过 100 篇。尽管总量不多，但增加速度较快，10 年的增加幅度超过 8 倍。同时，论文的研究内容和方法亦呈现出多样化的趋势，表明中国科技政策研究成果在国际学术界不断增长。一批中国科技政策研究者也取得了高质量的研究成果，在世界科技政策界产生很大的影响。

从国际研究热点看，中国科技政策研究热点与国际研究热点既有相同之处，也有不小的差异，反映了中国科技政策在努力追赶国际前沿取得了不小的成绩，但也表明存在着很大的差距。概括地说，中国科技政策研究在理论研究和实证研究上都存在着不足。在理论研究方面原创性不足，在实证研究中关注的问题比较窄、高信度的数据支持不足等缺点。中国科技政策研究在国际上有影响的大多是本国研究议题，这反映了中国科技创新实践的独特意义和研究价值，但若要在国际科技政策界产生很大的影响，需要超越这一阶段，更加关注国际科技政策更广泛的研究议题。

四、本学科发展趋势和展望

（一）科技政策学的发展趋势

在国内外发展的新形势下，科技政策学的未来发展将呈现如下几个方面的趋势：
1）重视科学技术前沿问题政策的研究；
2）科技政策研究与创新研究的交叉融合日益密切；
3）对社会、环境和可持续发展的关注将会占越来越重要的位置；
4）多科学交叉研究趋势增强；
5）定性与定量方法的结合增强；
6）面向中国实践的研究会进一步加强。

（二）科技政策学发展的未来重点方向

1）科技强国战略之下的科技政策发展模式；
2）科技投入的模式、管理及效益；
3）国家创新系统的建设和治理；

4）科技创新型人才的培养、使用及政策；

5）科技政策学的学科基础和方法论。

（三）发展科技政策学的措施

科学技术和创新在中国未来发展起日益重要的作用，对科技政策研究提出了更高的要求，必须大力推动科技政策研究的发展。为此，需要采取以下措施。

（1）从国家长远发展需求，研究和制定中国科技政策研究的长远规划和路线图

根据国家建成世界科技创新强国的"三步走"战略目标，研究国际科技政策发展的趋势和中国科技政策的需求，设置研究议程和主要科学问题，指导全国科技政策研究队伍开展研究。

（2）设立专门的资助计划，支持科技政策研究

在国家自然科学基金设立专门资助计划，长期稳定地支持科技政策研究。

（3）加强学科基础建设，促进学科交叉融合

加强科技政策的前沿问题研究。采取各种措施，促进更多的相关学科参与科技政策研究，促进各学科的交叉融合。

（4）加强中国实践问题研究，促进中国学派产生

中国科技和经济改革的发展提出许多新颖问题，为中国科技政策学的提供了丰富的源泉。加强中国实践问题的理论与实践研究，促进中国科技政策研究学派产生。

（5）加强科技政策研究队伍建设，加强科技政策教育

通过国家、部门、地方和研究机构等各种科技政策研究计划，凝聚科技政策研究队伍。促进科技政策研究者开展各种形式的学术交流。在高等教育中，加强科技政策的教育，设立专门的博士和硕士学位，设立专门的课程。

（6）加强与科技政策制定者的交流与合作

通过学术会议、学术报告与座谈等形式，加强与科技政策制定者的交流与合作，使科技政策学的研究更有针对性，提高科技政策学研究成果的可使用性。

（7）加强科技政策的平台建设

建设科技政策数据基础设施平台，系统地和连续地收集在各种调查、分析和研究中积累的数据，为国家科技政策学研究提供数据。

（8）发挥学术团体的作用

要积极鼓励学术团体积极承担科技政策研究任务，举办学术会议，促进学术交流和学术普及。

（9）加强国际交流与合作

加强与世界各主要国家和国际科技组织的学术交流与合作。通过参与和举办国际学术会议，促进学术交流和合作；举办科技政策培训班的形式，加强科技政策研究队伍的培训；加强与发展中国家科技政策研究的交流与合作，为发展中国家培养人才。

第三章

相关学科进展与趋势简介（英文）

1. Chemistry

This report summarizes the progress made by our country's chemical workers during the first half of 2015 to 2017, including five major branches of inorganic chemistry, organic chemistry, analytical chemistry, macromolecular chemistry, nuclear chemistry and radiochemistry as well as nine interdisciplinary and other specialized fields.

Chinese organic chemists continued their excellent performance in the past two years. New progresses have been made in the fields of synthetic methodologies and natural product synthesis; organofluorine chemistry and natural product chemistry have been already in the world's leading positions.

In the field of organic reactions and synthetic methodologies, a copper-catalyzed radical relay pathway for enantioselective conversion of benzylic C-H bonds into benzylic nitriles, affording with high enantioselectivity products that are key precursors to important bioactive molecules. This is the first research paper in *Science* magazine in organic chemistry field from China. In the field of natural product chemistry, more than 6000 new natural products were isolated and reported. Among the compounds isolated, 349 compounds have new skeletons and more than 1800 compounds exhibit bioactivities. Notable natural products include phainanoids A-F, a new class of potent immunosuppressive triterpenoids, and perforalactone A, a new 20S quassinoid with insecticidal activity against Aphis medicaginis Koch and antagonist activity at the nicotinic acetylcholine receptor of Drosophila melanogaster. Direct introduction of fluorine atom or fluorine-containing group into organic molecules: reagents and reactions for deoxyfluorination, monofluoromethylation, difluoromethylation, trifluoromethylation, difluoromethylthiolation, and trifluoromethylthiolation were developed.

In the field of inorganic chemistry, fruitful results in the following fields were achieved: (1) Basic scientific field, first extended the aromaticity concept to cubic metallic systems of $[Mn^I_8]$ and $[Zn^I_8]$ cluster-based MOF. (2) New energy utilization and storage, new materials, such as flexible artificial machine, super capacitor, self-repairing materials, ferroelectric materials, high efficient adsorption material. (3) Inorganic catalysis, a photochemical strategy to fabricate a stable atomically dispersed palladium-titanium oxide catalyst on ethylene glycolate-stabilized ultrathin TiO_2 nanosheets containing Pd up to 1.5% was reported. (4) Cancer treatment.

In the field of physical chemistry, chemical dynamics explores the dynamical nature of chemical processes: an oxygen generation pathway in the dissociative electron attachment process of CO_2, thus revealing the new mechanism of oxygen origin in the earth primordial atmosphere. For the photolysis of phenol molecules in the vibrational ground state of the first electronic excited state, the adiabatic model is not correct for describing the dynamics near the conical intersection.

Electrochemical still focused the fuel cells, water electrolysis. The general trend of the electrochemical power source is developing toward high specific energy and safe, it develops the high voltage, large capacity cathode materials, and high capacity anode materials, and on the contrary, it produces strong safety solid-based and water-based Li ion batteries. They rationally design high-performance electrode materials based on the atomic structure, molecular level and nanometer scale, the reaction mechanism and the structure-activity relationship of the electrode materials.

Studies on chemical thermodynamics and thermal analysis were focused on ionic liquids (ILs) and Supercritical CO_2 ($scCO_2$), including the physicochemical properties of IL systems, thermodynamic properties of CO_2/water involved systems and the non-equilibrium thermodynamics of the transfer-reaction process in IL aqueous solutions."IPE ionic liquids database" has been completed including 190000 database.

Biophysical chemistry is an emerging interdiscipline. It advances very fast worldwide and has made significant contribution to our understanding of life phenomenon. In China, we have also enjoyed the research progress and expansion of the scientific scope in biophysical chemistry. However, in comparison with the cutting edge of the world, biophysical chemistry in China still has a long way to go.

In the field of analytical chemistry, great progresses have been made in the fields of bioanalysis and biosensing, together with big developments in the fields of single molecular and single cellular analysis, in vivo bioanalysis, bioanalysis based on functional nucleic acids, biomolecular recognition, in vivo imaging, nanoanalysis and interdiscipline combination analysis. Chromatographic, as a branch of separation and analysis science, research in our country has made significant progress including sample pretreatment, chromatographic stationary phase and column technology, multidimensional and integrated devices, innovative chromatography instruments and devices, as well as the separation and analysis of complex samples (proteomics, metabolomics and Chinese medicine etc.).

An overview on the achievements of polymer chemistry, functional polymer including solar cell based on polymers, biomedical polymers, and polymer physics was provided. New methods for polymer synthesis such as amine-yne click polymerization, slow chain-walking copolymerization of ethylene with polar monomers, multicomponent polymerization, and COFs, polymer solar cell based non-fullerene electron accepter instead of C60. In the field of shape memory, discovered the selective actuation method for controlling multi-shape recovery using radiofrequency waves, proposed the concept of ultra-soft shape memory polymer and synthesized such a material. Based on the photoresponsive linear liquid crystal polymer (LCP), fabricated light-controllable self-driven tubular microactuators, in which complex liquids were precisely propelled under light irradiation without any aid of lines and pumps. This kind of tubular micro actuators presents a conceptually novel way to propel liquids by capillary force arising from photo-induced asymmetric deformation. It is notable that Chinese scientists have carried out the systematical research on crystalline CO_2-based polycarbonates prepared enantioselective copolymerization from CO_2 and epoxides and had made a great breakthrough.

Nuclear chemistry and radiochemistry covers several research fields such as nuclear fuel cycle, nuclear chemistry, radioanalytical chemistry, nuclear pharmaceutical chemistry and labeled compounds, environmental radiochemistry, management of radioactive waste, application of radiation chemistry, and so on. For example, preparing special adsorbing material to separate uranium from seawater and wastewater with high concentration. In the field of radiopharmaceuticals, the frontiers research in fundamental area are new PET drugs labeled with nuclides Ga-68,Cu-64, Zr-89 and new SPECT drugs labeled with nuclides Tc-99m and I-123 .The multimodal imaging medicines is also an important development direction, and accurate treatment of tumor by drugs labeled such as Lu-177, Y-90 always is hot area.

Environmental chemistry, as a discipline of studying the existence, behavior, effect, and control principle of toxic substances in the environment, researches the atmospheric fine particles and ozone, persistent organic pollutants, micro/nano plastics and resistance gene according to the study on analytical method, environmental fate, and risk assessment of organic pollutants. The recent advances in key mechanism of combined pollution formation in regional atmosphere, application and effect of environmental nanomaterials, environmental computational chemistry and predictive toxicology, environmental toxicology and health, effectively promote our understanding of environmental pollution and its remediation.

Green Chemistry focuses on the prevention of pollution from the molecular lever that can use resources reasonably and effectively: using non-toxic raw materials, under the condition of non-toxic reaction, few by-products or even "zero emission", and finally producing environment-friendly chemical products. the syntheses of a new platinum-molybdenum carbide nanostructure as the low-temperature and high-efficiency catalysts for aqueous phase methanol reforming has been reported. The synergy between the highly dispersed platinum center and the molybdenum carbide substrate accelerated the methanol-reforming reaction at the interface between platinum and molybdenum carbide substrate, giving rise to high hydrogen production activity in aqueous phase methanol reforming reaction.

Theoretical chemistry plays an indispensable role in molecular property prediction and reaction mechanism investigations. Prominent progresses are made in the areas of electronic structure theory, chemical reaction dynamics, and statistical mechanics, as well as their applications in life and material sciences. For example, the substrate preference of TET2 results from the intrinsic value of its substrates at their 5mC derivative groups has been demonstrated firstly.

The study of Metal-Organic Frameworks (MOFs), one of the frontiers of crystal chemistry, deepened the understanding of the relationship between macro-material performances and micro-structure, and developed some new systems and synthesis methods. At the same time, the design and structure of organometallic supramolecular materials, chiral metal organic macrocycles, cluster complexes, covalent organic framework (COFs), hydrogen-bonded organic frameworks (HOF) materials research and the preparation of novel non-linear optical (NLO) crystal materials have also made new progress.

In the field of colloid and interface chemistry, achievements can be generalized as follows: (1) Fabrication of novel amphiphilic ordered self-assemblies and their applications in biomedicine, especially with respect to the supramolecular assembly, one-dimensional ordered assembly, organic-inorganic composite assembly, ordered thin films with patterns on surfaces. (2) Application of colloid and interface chemistry in the preparation of micro- and nano- functional materials, including the morphological controlled inorganic materials, organic-inorganic composite materials, noble metal nanomaterials, and functional gels. (3) Novel application of colloid and interface chemistry in biosensors. (4) Novel methods of colloid and interface chemistry. As a theoretical and practical subject, the development of modern economics and society will provide wide space for colloid and interface chemistry.

2. Entomology

Insects are closely related to the food and clothing of human beings. Entomology is one of the important subjects in China. At present, about 11900 members of the entomological society of China academy of are engaged in entomological research. In recent years, the National Nature Science Fund of China, "973" project, focus on research and development plan, etc., under the support of entomology forefront of national demand and science in our country, in order to solve the major pest infestation and control the key basic scientific problems as the goal, highlight the characteristics of the combination of macro and micro biology research, integration of molecular biology, genomics, physiology, behavior and ecology, etc., among them in insect genomics, molecular biology, insect vector entomology, pollinators, invasion of entomology, global change, entomology, integrated pest management, and other emerging insects branch is especially rapid development, great progress has been made. Some areas are leading the world.

In the "insect genomics" branch,"i5k plan" (5000 insect genome project) and the "1 kite plan" (1k insect Transcriptome Evolution)(www.1kite.org) , our country important insect species in a variety of DE novo genome sequencing and analysis, functional genomics, Evolution editor in the areas of genomics, bioinformatics and genome got rapid development. In the "insect molecular biology" branch, the molecular mechanism of insect reproduction, immune regulation, molecular recognition and other fields has made great progress; In the "Vector entomology" branch, animal disease vectors insects omics, insect vector physiological and biochemical characteristics, transmission –meidatyed insect toxicology and control of diseases has made some progress. In the "pollination entomology" branch, the main progress in the investigation and analysis of pollination insect diversity, clarify pollination insects and plants adapt to the relationship between each other, reveal the farmland landscape, genetically modified (gm) crops, new neonicotinoid insecticides has been made. In the "invasive entomology" branch, many progress has been made in the invasion of insect biology and ecology characteristics, expansion and diffusion mechanism, host plant invasion of insect-associated microbial interactions and its invasion of insect natural enemies and the biological characteristics. In the "Global change entomology" branch, insects' response to global change is mainly focused on the study of the characteristics, mechanisms and laws of insect response under climate change and human activity. In the "Integrated pest control" branch, the scientists mainly developed the pest control for the prevention and control of agricultural pests. And new theory is proposed based on the service function of insect ecology regulation theory, developed to trap plants, ward off plants, plant source attractants, ward off agent, fan of sex pheromone to prevention and control technology. The current domestic development of entomology is insufficient and compared with the international advanced level.

In the past five years, 12996 research papers have been published, including 6589 papers in SCI journals and 40649 cited times, ranking second in the world. And the annual SCI literature shows that the overall trend shows a significant growth trend. In 6589 SCI literatures, the cited frequency was 40649 times, with the citation number 6.17, h-index = 57. From the citation frequency 303 in

2011 to 14751 in 2016, the citation frequency of 5 years increased by 47.68 times.

In recent five years (2012—2017), Entomology, great progress has been made in basic research in our country, a lot of research is published in journals at home and abroad, such as *Science, Nature, Nature Biotechnology, PNAS, Annual Review of Entomology journal* such influence. There are significant progresses have been made in basic research including the ecological effects of insect migration, analysis of the ecological risk of transgenic insect-resistant crops, molecular mechanisms of phase change in locusts, Invasion mechanism and comprehensive prevention and control of invasive insect pests, silkworm genome sequence analysis, and insect diapause hormone regulation, insects, fungi genetic evolution and adaptation mechanism, the host termites lignocellulose degradation mechanism and insect response to global change.

At the same time, in order to solve the major pest infestation and control the key basic scientific problems as the goal, there are several programs won the second prize of national science and technology progress such as "efficient use of key technology and application in biological targets of reducing pesticide", "development and application of new pesticide to prevent and control virus disease and insect vectors", "the middle and lower reaches of the Yangtze river rice planthopper outbreak mechanism and sustainable prevention and control technology", "leaf blight of rice stripe and black dwarf virus disease disaster rule and prevention and control of green technology", "highland barley and qinghai-tibet plateau green grass pests prevention and control technology research and development and application of ", which provide basis to ensure national food security and ecological security played an important role.

Entomology is a basic discipline, and it is also a practical subject, which is the basis for guiding pest management and beneficial insect protection. Entomology future developments will be guided by the modern omics and information science, surrounding and beneficial insects pests management protection as the main line, to represent the model insects, major health pests and the main resources of agriculture and forestry insect as the research object, and attaches great importance to the multidisciplinary cross and penetration, combination of macro and micro, in time by using the principle of the modern life science and biotechnology, technology and method of armed this major subject areas, promote the sound and rapid development of insects. The following direct need to be focused on in future: (1) Entomology and functional analysis. Choose major pests and insect resources in our country as the object, to carry out its genome, metabolic group, proteomics research and analytical functions, for further research on the innovation of the insect pests rampant harm mechanism, resource use. (2) Systematically investigate the insect groups in important areas of China, especially the insect resources of pollinators and natural enemies, and propose countermeasures for the protection and utilization of these insect. (3) Enhance the research of pollination insects, carry out the analysis of the diversity of pollination insects and provide the relationship of pollination services, and improve the ecological service function of pollinators.

(4) Strengthen the study of the relationship between the host and the vector insect-virus, cutting off the food chain of the media insects and reducing the transmission of the virus. (5) The research of plant pests - predators triple nutrition interaction relationship between physiological and molecular mechanism and chemical information, reveal the cooperative coevolution between plants-pests-predators, provide a new principle and method for sustainable control of pests. (6) To carry out the protection of the natural enemies use, natural enemy insects artificial breeding of nutrition and physiology basis, after the introduction of adaptability and its natural enemies and local competition, development of pest control methods. (7) Research on global change (including the global climate change, the landscape pattern change, land use change) response characteristics and mechanisms, put forward under the background of global change important pests of agriculture and forestry. (8) Using modern biotechnology means such as genetic engineering to find new ways of sustainable pest control.

Entomological research, beneficial insects protection in China and the harmful insect control technology overall level compared with foreign advanced level there is a larger gap, basic research is still relatively weak, the original innovation is weak, not closely combining study, lack of systematic research, intersection is not enough, and so on. The future will be the future development of entomology will be guided by the modern "omics" and information science, surrounding and beneficial insects pests management protection as the main line, to represent the model insects, major health pests and the main resources of agriculture and forestry insect as the research object, and attaches great importance to the multidisciplinary cross and penetration, combination of macro and micro, in time by using the principle of the modern life science and biotechnology arm this major subject areas, techniques and methods, focus on developing: insect proteomics and functional analysis, pollination insects, insect protection using natural enemy insects and decomposition, the host medium insects - analytical, the relationship among the virus insect response to the global change, sustainable control of pest new ways and new methods such as field, promote the sound and rapid development of insect science in China.

On the whole, entomology in China has great development in recent years, entomological research papers published in the world ranking second, reference rate is also ranked second, some research, technology or branch has reached the international advanced level and international leading level.

3. Psychology

In the new era of building socialism with Chinese characteristics, it is the specific requirements for the Party and the country to strengthen the construction of the psychological service system in our society, improve self-esteem and self-confidence, and cultivate rational, peaceful and positive social mentality. Meanwhile, these requirements are also the missions that Chinese psychological researchers must complete. In this background, Chinese psychologists have begun to pay much attention to the psychological problems faced by Chinese society in the new era, and explore the psychological and behavioral sciences closely related to social governance. "Social governance" is a process in which multiple subjects will participate. It emphasizes positive interaction and communication between different interest groups, so as to prevent and resolve conflicts. Therefore, social governance cannot dissociate itself from people in the society, nor can it ignore the mentality, demands, social behaviors and interactive processes of specific individuals, groups and social classes. Psychology is dedicated to exploring these problems and their internal mechanisms, so psychology plays a significant role in social governance.

The first aspect is the study of public attitude towards social governance. In recent years, domestic psychological research has paid much attention to the reality of the public mentality, and has described the general psychological characteristics of the whole society, involving the basic psychological processes of the public. The specific research topics discussed in this report include sense of fairness, social emotion and social trust. Research on sense of fairness reveals that people from the lower class usually feel more unfair than those from the upper class. They are more sensitive even if their injustice experiences are same or similar with the latter. The research on social emotion mainly discusses the internal mechanisms of the group emotional agglomeration and its influencing factors. The research on social trust focuses on the factors that influence trust as well as the impact of trust on society and the economy. It is also found that the social trust in our country has experienced the transitions from inner group trust to outer group trust, from interpersonal trust to institutional trust, and from lower social trust to higher social trust.

Then, there is the psychological research on social norms and values, where researchers not only explore the moral psychology advocated by the society, but also investigate the mentality and behaviors that violate social norms (such as crime) and its treatment. The studies of moral psychology mainly focus on the mechanisms of moral judgment, as well as the origin and influence of morality. In recent years, neuroscientists have explored the brain mechanisms of moral behavior and the mechanisms of gene-culture co-evolution. It is suggested that moral cognition and emotion are also physiological and biochemical reactions to a certain extent. Furthermore, issues such as the social conflicts caused by morality, the moral differences caused by social stratification, and the deterioration and progress of morality are the moral psychological problems closely related to social governance. The research of Chinese criminal psychology focuses not only on the psychological background or basic problems of the occurrence of criminal behaviors, but also the extensive applications. For example, scholars have made very prominent research in the areas such as the criminal psychological portrait, psychological tests, micro-expression in criminal detection, assessment techniques of criminal psychology and evidence-based correction, and so on.

In addition, there is also research on the organization and management issues in social governance, whose topics mainly include performance, change, emotion, leadership and so on. Through the analysis of these topics, we can see that the research in the field of organizational behavior focuses on the psychological and behavioral characteristics of individuals and groups in the organization, which involve individual, interpersonal, inter-group and other levels. Researchers synthesize various factors such as traits and situations, attempting to scientifically explain, predict and even control the feelings and behaviors of the subjects in the organization. The studies of organizational behaviors focus on results from two aspects: one is the achievement and performance of individuals as an "economic man", the other is the satisfaction and happiness of individuals as a "social man". Both of which require "managers" in the organization to pursue.

Finally, research on psychological problems in crisis management investigates people's psychological processes in crisis, such as risk cognition, sense of security, risk decision-making, and mass incidents. The social amplification framework of public risk perception, the ternary model of public psychological security and the psychological mechanism of mass events caused by management errors are discussed respectively. These analyses will provide insights and reference on how to manage emergent crisis with a good, orderly and efficient manner, especially when handling psychological and behavioral responses of the people involved in the emergency.

This report then compares the domestic and foreign studies in this field, analyzing them with examples from four aspects: concept and theory, cultural difference, research methods, and application. On the aspect of concept and theory, this report takes the content of public attitude as an example, and finds that there is a remarkable consistency in concepts, theories and research paradigms between domestic and foreign researches. In many conclusions, Chinese and western subjects also show roughly the same effect, so these conclusions can be verified by each other. However, in some specific research issues and approaches, there are still many differences between domestic and foreign studies. For example, foreign research on social trust gives more attention to the psychological process model. In terms of cultural differences, studies have found that Chinese people show the same motives of self-esteem and self-enhancement as the Westerners at the implicit and neural levels, but do not show the same obvious self-enthusiasm as the Westerners in the explicit level. In addition, Chinese people are very different from westerners in some classical moral cognition. This leads to different research positions on many issues in moral psychology between domestic and foreign researches. Moreover, in moral education, domestic research also has obvious cultural specificity. In terms of research methods, domestic researchers have begun to follow foreign researchers, using multi-layer model analysis, Bayesian estimation, and other new statistical methods, as well as experimental methods. Of course, most of the domestic studies are still based on the more traditional research methods, and in this regard, there is still a gap between domestic and foreign research. In terms of application, taking the Chinese application of crisis management as an example, there are two main strategies to deal with the problem: one is to "fix the problem", which belongs to the remedial response after the incident; the other is "prevention", which belongs to the proactive pre-incident response. This reflects that the domestic scholars conduct research related to the special conditions of our country. Rather than just following foreign research methods and results, they have more unique and realistic thinking, showing the research tendency of focusing on the practical problems.

The final part of the comprehensive report looks at the future directions of this field from several different angles. In terms of public attitude, in order to better promote psychological research of social class and social justice, we need to achieve more breakthroughs and innovations in the localization of the theories, the diversification of the methods, and the application of the conclusions. And it is necessary to conduct in-depth research on the occurrence and development of the group emotions in the network, and monitor real-timely the emotional changes of different groups with the help of the network, and quickly and effectively carry out interventions on group emotions. Future research also needs to meet the requirements of social development, focus on how to improve social trust, and investigate the various factors that affect the trust on the whole, so as to put forward a more integrated theory. And more research results should be applied to social development and governance. In the field of social norms and values, future research should address the issue of bias, prejudice and its restoration. In addition, more ecological research should be done, while new technologies should be introduced to explore new areas. In the aspect of organization and management, researchers should not only pay more attention to the reality of the organizational environment and management situation, go deeply into practice, and study the real problems, but also open the horizon, and listen to the theories and concepts of other disciplines in communications with them, so as to produce more creative and vital research. In particular, it should be noted that future research can promote the use of nudge instead of direct compulsory intervention. The so-called "nudge" is to change people's choice through the design of the choice environment and decision-making process, without the use of coercion or reward and punishment methods. At last, in the aspect of crisis management, future studies should keep track of the characteristics of people's risk cognition and their sense of security. And it is also necessary to verify and revise the risk cognition model, psychological security model and the iceberg model of mass incidents through empirical research.

4. Landscape Ecology

4.1 Introduction

Compared with the international research on landscape ecology, landscape ecological studies in China were linked closely to the social and economic development. The principles and methodology of landscape ecology were widely employed in decision-making, i.e., the framework of red-line on ecological protection, the revision of national ecological function zonation and adjustment of key ecological function zones, as well as the development of ecological protection at both national and provincial levels. The landscape ecological studies in China have made significant progresses in many fields, including development in trans-disciplinarily integration, deepened research in thematic fields, characteristic work in specific regions, and comprehensive assessment and practices.

4.2 Development in trans-disciplinarily integration

Landscape ecology and ecosystem services: (1) Innovative studies were carried out on the issues of double counting, formation mechanism, system analysis, and integrated management of ecosystem services. The systematic analysis and decision-making framework of ecosystem services and operation solutions were proposed to deepen and improve the ecosystem services cognitions and landscape ecological research. (2) In application-oriented theoretical research, the interactive relationship and its scale effect of ecosystem services driven by landscape ecological restoration and restoration practices were revealed, and the evaluation methods of improving ecosystem services were developed based on spatial-temporal variability. The spatial-temporal variations of ecosystem services at multi-scales of key areas were quantitatively evaluated with a variety of biophysical models. Independent R&D methods for regional ecosystem services had been developed, such as SAORES, a comprehensive evaluation and simulation system.

Landscape ecology and global change: (1) Through case studies on typical areas such as the Mongolian Plateau, the evolution characteristics and driving forces of the regional landscape pattern and ecosystem processes were analyzed. Driving mechanisms of landscape changes are studied by taking the invasion of shrubs and the change of lake area as a case. (2) The impacts of landscape pattern on vegetation dynamics in ecotone between forest and grassland, and on the soil moisture distribution at landscape scale were derived. The soil water status determines the vegetation landscape pattern, while the size of forest patches may be regulated by soil water availability, which is getting less with the decrease of precipitation. (3) The spatial resolution of global land use simulation data had been greatly improved, the combination of climate change scenarios and regional urban expansion models had been enhanced, and the land use models and socio-economic/ecological models had been tightly coupled together with climate change.

4.3 Deepened development in thematic fields

Urban landscape ecology: (1) A series of detailed characterization and quantification methods of urban landscape pattern and dynamics were developed in the research of fundamental theory, and the method of three-dimensional urban landscape pattern index and model was established, which initially clarified the three-dimensional model and driving mechanism of urban spatial growth. (2) In response to the ecological effects, the relationship between urban landscape structure and urban heat island effect as well as its scale effect were revealed, and the pollution environmental effect and mechanism of urban structure and green space pattern were also clarified. At the same time, the hydrological and water quality characteristics of urban landscape pattern were analyzed from the process. (3) A multi-scale urban ecological assessment method was improved, with rich case studies on the optimization of urban landscape pattern, in addition to the ecological security and ecosystem service oriented urban ecological land demanding and evaluation methods.

Wetland landscape ecology: (1) Based on the Yangtze River Mouth Salt Marsh Model (SMM-YE), the responses of coastal wetland structure and function to sediment reduction and sea level rise were simulated to reveal the landscape changes of coastal wetlands and their eco-environmental effects caused by the invasion of *Spartina alterniflora*. Combining historical data with field peat samples, it was found that climate-related net primary productivity is more important than the rate of decomposition for the long-term carbon accumulation. (2) Based on the long time series of multi-source data, a water level-elevation transformation method was proposed to explain the impact of river basin water conservancy projects on the vegetation changes of rivers and lake wetlands. By changing the model conditions to improve the simulation accuracy, a water level-area model based on MODIS image and Logistic regression was established. In addition, a monitoring method of wetland vegetation change based on the vegetation index and surface hydrological index was developed to improve the efficiency of wetland resource investigation. (3) A strategy of ecological restoration based on the theory of vegetation succession was proposed in the process of wetland restoration. By integration of traditional culture and modern landscape ecology theory, land/water use Reservoir wetland restoration and comprehensive utilization model named "modified pond-land terrace (MPLT)" was proposed. Meanwhile, some new perspectives were provided for the research and practice of wetland landscape planning and design after evaluating the effectiveness of wetland landscape protection and constructing the wetland park research system in China.

Rural landscape ecology: (1) As the principles and methods of landscape ecology were applied to the classification, evaluation and dynamic of rural landscapes, the ecological functions of rural landscape classification and evaluation were improved, and the characteristics of the spatial pattern and cultural connotation of the traditional settlement landscape were revealed. (2) The research on the impact of agricultural landscape pattern on biodiversity were carried out from different spatio-temporal scales. Related studies also investigated the impacts of spatial heterogeneity and other spatial patterns on the biodiversity of important functional groups, especially the natural enemies and pollination biodiversity and their ecosystem services. It was suggested that the integrated management of agro-landscape in agroecosystem and landscape mosaic is an important way to restore agro-ecosystem services. (3) Based on the effect of crop diversity and heterogeneity on the utilization of water, nutrients and pests control in traditional farmlands, landscape ecological characteristics of traditional agricultural production model was identified, which clarified the impact of agricultural landscape pattern on production and ecological processes, and promoted the protection of traditional agricultural culture landscape.

Forest landscape ecology: (1) The carbon stocks in different succession stages of *Larix gmelinii* forest were estimated after forest fire using the method of "space instead of time", and the non-linear regression method was used to fit the trajectory of forest ecosystem carbon recovery and quantify the turning point when carbon stocks are larger than the released. These methods can be used to assess the ecosystem restoration potential and reveal the time required for carbon stocks from a carbon source into a carbon sink and the recovery time to mature forest after the fire. (2) The ecological impact of long-term fire extinguishing on northern forests and changes in forest fire disturbances are clarified through the simulation of spatial and intuitive forest landscape model. It was proposed that under the existing high-intensity fire-fighting policies, a large-scale combustible management plan such as planned fire and crude fuel removal should be formulated. It was also clarified that climate warming will increase the carbon sequestration rate of broad-leaved Korean pine forest, broad-leaved Larch forest and *Betula platyphorus* forest while reduce the rate of Spruce-fir forest in Xiaoxing'an Mountain (the Xiaoxing'an District). (3) Based on field survey and ALOS/PALSAR radar image data, a prediction model of forest biomass was constructed combining with forest type, topography and stand information, which provided an important approach for simulating aboveground biomass of forest at landscape scale. In the meantime, using large-scale field investigation and sampling, along with remote sensing and geographic information system technologies, the carbon sequestration status and carbon sequestration rate of forest vegetation in China were estimated for the first time, providing important data for China's international negotiations on climate change.

4.4 Characteristic studies in typical regions

Landscape ecological studies in the Loess Plateau: (1) Based on location-weighted landscape index (LWLI), which can be used for pattern-process coupling research, location-function landscape index (LFLI), landscape capacity distribution index (LCDI), and flow accumulation index at catchment scale (FAI) were developed by integrating the impacts of function and spatial distribution of landscape on nutrient loss, flow direction, river network density, precipitation, ecosystem function, distance and flow path, and spatial heterogeneity of nutrient input. (2) We summed up the coupling methods and ways of the landscape pattern and ecological process, and proposed a multi-scale fixed-point monitoring and remote sensing monitoring technology system for pattern-process coupling research on the basis of direct observations of field fixed points.

(3) We carried out quantitative assessment and tradeoff analysis of key ecosystem services of the Loess Plateau systematically. The dependence of the social system on the natural system was sorted out from the natural system and social system perspective, respectively. A research paradigm of ecosystem composition–structure–process–service was proposed. A spatial assessment and optimization tool of regional ecosystem services was established based on GIS, ecosystem model, and multi–objective space optimization algorithm.

Landscape ecology research in the Karst area: (1) A spectral index based on the spectral absorption characteristics of mixed objects was established to extract bare bedrock information by analyzing spectral characteristics. The remote sensing classification method of Karst vegetation was modified and the accuracy of vegetation interpretation based on traditional remote sensing images was improved. (2) It was found that man–made disturbance was the main reason for the special phenomenon of "spatial inversion" in the nutrients and water regimes of Karst soil on the slope and small catchment scales. The special phenomenon of water and soil loss that surface loss and underground missing existed in Karst slopes was revealed. (3) A comprehensive assessment of changes in ecosystem services in Karst area in the context of ecological engineering was carried out systematically. The facts that ecological control measures improved the ecosystem service value and the functions of soil and water conservation and carbon sequestration were revealed.

Landscape ecology research in desert oasis: (1) The land use of multi–scale oasis, reconstruction of oasis spatio–temporal process, and influencing factors were analyzed. The quantitative analysis method for land use of arid area, spatio–temporal process of oasis and its driving factors was proposed. (2) Gradient analysis method on oasis urban landscape was developed, also in addition to the "clump–axial" expansion characteristics of the oasis urbanization along the traffic trunk axis. (3) The spatial distribution, composition, and structure of typical desert shrub landscape were studied. The "nutrient–island effect" and "salt–island effect" of desert shrubs, and the soil carbon accumulation effect of desert landscape were identified.

4.5 Multifunctional landscape and landscape sustainability assessment

In view of the complexity of regional landscape ecosystem service assessment, the quantitative assessment of landscape versatility was focused. (1) We enriched the evaluation index system of agricultural landscape versatility based on the assessment model of ecosystem services. The mapping capabilities of landscape versatility were enhanced, which can provide data support for landscape functional zoning, multi–functional hot spot recognition, and multi–functional tradeoffs.

(2) The extension of landscape sustainability assessment model from socio-economic system to socio-economic-environmental system was established, and the sustainable changes of key ecological transition zones and urban landscapes, especially urban environments was studied. (3) Landscape sustainability assessment was integrated into the planning and design of key ecological transition zones and urban landscapes. The optimization algorithm to clarify the impacts of urban green space on the sustainable planning of compact cities was improved.

4.6 Perspectives

Although great progress have been made in landscape ecological studies in China, there are still spaces for further development, such as innovations in theory and methodology. In the theoretical aspect, to establish landscape-information-graph theory, pattern-process-scale-effect theory and to develop landscape sensing ecology are particularly important. In the practical fields, to develop source-pathway-sink paradigm, pattern-process-design paradigm and the landscape-continuum-gradient paradigm will be highly imperative.

5. Resources Science

Resources science is a newly-developing interdiscipline, mainly studying the formation, evolution, quality characteristics, spatial and temporal distribution of natural resources, and its correlation with social development of human beings. This report analyzes the resources problems confronted by the social and economic development in China, reviews on the important progress and disciplinary construction achievements in resources research in recent years, and came up with the direction and major fields of research on natural resources in the future.

5.1 Resource problem confronted by social and economic development in China

After entering 21st Century, social development of human beings has faced with three global resources problems: the first is the earth system characterized as climate warming results in the reduction of biodiversity, declining of water resource and degradation of land; the second is that the water pollution, soil pollution and air pollution triggered by human activities reduces the quantity of renewable resources and causes the declining of quality; the third is that the regional conflict such as energy fighting and mineral exhaustion occurs frequently in strategic resource area and the geopolitics and economics risk increase.

China has a large population and the per capita occupancy volume of farmland, water, energy, mineral and other important resources is low. Not only is the global resource problem prominent, but the resource problem brought by industrialization and urbanization are more complicated. It mainly shows that: (1) The water and soil resources are highly utilized, but the effective demand is weak and the effective supply is insufficient, so the development and renovation are difficult and the optimized allocation of resource is complicated. (2) Energy structure is unreasonable and the resource flow cost is very high. (3) Management and high-efficient utilization mechanism of natural resource and asset is imperfect, the comprehensive utilization of resource is low and wastage is serious. (4) Resource region (or city) is transformed slowly, and social and economic development keeps depressed. (5) Some important strategic resources, such as energy and agricultural products and etc., depend high on the foreign trade and national resource has high security risk.

5.2 Recent research progress and disciplinary construction achievement

In the last several years, more than one hundred state key projects related with resources research has been initialized and completed, and tens of thousand copies of paper, research report, thematic map and monographs have been published. The research on resources science and the innovation of the technology for natural resource utilization have made far-reaching development.

(1) Natural resources investigation develops toward globalization, professionalization and normalization. In the last several years, the execution of some transnational projects of scientific survey for natural resources, such as Integrated Scientific Expedition in North China and Its

Neighboring Area, it is marks that the natural resource survey and research of our country has not been merely limited to domestic territory. Especially, the investigation supported by national science and technology key project, has been deepened to focus on some key areas and disciplinary fields, such as The Integrated Survey for the animal Resources in Southwestern Tibet Plateau,The Investigation of Changes of Salt Lake Resources in China, The Investigation of Oleaginous Microalgae in China and so on.

(2) Research on five traditional resources like land, water, creature, energy mineral and climate resources has made a remarkable progress. In several years, land exploitation, utilization, regulation and protection are systematically researched based on the new situation analysis of the urban-rural development transformation, and the results play an important role in solving resource problems in the industrialization and urbanization process. In the aspect of water resources research, it especially captured some major technologies of in-depth disposal of wastewater in some major industries and constructs the basic technology system and monitoring prewarning network of water environment treatment. Exploitation of medicinal animal and plant resources, utilization of crop residue resources, and the real time monitoring of forest resources have been become an important research field for biological resources. The efficient utilization of energy and mineral resources and its innovation of application technology are emphatically studied. And it is also a current attractive topic to study the impact of climate change on hydrology and water resources.

(3) Marine resource investigation and ocean resource exploitation technology realizes stepping development. The achievements were gained mainly in the marine resources exploration and mining technology, the development of marine biological resources and the development of marine equipment. Especially, 908 special project figures out the situation of marine environment offshore in our country, updates the basic data of offshore ocean and evaluates the development, utilization and management of ocean resource comprehensively. In addition, the design of seawater desalination project and the application technology development of marine biological resources also make prominent progress.

(4) Resource flow research, resource circulation research and ecological compensation mechanism research laid a solid foundation for establishing management and high-efficient of utilizing natural resources and assets. Especially, the trial compilation work of "Natural Resource Balance Sheet" completed recently provides a solid science and technological support for natural resource and asset audition of leaders upon dismissal.

(5) The disciplinary history of resource science has been analyzed, the construction resource science talent cultivation system and subject teaching material system is ongoing orderly. The professional research institute and professional team grows stronger and the position of subject is clearer.

(6) Research on hydrology and water resource, research on change in land coverage and land utilization, research on global climatic change and coping strategy, and research on biodiversity has been transformed from the past tracking orientation to participating in international cooperation plan and self-independent innovation. International science and technology cooperation ability and influence have stepped to a new stage.

5.3 The development trend and prospect of the discipline in the future

"13th Five-Year Plan" period is a victory-determining stage of building a moderately prosperous society in an all-around way and entering the rank of innovative nation. National scientific and technological innovation has listed "high-efficient development and utilization of resource" as a special field and proved special support for it. Nearly half of the prospective fields listed in the report Innovation 2050: Science & Technology and the Future of China prepared by Chinese Academy of Sciences are related to resources and the resource science has broad development prospect in the future. In the construction of disciplinary theory basis, traditional interdisciplinary synthesis walks toward cross, fusion and integration, and the independent theory system and methodology start emerging; and the international cooperation in resource science will be expanded and deepened further.

The research topic and technological innovation in the future will focus on the following aspects: (1) Coupling research of land and water resources. (2) Research on efficient development and utilization of ocean resource. (3) Research on efficient development and utilization of energy and mineral resources. (4) Technological method innovation of resource science in big data age. (5) Research on theoretical basis of natural resource and asset accounting. (6) Technological innovation of renewable resource utilization. (7) Research on national resource security strategy.

6. Photoimaging

This report defines photoimaging and outlines its relationship with the traditional imaging and digital imaging in terms of system architecture, technology core, openness and etc., and finally overviews the future trend of digital imaging and its important orientations and areas for research and development. This report and its contents are further extended and strengthened by 11 topics reports, which focus on and deal with specific branches or topics of photoimaging in a more systematical, professional manner and in more detail. This will allow the reader to have a comprehensive understanding of photoimaging about its history, current status of development, disciplinary attribute as well as the related industries which are spun off or/and supported by photoimaging.

Photoimaging is an old but at the same time new area, an intersection between photosensitiveness and imaging.

As the entrance, photoimaging is responsible for color-separation, A/D transfer, storage, profile description and even visually rendering of the captured image. All these functions are carried out in one system and completed without considerable time-delay as what we experience with a digital camera. Among a number of important technologies, photosensor is the key technology. Unlike the sensor used for silver-halide photography or the traditional imaging system, photosensor used for digital imaging works on the basis of photophysically reversible change or process when struck by photon, such as photoconduction, electron photoemission and etc. For example, digital camera uses either an area-arrayed CCD or CMOS as the sensor, which transforms the photons entering the aperture and arriving at the sensor plate proportionally into electrons, that will then be counted and transformed into digital data by the installed computer scanning system. In short, digital camera or any other digital imaging capturing system is responsible of transferring the captured image, usually in its analogue form into digital image which can be directly placed into and stored and flow in the networked digital image resources. Later access, manipulation and output by application system or equipment are allowed.

As the exit, photoimaging is responsible, in most cases, of transferring digital images or contents to visible images which can be directly recognized by human visual system or to functional patterns which can be used to build mico- or even nano-sized structures or functional elements on suitable substrate, like silicon wafer.The first path is especially for applications of graphic communication and the second for manufacturing, for example, large scale integrated circuit. For the purpose of graphic communication, digital image or content can be output as visual images rendered by light emitting displays (such as LCD, OLED, PDP and etc.) , paper-like or e-paper displays (such as those developed by E-Ink, Gyriconandand etc.) , traditional/digital printing systems and other hardcopysystems. Electrophotography, inkjet, iconography and magnetographyare typical hardcopy systems, which transfer the digital image/content directly into inked or visual image/content. Taking as example inkjet or digital printing system which is based on inkjet technology, digital image or content is first interpreted or transferred into the corresponding bitmapped image or the table of On/Off which is then converted into the actuation waveform to control the scanning On/Off states of the printhead or the placement of jetted ink droplets on substrate. This allows a direct transfer of digital images/contents into inked or visible ones on substrate. For the purpose of manufacturing, digital images/contents or more precisely speaking patterns described in digital form can be used to control the scanning On/Off states of laser-or electron-beams for exposing patterns directly on silicon wafer in the case of direct writing system or on an substrate which will then be used as mask for transferring the patterns onto silicon wafer.

Being the entrance and exit of the networked image resources or the core of digital imaging, photoimaging in the digital erais, however, more divergent or decentralized, and its R&D and other related activities are distributed in much wider and diversified fields. So do the research platform, resources and the related expertise education and personal cultivation, which are characterized by a more decentralized and opensystemarchitecture. Finely segmented expertise resources and their coordinated collaborations are the reality and theme of research and development of photoimaging, and its success in digital era, too. Photoimaging in digital era is a typical interdiscipline, characterized by diversified disciplinary branches and their integration. Each branch may correspond to an independent industry or a certain market segment. CCD, CMOS, flash memory, photoconductor, toner, internet and internet of things originates and belong to other industries or fields than photoimaging, but become its necessary or key elements when integrated in or combined with digital image system. For example, if well designed and manufactured, area-arrayed CCD or CMOS is the key image capturing sensor of digital camera, belt- or drum-shaped photoconductor and charged toner will be the key imaging and visualizing components of any electrophotographic systems, such as laser printer, multifunctional photocopier and electrophotographicdigital press. Any company, no matter how big or small it may be, can make success stories in photoimaging if it relies on innovation and possesses the proprietory technologies. It is extremely difficult, if not impossible, and unnecessary, too, for an individual company or organization to monopolize all the core technologies of photoimaging and therefore, hardly to create a giant, again, who dominates the photoimaging market, like Kodak or Fujifilm in the traditional imaging era. This brings all technological runners including those from China to the same start line of a new territory, the business and market for digital imaging and creates possibilities and opportunities for China to catch up and even lead the world in many areas of digital imaging.

7. Rock Mechanics and Rock Engineering

Rock mechanics and rock engineering is an applied science discipline which use mechanical principle and method to study rock mechanics and engineering problems. In recent years, with the rapid development of our society and economy, the rock mechanics and rock engineering obtained a great progress. In general it has the following features.

7.1 Introduction

7.1.1 China is gradually transforming from a big power to a great power of rock mechanics and rock engineering

Our country has built and planned huge amounts of rock engineering projects, which attracted worldwide attention. Numbers of large-scale hydraulic and hydropower project d, mining and energy exploitation, underground energy storage and nuclear waste disposal, the ground and underground transportation construction, and other rock engineering construction have started: whether Baihetan hydropower station, China's second hydropower station, or The world's largest radio telescope that named "Five-hundred-meter Aperture Spherical radio Telescope" (FAST), whether "12 · 20

Shenzhen Landslide Accident", or the programme and construction of Xiong'an new district, were active in the field of rock mechanics and engineering scientists and engineers. The achievement on theory and engineering practice of rock mechanics by Chinese scholars has received the worldwide recognition. Eda Quadros, who is the current chairman of the International Society for Rock Mechanics and Rock Engineering, said in GEOSAFE 2016, "the barycenter of rock mechanics and engineering is now on China".

7.1.2 Adhere to serve the main battlefield of the economy, and engineering practice and technological innovation promote each other

As the "One Belt One Road" national strategy boost, that involved in a massive rock engineering.Our scientist or engineer of rock mechanics and rock engineering adhere to serve the main battlefield of the economy, closely around the engineering practice and practical problems on scientific research and technological innovation. The technological innovation and the engineering practice of transportation, energy, water conservancy and hydropower, mining and other fields, promote each other.

7.1.3 The interdisciplinary integration is getting higher and higher, the computer network technology brings new vitality to the discipline development

In recent years, the main scientific and technological achievements in the field of rock mechanics and engineering, the interdisciplinary integration presents the trend of higher and higher.In addition, with the rapid development of Internet Plus, cloud computing, big data, the traditional rock mechanics and engineering system combined with the modern computer technology, promotes the mart management and data sharing, promotes the information level of the overall increase of civil engineering, transportation, water conservancy, energy, national defense, and other infrastructure areas in China.

7.1.4 Young scholars are active in research and innovation, and the echelon construction has been effective

Related research papers in the domestic field of rock mechanics and engineering steadily improve on quality and quantity, particularly for young scholars.In recent years, China has improved its international academic standing on rock mechanics and rock engineering.In addition to take the active part in abroad academic conference, the large international academic conference were organized in China frequently, which provides the display platform for young talents, and promotes the young and middle-aged academic echelon.

7.2 Advance in Rock Mechanics and Rock Engineering

In recent years, due to the rapid development of our economy and with the rock mass of the project construction, the reservoir dam, railway tunnel, bridge construction, underground mining engineering, civil air defense project and the utilization of urban underground space, promoting the rock mechanics and rock engineering disciplines obtain a lot of achievements in research and innovation on the theory and numerical analysis, survey technology, the experiment equipment, the excavation and reinforcement technology and disaster warning.

7.2.1 Rock mechanics theory and numerical analysis

(1) The rock strength and the strength criterion: According to the traditional Hoek Brown – strength criterion, combiend with the engineering practice, proposed the corresponding correction criterion.Among them, the nonlinear three-dimensional H-B strength criterion (GZZ criterion for short) has become one of the international society for rock mechanics (ISRM) proposed method. The criterion considers the intermediate principal stress effect on the strength, and fully inherited the advantages of the H-B standard, that can be directly used test and engineering experience accumulation of H-B criterion the same parameters.

(2) The deformation of the rock and the rheological properties: Focus on the nonlinear creep damage model, the scholars put forward respectively function beween elastic modulus and long-term damage variable expression; established the initial damage evolution equation of rock damage variables.

(3) Rock fracture and damage mechanics: Reveals the rock fracture, fracture cracks initiation and propagation, linking mechanism, and the mechanism of rock crack extension study from two-dimension to three-dimensional.

(4) Rock dynamic response: With the improved Hopkinson pressure bar (SHPB) test equipment, studied the different loading rates, a combination of dynamic and static loading, axial static load, impact velocity and different fissure Angle under the condition of rock dynamic characteristics test.Developed the technology of single wave loading, waveform shaping technology and dynamic experiment of new technology. Proposed and developed three sets of dynamic fracture experiment method, DSDM discontinuous displacement and force model.

(5) Multi-field coupling model and application: Based on equivalent seepage resistance of seepage and stress coupling model, the property of fracture dilatancy under the action of compression-shear stress, under the action of seepage flow stress coupling model and the dissolution of the rock, rock

elastic-plastic dynamic model of stress and seepage damage coupling model, layered salt rocks of the consideration of seepage and stress coupling action section crack extension model and evolution of shale permeability mathematical model and rock mass under condition of low temperature freezing and thawing THMD coupling model.

(6) Zonal disintegration of deep rock mass: Established the mechanical model with dynamic excavation unloading in deep partition isotropic circular cavity surrounding rock burst influence under the condition of hydrostatic pressure and non-hydrostatic pressure.

(7) Rock mass numerical methods: Based on the research of multi-scale constitutive model and the rock creep constitutive model, developed the numerical simulation method of rock mass with continuous method (Finite Element Method, Boundary Element Method, FLAC, Rigid Body Interface Element Method) and discontinuous method (DEM, DDA, NMM, Key Block).

7.2.2 Rock engineering survey, test and design

(1) Rock engineering geological survey: Based on the widespread application of Total Station, GPS technology and three-dimensional scanning technology, put forward the measuring the parameters of rock mass structural plane, monitoring of slope surface deformation, identification of dangerous rock mass, and rock mass properties in-situ test methods. Based on survey of drilling technology and borehole in-situ survey, puts forward a variety of new methods on rock mass parameters. Based on large hydropower engineering investigation, present the hydropower engineering geological investigation and analysis of information integration design method and its solution.

(2) Indoor testing instruments and equipment: Developed the CT examination of rock fracture process, soft rock shear test and rock creep testing technique, field deformation test, etc. Developed the microcomputer control electro-hydraulic servo flexible triaxial experimental machine, the medium-sized rock true-triaxial test system, deep tunnel frame type true triaxial physics test system, etc.

(3) In-situ rock test: Developed the high pressure load and complex stress path servo control of the rock true-triaxial test system, and large-scale triaxial rheological test system. Developed the method and technology of stress test of the hollow core of deep hole underwater cavity.

(4) Dynamic design and feedback analysis: Presen the analysis method on hydraulic rock mass excavation and supporting dynamic feedback, the Spatial-temporal prediction and analysis evaluation on complex environmental the safety (stability) of rock engineering.

7.2.3 Rock excavation and reinforcement

(1) Coal mining: 110 Method provides a new effective solution for the safety of coal mining, called mining technological changes in China's third exploration. This technology makes an important

contribution to our energy security.

(2) Water-power engineering: For the problem of excavation of large hydropower engineering foundation, a conjugate deformation mechanism is proposed to explain the stability of large deformation and slope.Based on complex dam foundation excavation and protection, developed the digital management system for whole process of monitoring and evaluation, prediction and decision support.

(3) Engineering reinforcement: According to the traditional bolt anchorage mechanism and characteristics of joint rock, puts forward the concept of main crack, and study the anchorage effect of anchorage rock block on crack angle and lateral stress, and established the joint deformation of the anchor bolt under shear load model.

7.2.4 Rock engineering monitoring and early warning

(1) Slope (landslide) monitoring: Deep sliding surface based on the slope sliding dynamic change remote real-time monitoring system has a set of reinforcement and monitoring, control, and forecast the distinct advantage of the multiple functions.

(2) Underground engineering monitoring: In mechanism and characteristics of rock burst, developed a strain of rock mechanics test system and impact rock mechanics test system.In the technology of monitoring, developed a set of the domestic first with independent intellectual property rights of kilometer-scale destructive rock-burst monitoring and orientation system for mining area.

7.3 Further Development and Outlook

Rock mechanics is a science which related to the engineering application closely. The problems tend to have the feature of large scale, long construction period, engineering geological conditions complicated and so on. In recent years, our country's scientists and technologists made outstanding contributions of rock mechanics and rock engineering for the discipline development with fruitful work.

To address the new problems and emerging challenges, the research focus of the future is mainly reflected in:

(1) Cross-scale fracture mechanics: Development field in situ experiment platform construction, at the same time, accelerate the construction of indoor small size platform. And build rock structure micro-fine-macro multi-scale experiment and numerical research.

(2) The comprehensive analysis of the prototype monitoring, geological mechanical model and numerical calculation: Research and development with independent intellectual property rights at the high performance monitoring equipment and large structure geomechanics model test, and

promote the basic theory, model and key technology problems in the further research.

(3) The research of mutual fusion between rock mechanics, rock engineering and modern computer technology: Based on analysis of large data of rock engineering design, analysis of construction and operation of monitoring, simulation and engineering decisions, seize the opportunities brought by artificial intelligence, and make the leap-forward development for rock mechanics and rock engineering for the future.

(4) Deep underground engineering will be the rock mechanics and rock engineering strategic frontier in the future. The safety of the deep geothermal energy and mineral resources, development of effective and reliable storage is related to the national economic strategy of sustainable development and energy security. Sustainable, safe and green energy supply is a basic guarantee and rigid demand of rapid economic growth in China even all over the world.

8. Mechanical Engineering

Design, as the starting point and precursor of the formation of mechanical products, determines the functional quality of the product as well as the value of the manufacturing and service. The product's structure, performance, cost, maintainability, man-machine environment, style and other key factors which affects the product competitiveness are all determined by the design stage. The competition of modern manufacturing industry is often the competition of product design. So, the innovation ability of product design has become the core and key to keep the competitiveness of enterprises. Under the current development trend of digital, information and intelligent, different disciplines are mutually and strongly integrated. This brings both great opportunities and great challenges to the mechanical design, and it will also lead to profound changes in modern design methods. This report mainly summarizes the innovations and breakthroughs in the research of the theoretical basis and different industrial applications of the mechanical design, and by comparing the research progress at home and abroad, the research trend of mechanical design theory and application technology in the future is prospected.

8.1 The basic theory of mechanical design

On the basic theory of mechanical design, it will be illustrated mainly from the several aspects of industrial innovation design theory, digital design theory, the structure optimization theory and some new design theory.

(1) Industrial innovation design theory. Industrial innovation design is an advanced design formed by the organic combination of industrial design and innovative design. It covers three distinct phases of the concept fuzzy front-end of industrial products, product development process and commercialization. With the continuous development of computer technology and the arrival of

the digital information age, the industrial innovation design industry around computer technology came into being. For example, the industrial design, the computer aided industrial creative design, the computer supported interactive design, the computer supported collaborative design and the computer aided innovation design. The realization of industrial innovation design not only depends on the professional knowledge and innovation ability of enterprise R&D personnel, but also involves the new technology of interdisciplinary combination. It combines multicultural design technology, personalized configuration design technology and innovative collaborative design technology based on network large data and so on. Meanwhile the application of technological innovation method can improve the innovation ability and innovation efficiency of research staff and help them to develop innovative design schemes for industrial products quickly with high quality. The future development of industrial design innovation theory can be briefly summarized as three basic properties, five kinds of development characteristics and five major trends. The three basic attributes are green, virtual and humanized. The five characteristics of development are global product integration, green low-carbon sustainability, intelligent mechanical products, ergonomics and design capability. The five trends are international globalization trend, low consumption and low pollution trend, innovation transformation design trend, rich and poor balanced people's livelihood design trend and the multi-element and co-win design trend.

（2）Digital design theory. Digital technology is not only the key to promote industrial development and enhance economic competitiveness in the world, but also the main means of information technology, independent innovation, design and manufacturing. In recent years, the development of digital design theory mainly focuses on computer aided design theory and numerical simulation based design theory and other fields. In the field of computer aided design, computer aided design integration technology, computer aided design standardization technology, computer aided design intelligence technology and computer aided design virtual design have made great progress. However, compared with the developed countries, there is still much room for improvement in the independent R&D capabilities of 3D CAD systems, such as data exchange and visual integrated CAD technology. In the field of design theory based on numerical simulation, the information technology is combined with the manufacturing technology deeply, which leads the digitization of mechanical equipment to develop in the whole life cycle of mass data integration direction. This realizes the complicated modeling, simulation, analysis and excavation for each phase of creativity, design, processing, assembly, testing, service, maintenance, scrap and recovery. Around the numerical simulation, the integrated modeling technology, the life-cycle oriented product digital design representation, modeling and analysis technology, the accurate identification of digital modeling parameters based on computational inverse and the evaluation system of uncertainty in modeling and its spread and so on have achieved great progress. The achievements in the field of digital design will help China's design and manufacturing industry to lay a solid foundation for the development of the industry.

(3) Structure optimization theory. The structural optimization design is to use scientific and target orientated process and method to replace the traditional design. The aim is to find the structural form that is both economical and applicable, and to achieve the best performance of the structure with the least material and the lowest cost. In the last two decades, the structural optimization design, which aims at improving the structure economy and the structure performance, has made remarkable progress in theory and method, and has been widely used in the design practice of complex equipment. From the current situation and development trend of the research, the multi-disciplinary and multi-objective structure optimization design, and the structural topology optimization design are still the hot spots in this field. Meanwhile, aiming at the uncertainty of the technical difficulties in the project, the uncertainty optimization design based on the related research is increasingly rising. In these fields, many scholars have done a lot of work and have obtained corresponding theoretical and technical achievements. For example, the multidisciplinary design optimization method based on collaborative design, the topology optimization design method based on variable density method, the reliability based optimization design method and the sequence uncertainty based hybrid uncertainty analysis method. Some of these technologies have already been applied to engineering practice. In complex structure and product digital design, the global and local cooperative optimization, the static and dynamic cooperative optimization, the multidisciplinary robust optimization, the multi-objective cooperative optimization, the topology optimization design under multi-physical field, the multi-reliability approximate optimization for large scale systems, the dynamic or time-varying uncertainty optimization and the life-cycle oriented product reliability optimization are still the urgent problems to be solved in the field of structural optimization.

(4) New design theory. In order to meet the demand of market competition of domestic and international machinery products, many new theories and methods of mechanical design have been put forward and widely studied. There are mainly material-structure-function integrated design theory, green design theory, bionic structure design theory and 3D printing design theory. The domestic material-structure-function integration design mainly studies the new method of high performance precision forming from the angle of equal material, material reduction, material increase and compound manufacturing. Many foreign countries are devoted to the research and development of lightweight polymer composites to replace heavy metal components in equipment. The development of green design is mainly reflected in the control and reduction of harmful substances in the product manufacturing process, and lightweight design of products, so as to save material usage. Bionic structure design is intercrossed with mechanical engineering discipline, nanoscience, information science and biological science, so as to imitate the new structure with directional

function. 3D printing technology is constantly applied to many fields, such as manufacturing, aviation, biomedicine and other fields, such as 3D printing automobile, 3D printing blood vessel, 3D printing and regenerating articular cartilage and other high-performance machinery or structure. Material-structure-function integrated design will conduct multi-scale structure and construction performance research, reveal the interaction mechanism between material organization evolution and structural deformation, and explore the principle of composite manufacturing, so as to achieve the overall manufacturing of high performance complex structure. Green design will turn to the direction of low carbon economy and the direction of manufacturing and providing life-long service around high value-added products. Bionics, along with the development of biological manufacturing technology, produces bionic organisms or bionic intelligent robots to serve human life. 3D printing design from R&D to industrial application, and integrated with the information network technology, will bring a new round of industrial change.

8.2 The machinery industrial design

For the machinery industrial design, it can be introduced from the main five aspects, structurology and robot design, mechanical transmission design, mechanism strength and reliability design, machine tool design and vehicle engineering design.

(1) Structurology and robot design. Modern mechanism is one of the important branches of mechanical engineering, and it is the basis of designing and developing various modern mechanical equipment and mechanical and electrical integrated products. In recent years, the research on mechanism and robot has been developing rapidly and has been expanded in many fields. In the field of parallel mechanism, the research contents mainly involve workspace analysis, scale synthesis, dynamic modeling, singular pose analysis, design of less freedom parallel mechanism, rigid flexible coupling and so on. In the field of robotics, the GF set theory of parallel robot type synthesis is proposed. The whole area quantitative evaluation and scale synthesis visualization method of parallel robot has been formed, which laid the theoretical foundation for independent research and development of parallel robot equipment in China. At the same time, on underactuated robot dynamics and motion control, low DOF parallel robot in space (the number of degrees of freedom < 6), comprehensive evaluation and configuration parameters design and mechanism theory and control of parallel robot and bionic robot, a lot of research has also been done, and very fruitful results have been obtained.

(2) Mechanical transmission design. Mechanical transmission is the foundation of the equipment manufacturing. The development of China's energy, transportation, aerospace and

marine equipment has raised higher requirements for the precision, efficiency, power density, reliability and environmental adaptability of transmission parts and systems. In recent years, the research of mechanical transmission design focuses on the three main aspects, i.e. the multi-media multi-medium and high efficient transmission design considering energy conversion and transmission efficiency, the precision hybrid drive design considering the transmission accuracy, and the new type of driven design based on functional materials. In detail, the three aspects include the principle of the scale criterion and shaping of the micro gear, the LIGA and quasi-LIGA technologies of the micro gear, micro etching techniques, micro-spark cutting technology and micro-cutting technology, the contact and adhesion analysis, friction analysis and multi-physical field coupling analysis of the micro-gear, the traction-type non-stage transmission and pulse-type non-stage transmission of the cone-wheel, the hydraulic transmission and hydraulic transmission, the multistage hybrid transmission and the hydraulics-mechanical multistage hybrid transmission of planetary gears, the improvement of the shunt transmission system and the transmission based on gradient material, memory alloy, composite materials, intelligent materials, etc. With the development of materials, machinery, sensing and control, and the progress of manufacturing technology, the design of mechanical transmission will be developed in three main directions, the design of the transmission parts and unit with high performance and high efficiency, the design of the precision regulation of the hybrid transmission, and the driving mechanism of intelligent structures and new actuators. For example, the compound modulus and parameter match of the hybrid transmission, the movement, force, energy and control strategies of the precision hybrid transmission, the integrated optimization design of the drive, test, control and new principles of the intelligent structures, and the new transmission principle based on multi-field multiphasic coupling effect.

(3) Mechanism strength and reliability design. Due to the varieties and complexities of the modern machinery, the strength and reliability design is undoubtedly the concern of structural design. But in modern mechanical systems, new techniques, new materials and new processes are widely used, therefore the lack of practical experience makes it difficult to adjust to the safety factor under the different conditions. Meanwhile, the material strength, the external load and the sizes of the mechanical parts exist various deviations, and those parts may fail to achieve the designed effect. Therefore, the safety factors cannot be regarded as a constant but a probability variable for the reliability design. The domestic strength and reliability research on the mechanical products is relatively late. In the 1980s, the Institute of Mechanical Product Reliability Design Research is established and set a strength and reliability design criteria. After that, the strength and reliability design of mechanical products attracted more and more attention. Although the domestic scholars had established the system strength and reliability rating and the design basis, but the practical

applications are relatively less. Compared with developed countries, the distance is relatively large. However, the research and development of the domestic strength and reliability research is fast, and the numerous failures of aircraft landing gear agencies, satellite dishes, weapons manipulations have attracted many scholars. With the application of new technologies and new methods, the research trend of the institutional strength and reliability is expressed in the following aspects, the in-depth application of the computational simulation, the breakthrough of professional reliability analysis software, and the integration of multi-disciplinary theory and technology.

(4) Machine tool design. With the development of national science and technology major project-"Advanced numerical control machine and fundamental manufacturing equipment", the level of middle and high grade machine tools has been continuously upgraded, the advanced numerical control (NC) system and the key technologies of the functional components have been also gradually conquered. The machine tool design is developing rapidly towards the direction of "High precision, high efficiency, complex, intellectualization, green, high flexibility and modularization". Meanwhile, the researches of recent years mainly focused on the dynamic simulation and analysis technology, the error analysis and compensation technology and the reliability analysis and optimization design technology of NC machine tool. The independently researched and developed NC machine tools are also gradually moving towards the international market. High-precision NC machine tool, high-speed NC machine tool, compound NC machine tool, curved NC machine tool, heavy load NC machine tool and grinding NC machine tool have been achieved significant breakthrough due to the achievements of the major project. However, the main development and progress are all concentrated in medium and low grade machine tools, the market of high grade machine tool is mainly occupied by foreign enterprises. Nowadays, our machine tools, functional components, control systems, tools and measurements still have disparity in accuracy, reliability, stability and durability comparing with foreign advanced level. With the continuous development of high-end technology represented by the aircraft industry, the requirements for the performance of the product and the processing efficiency of the components are getting higher, and the control of precision and cost are demanded, so that the performance of machine tools needs to be further improved in the future. High speed, high precision, compound, intelligent and environmental protection is the inevitable development trend of machine tools in the future.

(5) Vehicle engineering design. Vehicle engineering, an important research field of mechanical design, is the concentrated expression of the latest progress of mechanical design. The development of vehicle engineering and technological progress are the important indexes for evaluating the development of the mechanical design discipline. The rapid growth of sales of the national brand automobile of China and the rapid development of rail transportation industry reflect the overall technology progress of China's automobile industry. As new materials, new structures, new

calculation and analysis methods are constantly emerging, the massive data analysis and design have been achieved, and the safety and reliability of structures have reached the best level of history. Lightweight technology based on new materials, new techniques and new structures has been widely used in vehicle design. The active and passive safety technologies have been considered as the core technologies of competition. At the same time, in order to alleviate energy pressure and reduce environmental pollution, some new energy techniques such as hybrid power, fuel cell have become one of the core techniques of technical innovation and industrial upgrading. The optimization of comprehensive performance of the vehicle will become a hot and difficult point in the academic and engineering fields with the rapid development of the vehicle industry. In addition, the scope of automobile safety technology is becoming more and more wide, and gradually towards the direction of integration, intelligence, systematization and full membership, etc. The new energy technology is recognized as the future development direction of the next generation of vehicles. Meanwhile, it is significant for promoting the transformation and upgrading of the automobile industry, enhancing the international competitiveness of the industry, and building an environment friendly society.

9. Automation Science

Automation science has been an important critic to evaluate the technological advancement and general power of a certain country. Meanwhile intelligent automation technology, with automation control and information handling as core, has become a main power to promote productivity advancement, improve human life and accelerate social development. After development of several decades, basic theory of automation science has been matured and widely applied in the fields of industry, agriculture, military, transportation, commerce and medicine. In recent years, the development of information technology, such as computation, internet and communication, has provided more development opportunities, and promoted the emerging new technologies.

9.1 Current Research

According to our undergraduate education system, automation is the first-level discipline with five coordinate disciplines: "control theory and engineering", "pattern recognition and intelligent system", "system engineering", "GNC", "detection technology and automation equipment".

Control theory and engineering is the base of control science since it came out. Nowadays, intelligent control theory and control theory, targeted to complex system, has been attached with more attention and research. Firstly, research about classic PID control has made considerable achievements, which do not only answered the question why the PID control has been widely and efficiently applied in the system engineering but also provide theoretical guidance in the design of PID parameters for engineers. Intelligent control was born in the background in which there are more and more requirements with the highly complexity and uncertainty for control property. Under big data background, research and development of control driven by data, study and optimization, which meets the general trend of technological development and strategies that is advocated by our country along with big data and artificial intelligence. Data-driven has freed the system under control from the reliance on mathematical mode. With classic dynamic programming as theory base, Automatic Dynamic Programming has adopted the advanced technology of AI and provided new methods for the large-scaled systematic control with more complexity and optimization. China has made considerable achievement on the research of Automatic Dynamic Programming, which has been applied in the Smart Grid, intelligent vehicles, industrial processing, and application demonstration. Along with the rapid development of China's economy and society, industry has brought great economic benefit along with some environmental problems. In recent years, domestic scholars have serious researches on the theory basis and application in the fields of environmental protection. There are new research achievements in modeling, optimized control, parameter evaluating and forecasting, and fault diagnosis. The application of automation technology in the air quality monitoring, big data, source apportionment and general comprehensive decision, has accelerated the progress of solving air pollution problem in China.

Robot is an important carrier to realize intelligent control of complex systems and its technology also plays an increasingly important role in the development of modern society, among which the humanoid robot technology has made continuous breakthrough and reached the advanced level internationally. China's superiority in intelligent architecture is in terms of the application of automation and robot technology in the construction of large infrastructure projects such as bridge and tunnel. Underwater robots have made great achievements and been successfully applied in Marine research and other fields. Intelligent engineering robots have led the technology upgrading of engineering machinery industry. Specialized robots have been developed from scratch and grown in variety and application. Integrated with computer, communication and artificial intelligence technology, the simulation science and technology, which especially has been attached great importance to and widely applied in China's defense and military field, has been born and developed in control science, system science, computer science and others.

Under the guidance of major national strategic projects such as China Manufacturing 2025 strategy and in formalization of manufacturing industry projects, China's control technology of manufacturing system has made some research achievement in the field of equipment, management, mode, information and so on. Also advanced intelligent controllers and control software have been developed. Digital sample workshops and the benchmarking enterprises of intelligent factory have come out. With the rapid development of internet technology, internet and manufacture have been closely linked by" Internet plus" collaborative manufacturing, which has given birth to a new model of manufacturing.

In terms of pattern recognition and intelligent system, China has development well in the basic theory and method of mode classification and machine learning. In terms of computer vision, China strives for the commanding height of theory and technology on frontier issues. Bioinformatics extends the study of control science and engineering from mechanics and electronics, physics, chemistry to the life system based on molecules and cells. In the field of multimedia analysis, Chinese scholars greatly favor the feature analysis of geographical distribution, social events, description generation, information countermeasure and multi-source heterogeneity across communities. Brain-computer interface technology has come a long way and its research findings cover many aspects such as limb auxiliary, neural rehabilitation, military affairs and recreation and so on. In the field of smart buildings, the method of information technology and intelligent control makes the management of buildings easier and efficient day by day.

In system engineering, with the development of science and technology and great support for scientific research by China, there are endless researches of system science. In 2006, the scientific research institute of mathematics and system science of Chinese academy of sciences gives a historical source definition and outer edge of systematics, which has generated a great response in the field of system science and engineering and shown its future development trend.

In system theory and method, China has made much research achievements in complex network, multi-agent system and system engineering. Complex systems, as basic research, have also been included in the Outline of the National Long-term Science and Technology Development Plan (2006—2020). In process industry automation, along with deepening integration of IT application with industrialization, continuous breakthrough has been made in the new theories, new methods, new technology of the production process detection, modelling, control, optimization and policy support, as a result of which multi-level integration automation system has gradually formed in enterprise resource planning, production execution system and production process system. In smart traffic system, great long-term progress and much international leading research achievement have been made in rail transit, ground mass transit, urban parking, traffic flow theory and traffic signal control, traffic planning and design, and traffic big data. The research level of China in the field of intelligent transportation is in synchronization with the world. The automation of urban rail transit in China, which has been developed from manual driving to autonomous driving and further developed to the unmanned one, has been greatly improved. High-speed railway has developed rapidly. A high-speed railway network with four vertical and four transverse frames has been built now. In network information service, P2P service, grid service, oriented architecture, cloud computing, internet of things, big data and other information service technology brings great opportunities for network information service.

With regard to navigation, guidance and control, our scholars, as the positioning and navigation technology matures in the past 20 years, focus on such aspects as strong nonlinearity, high uncertainty, rapid changing parameters, and strong coupling of aircraft and spacecraft. Paying attention to the control of aircraft and spacecraft in complex environment, analyzing and discussing the intelligent control methods of aircraft and spacecraft, our scholars have pushed China's aerospace industry into an era of intelligent autonomy.

In testing technology and automatic device, we have made systematic and pivotal progress in micro-fault diagnosis, intermittent fault diagnosis and fault diagnosis of closed-loop system for dynamic system fault diagnosis and fault-tolerant control. Hot issues, such as adaptive control of new energy sources, advanced measurement and intelligent control technologies, and exploration of machine learning and artificial intelligence applications, have promoted the development of power generation and automation technology with the escalation of energy consumption and the transition of economy of China. Power generation and automation technology adapts to the development of new energy sources, thus improving the reliability of UHV interconnected power grids and promoting the construction of smart power plants. Progresses in distributed energy grid-connection have been made mainly in the control strategy, seamless switching technology, automatic frequency, voltage and black start technology, energy management technology, collaborative control technology, self-healing control of energy systems with distributed power and research in fault diagnosis. Electric power companies and many universities and research institutes have carried out research on key technologies such as micro-grid system and energy Internet and have made some progresses. For the smart city research, Chinese scholars have conducted various researches on the origins and development of smart cities, their architecture, key technologies, standardization and evaluation systems, and typical cases among them by focusing on the development of smart cities since 2013.

In addition to important developments in traditional fields above, automation has a strong vitality in interdisciplinary and emerging applications. In recent years, the popularity of the Internet and mobile smart devices have also promoted the explosive growth of social media and social computing research. With its rapid development in the past ten years, information-physical fusion system has been one of the most promising and cutting-edge technologies for the life of human beings. It has been widely used in fields such as manufacturing, smart grid, intelligent transportation, smart medical treatment and intelligent network. Information-physical fusion system is becoming an unprecedented way to reshape more research areas.

Being important nodes in a network-based, autonomous system, UAV, unmanned vehicles and unmanned aerial vehicles are of great significance to the development of science and technology in military, civilian and defense industries. The UAV is an aircraft that can automatically complete the entire flight process with the control of electronic equipment. It has made some achievements in navigation, structural design, modeling and flight control, networking and multi-aircraft coordination. Unmanned vehicles are smart cars that can automatically plan their driving directions to achieve their intended driving targets. Currently, the research hot spots in the field of unmanned vehicles are mainly the perception, decision-making and control of unmanned vehicles. Unmanned vessel is an intelligent platform for autonomous or semi-autonomous sailing on the surface of the water. At present, China has made some technological breakthroughs in the technical fields of unmanned aerial vehicles, unmanned vehicles and unmanned ships, which will have far-reaching effects on military and civilian.

With the development of modern information technologies such as Internet of Things, the Internet, big data and cloud computing, smart agriculture has emerged for a modern and new type of agriculture with innovation and sustainable development. A number of demonstration projects have been implemented with technological elements with the support of the Ministry of Science and Technology, the Ministry of Agriculture and local governments. With the implementation of agricultural policies such as "Separation of the Three Powers" of the country, smart agriculture will be promising and will change the current mode of agricultural production profoundly.

9.2 Current development and future trend both at home and abroad

Considering development status in China and abroad, Chinese scholars have been an unneglectable force in the international control theory field. But in the theoretical research based on practical application and new research of multi-discipline, we still need more efforts. In the field of nonlinear control and robust control, the domestic research enjoy the internationally advance in both theory and industrial application. The achievements of intelligent control theory are mostly put forward by foreign scholars. Parallel control and parallel management have entered a period of rapid development in China. In foreign countries, the research on parallel control and parallel management is still relatively preliminary with no focus. There is still a gap between the level of China's visualization technology and foreign countries, especially in the field of manufacturing. However, our country has made obvious achievement in aircraft design simulation, vehicle new energy power system modeling and state estimation. The research content and scope of Chinese robotics are not much different from that of foreign countries. At present, there are significant differences between domestic and foreign industrial robot products, and domestic enterprises have not yet started large-scaled production. Foreign robots occupy an absolute advantage in high-end applications. As far as the density of the robot is used, the density of industrial robots in our country is still far below the global average. The development of architecture robots in China falls behind the advanced level of foreign countries in general, and the advantages are concentrated on the construction of large infrastructure such as tunnels and bridges. The application of new technology and new ideas in the field of architecture is insufficient. In the field of unmanned systems, there is still a big gap compared with the advanced countries, especially the United States. We need to carry out autonomous sensing and research of unmanned system under complex environment, put forward more effective control methods and strategies, and do experiments on unmanned system, so that the unmanned system can be better used in national defense and civilian. In the field of the theoretical research of the brain machine interface, the domestic and international standards are fully integrated. In the application of brain computer interface, the gap between domestic and foreign countries is larger. In the basic theory of pattern recognition and machine learning, our country is in the international front level in low rank learning, classifier integration and multi example multi label learning. In the field of in-depth study and application, domestic scholars have to take a gratifying result. In China, the analysis of multimedia social events started late, and has achieved fruitful results in recent years. In the aspect of Multimedia description generation, the whole country needs more effort to keep up with the world. The development of intelligent buildings in China is late. In general, the development of intelligent buildings in China is closely related to the world trend. In the field of computer vision, domestic researchers have breach-through in many aspects while there are a lack of pioneering achievements in the hot topics in recent couple of years. Bioinformatics in China is at the forefront of the world in the aspects of group data processing and analyzing, multi-type data integration, quantitive systems biology and synthetic biology. We are unique in the fields of Chinese medicine system biology and network pharmacology. While many key new technology are originated from abroad and behind the world. In the field of systematic theory and methods, research on complex system has slowed to a smooth upward-state from its high-speed period. In terms of industrial automation, there are world-state research achievements in the aspects of technical process model, advanced control, production optimization and support of production and management decision-making for enterprises. In industry-university cooperation, there is still a certain gap between our process automation and the foreign counterparts. In aspect of intelligent transportation, the research in our country is started relatively late, but it shows a strong momentum of development. All the relevant researches and applications have made breakthrough progress, and a number of leading international research results have emerged. Our research in intelligent transportation has achieved the synchronization with the world. In aspect of network information services, domestic scholars and enterprises have made a series of achievements in the research of the application of network information services. However, we should propose more mainstream methods and core algorithms in basic theories and technical issues with important application prospect. Our sustained and substantial investment in the field of aerospace has greatly promoted its development. However, the width and depth of the application of advanced control methods still remain in the infancy stage. In terms of testing technology and automation devices, significant progress has been made in the research and development process in China. And we have got some research achievements in the aspect of fault-tolerant control of dynamic systems, clean combustion of conventional power supplies and grid adaptability, micro-grid, smart city structure based on information and communication technologies, in a leading international level. In the research of power generation automation, the demand for the development of renewable energy, coal-fired power generation and nuclear power generation is not satisfied in home and abroad, so that the world is facing with the huge challenge of integrating the control of thermal power plants into the smart grid; China's development of new clean power generation technology is relatively slow, and also lack of the study on storage-assisted power generation flexibility control. In the aspect of distributed energy grid, the international community is concerned about issues such as distributed generation and energy Internet technologies. In the research of smart city, our country is currently trying to realize the construction of smart city from such perspectives as information and communication technology, urban "wisdom growth" and knowledge economy and creativity economy. The international community however has made great development in the investigation and research on the target-, dynamic- and application-type components of smart city.

9.3 Strategic Needs of Corresponding Disciplines in China

In order to promote the development of control science and engineering, the following strategic needs are proposed: (1) Pay attention to interdisciplinary. The new scientific theories, new inventions and new engineering techniques often happen at the edge or intersection of disciplines. Emphasis on interdisciplinary science will make science itself develop to a deeper and higher level, which is in line with the objective law of the nature. (2) Industry-university cooperation. Gradually promote from easy to difficult to form a complete industry chain including project research, manufacturing and use management. (3) Training and support of basic personnel. Personnel training of all ages plays a crucial role. At present, primary and secondary education in our country lacks relevant courses in science and technology practice, and the students are lack of practical ability. So it is necessary to increase exchanges at home and abroad and to study personnel training mechanisms in more developed countries. (4) Funding for basic and critical research. At present, China's economic investment in science and technology has exceeded that of many countries. However, catching up with developed countries, even leading the world's advanced level requires larger capital investment in basic research and major application methods. (5) Improve the social status and economic income of researchers. Research is now a profession, which is hard for good young people to choose it as a lifelong career if the research positions are not attractive enough, except the driving of pure interesting. If state can raise the social status and economic income of the researchers, more and more people will devote themselves to scientific research so that a virtuous circle will be formed that will promote the development of science and technology in our country.

10. Standardization Discipline

From the end of 20th century, with the development of information technology and the emergence of standard essential patents, the standards play a more and more important role in market competition, industry development and public management. WTO/TBT agreement further improves the status of standards and standardization in international trade and global governance. As a comprehensive interdisciplinary among Science and Technology, Economics, Management, Law and Social Science, Standard Learning presents a trend of comprehensiveness and differentiation, mutual pervasion cross with other disciplines. In recent years, along with the research and practice of standardization science and technology continuously thorough, new theory and method emerge continuously, the field of standardization discipline continues to enrich and develop. In addition to traditional Economics and Management Science, Network Economics, Law, Industrial Innovation Theory, the Theory of Public Governance, Social Science and other fields are launching the interdisciplinary research on the connotation and extension of standard and standardization, which gives new characteristics to standardization discipline development. With the efforts of international organizations for standardization, some national standardization organizations and the government agencies, the level of global standardization school education and the occupation training has been improved to some extent.

From the end of 20th century, our country began to implement the standardization strategy, vigorously supported the independent innovation technology to form the national standard, promoted our country independent innovation technology to transform into the international standard. After 21th century, our government began to initiate standardization reform, encourage industry alliances to set standards, and then promote social organizations to set standards. In 2015, China started to amend the *Standardization Law of the People's Republic of China*. China's reform and opening up, China's accession to the WTO, the implementation of standardization strategy and the practice of standardization system reform provide a broad field for the development of standardization scientific research, and the academic circles in our country have gradually formed their own unique academic characteristics and research methods.

This report has four parts, including introduction, progress in the field of standardization discipline, characteristics and comparative analysis of the construction and development of standardization discipline at home and abroad, and prospects for countermeasures and suggestions. More details as follow.

(1) The progress of standardization discipline, including the research on body of knowledge of standardization, concept and classification, standard economics, standards and technology innovation, standard essential patents, formal and informal standardization organizations research, standardization management system research and practice.

Research on body of knowledge of standardization: Before the 21th century, the research on body of knowledge of standardization discipline is still in a starting phase, and the standardization discipline is more centered on the theoretical basis, discipline status and so on. Since 21th century, the research on body of knowledge of standardization discipline is rising gradually at home and abroad, many scholars have put forward the framework of body of knowledge of Standard Learning from different angles.

Concept and classification: From the Sociology dimension, the Economics dimension, the Philosophy and the Terminology dimension carries on the analysis research.

Standard economics: From micro-economic and network externalities, standard economic benefits carries on the analysis research.

Standards and technology innovation: Standardization and technology innovation are attracting more and more scholars' attention. These researches are mainly in the fields of management, economics, commerce, engineering management, computer and communications.

Standard essential patents: Academic circles research on standard essential patents can be divided into two categories: the research of industrial innovation and strategic theory, and the research of Law circles.

Formal and informal standardization organizations research: The investigation of standardization organizations involves the fields of Management, Sociology, Economics, Social Governance Theory, etc. The object of investigation includes all kinds of organizations such as the traditional standardization organization, the standard consortia and so on.

Standardization management system research and practice: In recent years, with the increase in the number of management system standards and the number of organizations applying multiple management system standards, ISO has realized that these standards need to be more consistent. Meanwhile, it is also recognized that these standards need to be revised in the light of the functions and objectives of these organizations and the changing business environment in which they operate. This enables the ISO management system to have the basis of a new architecture and common terminology and core definition.

Through the analysis we can see that the main contribution of Economics in Standard is economic benefit, standard essential patents and technological innovation and so on. The main contribution of Law in Standard is standard essential patents and anti-monopoly, patent policy (RAND/FRAND) and so on. The main contribution of Sociology in Standard is mainly in basic concept and classification, standardization organizations investigation and so on.

(2) Comparative analysis on characteristics of the construction and development of domestic and international standardization discipline, mainly consisting of two parts: the theory of standardization and standardization education.

The research on standardization theory are introduced and expounded in an all-round way from the development characteristics of the study of foreign standardization theory and the development of theoretical research and comparative analysis in China.

The standardization education is introduced and elaborated from three aspects: The current situation of international standardization education , the standardization education and its comparative analysis , the development and trend of standardization education in China.

Through the analysis we can see that the theory research on international level has been developed to a comprehensive intervention, multidisciplinary research stage deeply with Economics, Industrial Innovation, Law, Public Management, Sociology and other fields. Characteristics of standardization research in China is mainly to deal with the problems and challenges that China is facing in the process of economic globalization and market reform, including the research on body of knowledge of the standardization , the research on standard economic contribution rate in the macro, meso and micro level, the research on the contrast of the standardization system in different countries and the reform of our country's standardization system and so on. This part also includes the summary and comparison of the latest development of the domestic and foreign standardization education.

(3) Prospects and countermeasures parts are mainly based on the analysis results of this report, forecasts the development of standardization discipline in the future, and puts forward some countermeasures and suggestions.

We must strengthen the training of standardizaton talents, promote the construction of standardization discipline, and support more universities, research institutions to offer standardization courses and education with record of formal schooling, set up standardization professional degrees, and promote standardization universal education. We must strengthen the international standardization of high-end talent team construction, and strengthen standardization professionals, management personnel training and enterprise standardization personnel training to meet the different levels of different areas of standardization talent needs.

The relevant body of knowledge of standardization in this report, the concept and classification, standard economics, standards and technology innovation, the standard essential patents, the latest research of formal or informal standards organizations are the important supplements to the *2011—2012 Report on Advances in Standardization Science and Technology*.

Research on China's standardization theory has achieved some results. However, as standardization discipline is a new and interdiscipline, it needs to be filled a lot. On the whole, with more researches added, the basic theory is gradually deepened, the definition of standardization discipline itself is more clear, the discussion is more and more focus on the standard and standardization connotation and extension. We believe that with the support of the government, in the joint efforts of the majority of standardization experts and educators, the construction of standardization discipline theories and the body of knowledge will be further developed, making greater contributions to the construction of the national innovation system, the standardization reform and development of the national economy.

11. Surveying, Mapping and Geoinformatic Science

A rapid progress has been made in the surveying, mapping and geoinformatic science in China nowadays. The development of China's Surveying and Geoinformation technology is stepping into a critical period of building a "wise" China in a more comprehensive way. Presently, the surveying products and services are in great demand, as well as the opportunity for the development of the geoinformatic industry. Meanwhile, we have entered a difficult period of time where accelerating the speed of building a strong country with surveying and mapping capabilities has become more technical. We have completely transformed and upgraded the pattern from a conventional surveying and mapping data collector to an information and service based provider for surveying, mapping and geoinformation. This report reviews and summarizes the strategic position, the transformation and its upgrading of surveying, mapping and geoinformation in recent years, mainly from 2016 to 2017.In addition, we also retrospect, summarize and scientifically evaluate the current situation of the new viewpoints, new theories, new methods, new technologies, new achievements on the transformation and upgrading of surveying, mapping and geoinformatic science of China and condense some key technological progresses and important research results of the transformation and upgrading.These mainly reflect on the subjects of the geodesy and navigation, gravity survey and earth gravity field, photogrammetry and space mapping, cartography and geographical information engineering, surveying and deformation monitoring engineering, mine surveying and subsurface space surveying technology, marine and rivers and lakes geodetic surveying and so on.We have made remarkable progress in geodesy, temporal and spatial reference, navigation and geodetic reference systems and data fusion, the earth's gravity field and other related areas of basic and applied basic research, which keep pace with international advanced level, or even surpass and lead the international level in some areas. National modern surveying and mapping benchmark system infrastructure construction project successfully completed and 2000 National Geodetic coordinate system are promoted by most national ministries department. Beidou navigation satellite system has the ability of navigation and positioning, and it will cover the global range of services in 2020. The report reviews the important sensors used in photogrammetry and remote sensing, and introduces the recent progress in very high spatial resolution (VHR), Synthetic Aperture Radar (SAR), Light Detection and Ranging (LiDAR), and other thematic data. Meanwhile, this report summarizes the latest advancements in data processing methods and applications of photogrammetry and remote sensing in terms of survey adjustments, mobile mapping, three-dimensional Geographic Information System (3D-GIS), and high spatial resolution image processing. According to the development of Cartography and technology of geographic information since 2015, the theory of modern cartography, digital mapping, the technique of GIS, the production and renewing of geographic information, mobile mapping and internet mapping, the application and service of geographic information are described in this paper. The innovating of technical theory and system accelerates the development of engineering surveying discipline. The development of geodesy, photogrammetry and other disciplines, and the application of new technology such as space positioning technology, geographic information, laser, wireless communication and computer technology have greatly promoted the progress of engineering surveying which has undergone profound changes. Many advanced technologies and equipment such as 3D laser scanner, smart total station, magnetic gyroscope, geological radar, pipeline 3D reconstruction measuring instrument, UAV and so on are emerging. Based on spatial informatics and system engineering theory, the geometrical and physical properties of all-round objects are observed and sensed by means of mapping, remote sensing, geophysics and internet of things, meanwhile, dealing with and solve the mineral resources protection, mine development optimization, production environment safety, mining subsidence control, the mining ecological restoration and other scientific and technical problems, mine surveying has achieved remarkable achievements, and made great influence on the social and economic development. The national geographic conditions monitoring has been proposed as a new requirement and demand of the surveying, mapping and geoinformation industry under the new situation of economic and social development, as well as the key responsibility and strategic task of the departments of surveying, mapping and geoinformation to active service for the science development. The first national geographic conditions census in China was completed in 2015 and the acquisition of geographic conditions information has been changed over to the normalization stage since 2016. Based on the national geographic conditions census, through the fundamental and thematic monitoring, building a system with complete functions of dynamic monitoring, comprehensive information analysis and publishing, forming normalization mechanism of geographic conditions monitoring, to provide business and normal services of national geographic conditions information. The research progress of cadastral and real estate surveying and mapping was described on such terms: land survey, land use / cover change, investigation of real estate rights, surveying of real estate, real estate registration, and the construction of information system. Marine surveying and mapping is a science and technology to study the collection, processing, representation, management and application of geospatial information related to marine and terrestrial waters. In recent years, marine environmental information survey, marine multi-source data comprehensive processing and other aspects have made great progress. Integrated marine survey and measurement platform has been put into use. Marine navigation and mapping benchmarks tend to be fine. The sea and land vertical reference conversion system of the whole sea area is established initially. Marine navigation and positioning technology has been researched in-depth and applied effectively. Off shore area positioning accuracy is up to centimeters level. The land, sea and air integration maneuvering coastal terrain measurement scheme is formatted. Comprehensive utilization of unmanned aerial vehicle aerial survey technology, three-dimensional laser scanning technology and multi-beam sounding technology can achieve accurate measurement of coastal terrain. The chart mapping is oriented towards the development of intelligent and automation. The platform of marine geographic information sharing is basically completed, and the integration and sharing of multi-source and heterogeneous marine environment data are realized.

This paper reviews the social applications and services of geodetic surveying and geographic information technology after transformation and upgrading. The main progresses include basic framework construction and services of geographic information, general investigation and monitoring of geographical conditions, public service platform for geographic information of "Tianditu", the Beidou II satellite project, the reliability of the great projects in spaceflight and the key technologies of remote sensing information, the key technology and its applications of the geodetic information system at provincial level, theoretical methods and software of measurement engineering space informationacquisition, cadastral and house property surveying and mapping, and so on.This paper also summarizes the major trends in the transformation and upgrading of surveying, mapping and geoinformatic science and reviews the transformation and upgrading of surveying and geoinformatic technology, rapid processing and analysis of technological upgrading, expertise and technical level of the key technological upgrading. This report briefly introduces the progress of disciplinary construction and personnel training, research platform, important research team, and combines major international research projects and major research projects of the subject, analyzes and compares the latest research hotspots,leading edge and trends of the subject, evaluates the development of this subject both at home and abroad,compares the differences between domestic and international mapping technology and geographic information, analyzes the development strategy and key development direction of surveying and mapping and geographic information in the next five years in china, and puts forward the related development trend and development strategy.

12. Mineral Processing Engineering

Mineral processing engineering, which ever been called as ore beneficiation engineering for a long period in the past, is also been referred as mineral engineering in some literatures. In most of the domestic university's subject classification, mineral processing engineering is been recognised as a branch subject of mining engineering. In some other countries, mineral processing engineering often been categorized into metallurgical engineering, chemical engineering or materials science and engineering.

Mineral processing engineering is a method to effectively separate target minerals (valuable minerals) from useless minerals (usually referred as gangue minerals) or harmful minerals, or separate a varieties of useful minerals. In the industry chain of mineral resource exploitation and its comprehensive utilization process, mineral processing is a vital part between geology, mining, metallurgy and chemical engineering. The raw ore, though mineral processing (beneficiation) treatment, could significantly increase the grade of the target minerals, which will greatly reduce the transportation cost, and also reduce the cost and difficulties of subsequence processes, such as metallurgy, chemical treatment, reduce the overall production cost. Hence mineral processing engineering plays a important role in the national economic development.

In recent years, the discipline of Chinese mineral process has made considerable progress, especially in the area of mineral processing theoretical study, processing techniques, equipments, reagents, plant automation and process control, providing theoretical, technical and talents supports for the development of Chinese mineral processing, playing an important role in the global mineral processing development. To promote the further development of mineral processing discipline, and to increase the role of mineral processing played in the resources utilization and environmental quality, the investigation and evaluation of the mineral processing discipline development is conducted. By the actively cooperation of all parties, the investigation team have found out the general information of mineral processing discipline construction, talents training, scientific research, systematically illustrated the achievement and cornerstone of the discipline development, comparatively analyze the general trend and future development of both domestic and international mineral processing discipline, uncovered and discussed the potential problems and challenges during the future development.

(1) Chinese mineral processing subject construction has been made significant achievements, formed a complete, multi-discipline system including teaching, scientific research, engineer design and application with high academic level research team, established a completed research platform that serves and promotes basic scientific research, technical development and transformation of scientific and technological achievements.

The subject of mineral processing discipline has been continuously constructed and developed by all the involved Chinese universities, colleges and research institutes, and hence has strong international competitiveness. Domestic mineral processing education has undergone rapid development over the past few years, up to Sept 2017, a total of 37 universities or colleges have opened mineral processing subject discipline in China with two secondary key subject discipline. More than 600 full-time lectures or teachers are currently listed in the subject of mineral processing, with more than 15000 undergraduate students, 3600 master students and 650 doctoral researchers.

There are more than 20 mineral processing or related research institutes in China. These research institutes have professional research teams, advanced technical platform and comprehensive research apparatus, which has been becoming the important bases for researching the mineral processing basic theories, application basis, transformation of scientific achievements and promote scientific research cooperation as well as international academic talks.

China has owned totally 22 mineral processing related national key laboratories or engineering technology research centers, and 58 provincial or state key laboratories or research centers. These high-standard research platform has played a significant role (both scholar and apparatus) in the development of the mineral processing research.

At present, there are more than 400 thousand people working in the area of mineral processing field in China, including 6 academicians, more than 350 national outstanding contribution awards owners or state council special allowance, 10 or more Yangtze River Scholar Professor. 10 or more National Science Fund for Distinguished Young winner, and 20 cross-century or new century talents project experts. Currently, in China, there are dozens of distinctive professional mineral processing science and technology research teams, these teams have played a vital role in promoting the construction and development in national mineral processing technology. In recent years, with the constantly support of national vertical projects as well as mining enterprises horizontal research projects, large quantities of important academic literatures and scientific achievements were made, and hence will provide strong technical supports for the current and future development of appropriately exploited national mineral resources.

(2) The progress of mineral processing technology provides a solid technical support for the

development of mineral resources in China, formed a variety of specialized basic theories research directions. Some of the techniques formed and the apparatus equipped has international leading level. A number of modern mineral processing plant was built with cutting-edge technology and excellent processing equipment.

In terms of basic theory research, new research directions and ideas were formed, such as Genetic Mineral Processing Engineering, chemical modeling and molecular design, fluid and inclusion effects in flotation, magnetization roasting and depth reduction, metal ion coordination control and molecular assembling, which has been attracted global interest in the field of mineral processing.

In terms of technology development and scientific achievements transformation, the research work is mainly focused on the conspicuous problems or issues occurred during the mineral resources utilization, and breakthrough progress has been made in the aspect of mineral processing techniques, reagents, equipments development and manufacture, mineral resources comprehensive utilization, processing automation, mineral material research and its downstream processing, formed a series of new techniques, new achievements with global influences, and have already been put into industry applications. The beneficiation technology of low grade, complicated refractory ore in China is in a leading position. The manufacture and industry application of large-scale autogenous mills, semi-autogenous mills, ball mills, large flotation machine as well as magnetic separators is also comparatively superior, built large modern processing plants with advanced techniques and equipments, such as Luming Molybdenum mine, Wunugetoshan Copper-Molybdenum mine, Jiama poly-metallic mine, Yuanjiachun iron mine, Anqian iron mine and etc.

(3) The mineral processing research focus of China is different from other countries in the world. China is mainly focused on the development of cutting edge techniques and their applications, while other countries paid more attention on the basic theory study in the micro base. In the global scale, the Chinese mineral processing can be described as "Leading in processing technique, on par with in equipment automation and being leaded in basic theory study".

Chinese mineral resources is "lean-fine-miscellaneous" compared to other countries in the world, and many sophisticated problems or fresh topics that have not yet attracted attention of the researchers from other countries. Therefore, a vast of research forces and research funds has been already put into use in both government-macro side and enterprise-micro side. Within the research efforts, the world leading cutting-edge technologies have been successfully developed, such as fluidizing magnetizing roasting-magnetic separation; low-grade copper green-recycling biological extraction technique; high concentration and multi & rate flotation technique for complicated copper-zinc sulfides; high-efficiency flotation technique for complicated refractory low grade wolframite-scheelite intergrowth poly-metallic, specialized technique that combined mineral process with metallurgy to process low-grade sulfides-oxides minerals. The mineral processing techniques is in the world's leading position.

The researchers outside China are more focused on the mineral processing basic theory study, especially in the area of mechanical analysis of mineral particle crushing-grinding process, motion trajectory and force field analysis of the separation process, interaction between bubble and particles, mechanisms of reagents to the surface of the minerals. Foreign researchers also pay more attentions on developing new mineral processing method, such as wet leaching, microbial beneficiation, theoretical study of fine particle three-dimensional flotation techniques, with new separating machine to enhance separating process. China has also conducted a lot of researches on the basic theory study of mineral processing, but with different emphases, generally focused on basic application studies on a specified mineral and generally has more link and interaction with practice. However, the domestic research force is generally weak in micro level research as well as computer modeling and simulation.

The mineral processing equipments outside China relatively developed earlier, and has already formed relatively larger scale production, the technical performance and automation is superior compared to domestic equipments. The key techniques are mainly belongs to developed countries, such as United States, Russia, Britain, Germany, Japan, Australia and Sweden. Chinese mineral processing equipments production started late, but developed rapidly, the development and application of large size flotation machines and magnetic separation is in the leading position in the world. However, in terms of the overall manufacture and techniques, there is still a big gap compared to developed countries.

(4) The mineral processing develop in the future, should meet the needs of the society and the industry with better understanding, should pay more effort on intelligentialized scientific research and teaching, should also pay more attention on the basic theory study in the micro level, and need more diverse multi-discipline talents to support innovative research works. The following are the main trends of the mineral processing development.

(a) The mineral processing development should have better understanding to the needs of the industry and the society. The general need of the industry and the society includes: increase the resource recovery and percentage of utilization, reduce the production cost, reduce the water and energy consumption per ton, reduce pollutant emissions, and improving process reliability. Based on the needs listed above, the general trend of the mineral processing development can be described as: Green, High-efficiency, energy saving, low consumption, low emission, highly automation. In addition, the development of mineral processing should also considering energy and resource consumption, establish "Mining-mineral processing-metallurgical" process concept, emphasized on the academic communication and maximize economic and resource utilization.

(b) Intelligent mineral processing research and education system. Currently, the numerical test method system, such as CFD, FEM, DEM, have been widely applied in the mineral separation theory study and equipment development, the role of information technology in mineral processing technique research and teaching is become increasing prominent, and it will become a trend of future mineral processing development and teaching method. "Internet + Education" and : Internet + Mineral processing" will play an important role in the future and the "big data" will provide a lot of benefit to the research work, modeling virtual processing plant will also play a significant role in the mineral processing teaching and production management.

(c) Advanced detection method and experiment methods will promote micro-level basic theory study. During the past hundred years of mineral processing development, theory driven by technological progress has always been the main feature. Due to lack of research method, the mechanism of many mineral processing separation has not yet been fully revealed. By the progress of the detection technique and test method, the basic theory research will be promoted, and eventually guide the real plant practice.

(d) The application field of the mineral processing technique will become wider than ever. Green recycling is the most important develop model in the near future, in the formal report of the 18th CPC National Congress, "Green development", "Recycling development" and "Low carbon development" were firstly been pointed out. Developed countries, such as EU and Japan has already built recycling economic development model, the recycling economy is still in this infancy. Based on the difference in physical or chemical properties, mineral processing technique could used different processing method to conduct the material separation and enrichment, thereby broaden the research field.

(e) Mineral processing discipline needs more composited talents for the future development. For a long time, because of the influence of education concept and education system, the universities in China has more emphasize on the specialized graduates without interaction with similar subject. The future development of mineral processing requires composited talent with multi-discipline knowledge.

(5) The development of Chinese mineral resources is currently growth rapidly with confronting many challengers, confront the contradiction between the insufficient resources supply and the increasing needs of the mineral resources, confront the contradiction between excessive exploitation and the limited capacity of the ore body and the environment, confront the contradiction between the backward industry structure and the rapidly growth of the techniques.

The needs of human race to the natural mineral resources continues to increase while the high-quality resources is depleting, the limitation of current processing strategy can not satisfy the green and healthy development of mineral processing. Firstly, the contradiction between the increasing needs of the resources to the depleting of the mineral resources, a comprehensive and smart way to utilize the mineral resources is necessary. Secondly, the high-quality mineral resources is depleting, and a effective way to process lower grade and refractory mineral resources is necessary and urgently; Thirdly, develop a green and sustainable processing system with much lower environmental footprint is the mostly urgent need for our future development. Fourthly, the national reform of supply front and industry upgrade will need improved techniques with equipment with better performance and low consumption. Fifthly, the development of the industry need to have more interaction with the information technologies. Sixthly, the upgrade of downstream products will need higher quality mineral processing products.

Digitalization and intelligentization is the general trend of development. The development of intelligent mining is the key to improve the product quality, improve production efficiency and reduce the production cost. Focused on the application field of digitalization and intelligentization, through technique innovation, realize industry upgrading, and seek a bright, positive and smart development for mineral processing in 21 century.

Directed by "The great rejuvenation of the Chinese national", the government has remarked the proposes of " The belt of road initiatives", "Made in China 2025", "Internet plus". In terms of mineral processing development, the continuous development of resources conservation, efficiently utilization of mineral resources, green and low carbon emission techniques, high efficiency techniques will become more important in pushing the technology upgrade, industry upgrade, and sustainable development of the industry.

13. Rare Earth Science

In the past few years, great progresses in rare earth science and technology have been achieved. The main contents and some examples are described briefly as follows.

(1) Rare earth minerals. Weathered crust rare earth ore is rich in medium and heavy rare earths, and it is the main source of medium and heavy rare earths in the world. This paper reviews the occurrence and evolution behaviors of rare earths in the weathering system during the migration and enrichment processes, the mineralization characteristics of the deposits,

and the variation of rare earths partition during the metallogenic processes. Technology development for the leaching of rare earths from weathered crust rare earth ore, leaching kinetics and mass transfer mechanism of rare earths are also summarized. Furthermore, this paper mainly summarizes the following points. Development of the injection well location, the injection methods, the seepage process and the stability of ore body during leaching. Based on the analysis of the problems raised in the in-situ leaching process, the future directions are pointed out. Further researches should be focused on the metallogenic law of rare earths during the migration and enrichment processes, the seepage mechanism during the in-situ leaching process, and the reduction of the stability ore bodies and the landslide mechanism, thereby promoting the development for the green and efficient mining of weathered crust rare earth ore and providing rare earth resources with higher quality for the world.

(2) Rare earths metallurgy. Resources and environmental issues have become increasingly obvious during the large-scale exploitation and utilization of rare earth resources. Therefore, to develop green chemistry for rare earth separation and industrial technologies for rare earth metallurgy and separation and improve the resource utilization efficiencies and environmental protection level are the only way to fulfil the balanced utilization of rare earth resources with high value-added and consolidate the international status of the rare earth industry in China. This review gives an overview of the important progresses in the environment-friendly separation of rare earths and industrial application in China in recent years. These achievements have already incurred worldwide impacts, and possess profound significance in supporting and further directing the rare earth industry in China. Based on the development of the rare earth industry, comprehensive utilization and cleaner production should still be on the main directions in the future, to support the sustainable development of the rare earth industry.

(3) Rare earth chemical analysis. In the last five years, the rare earth analytical chemistry has made progresses in the following fields: the application of advanced mass spectral technology in rare earth metallurgical analysis, the rare earths on-line analysis, the rare earth waste analysis, the analytical standard methods, the pre-treatment and rare earth analysis of biological & environmental samples, the relationship between rare earths and mental sickness, and the rare earth labelling of bio-marker to sickness.

(4) Rare earth permanent magnet materials (REPMs). In recent years, with the rapid growth of production capacity of China's REPMs, as well as the increasing demand for REPMs, the demand and supply of rare earth resources becomes imbalance. Reasonable and balanced use of rare earth resources is a sustainable development strategy of China's rare earth industry, and the development of resource-saving super-high performance rare earth materials has become the key point for the rapid development of China's REPMs industry. According to the advantages of China's resources,

the development of high abundance of rare earth elements instead of neodymium has also shown rapid development.

With the application requirements of sintered Nd-Fe-B magnets in the wind power generation, hybrid cars / pure electric cars and energy-efficient appliances and rapidly increased low carbon economy, double-high magnets (high energy product and high coercivity) and low cost are the main direction for the development of the magnet, to meet the new situation application of REPMs in emerging areas and the rising prices of raw materials, but also to promote efficient use of rare earth resources. In recent years, great progresses have been made in the fields of high-performance REPMs, heavy rare earth reduction technology, hot pressing and deformation technology, waste magnet recycling, etc.

(5) Other rare earth magnetic materials. In recent years, magnetostrictive materials have been widely concerned, as a key energy conversion material used in sensors, generators, linear motors, actuators, valve components, displacement devices and underwater scanning sonar. Researches should not only continue to pay attention to improve the performance of the existing magnetostrictive materials, but also focus on the development of novel materials with large low field magnetostrictive effect, low price, high mechanical strength and narrow hysteresis.

It has become an urgent problem to find a new type of refrigeration technology, which is environmentally friendly and energy efficient. Magnetic refrigeration technology, which is based on the magnetocaloric effect of the magnetic materials, is a new technology with important application prospect, which captures extensive attentions of the researches around the world.

High frequency soft magnetic materials are a kind of important functional materials, which are widely used in the fields of information and energy technology. With rapid development of information technology, soft magnetic materials are required to work with higher frequency from kHz to MHz, even over GHz. The requirements for materials are also extended from traditional thick and heavy blocks to light and thin structured composites. However, most of the existing materials are still ferrites since the high frequency soft magnetic material has just been discovered, which could not meet the requirements of the information industry. Therefore, it is urgent to develop high performance soft magnetic materials with high resistance and low power consumption, to solve the key problem in the development of soft magnetic materials.

(6) Rare earth catalytic materials. Rare earths are excellent catalytic promoter in petroleum cracking process and purification of vehicles exhaust due to their unique 4f electrons, which help to reduce the use of noble metals and enhance the catalytic performance. Putting an emphasis on the investigation of rare earth catalytic mechanism and expanding the applications of light rare earth such as La, Ce in the petrochemical industry and environmental governance are beneficial not only to the efficient use of rare earth resources and the structural optimization of rare earth industry, but

also to the alleviation of increasingly striking contradiction between supply and demand of noble metals, which brings prospects for the new catalytic materials. This report illustrates the latest technical progresses in rare earth catalytic materials and industrial development in the years of 2015—2016, including vehicles exhaust purification, VOCs, catalytic combustion of natural gas, petroleum refining, de-NOx in stationary emission and so on. The development of the fundamental research of rare earth catalytic materials, catalyst products and industrial development are analyzed, and all the three aspects are compared with abroad, and a development target for rare earth catalytic materials in the future is proposed.

(7) Rare earth optical functional materials. Rare earth optical functional materials is the core or important materials of the current strategic new industries such as advanced lighting and displays and the hot issues such as solar spectrum conversion materials and bioimaging nano-probes. This topic is first focused on the problems or technology bottleneck of semiconductor lighting and display including white light LED and laser LD as well as solar spectrum conversion material and biological persistent nano-probes. Then the main progresses on rare earth optical funcational materials for the above applications are contrastively summarriezed in detail from materials design, synthesis, new research methods, optical performance optimization and new application exploration, etc. Finally, the future development trends of rare earth function materials are proposed.

(8) Rare earth hydrogen storage materials. The research of hydrogen storage material with high efficiency and low cost is the key point to solve the large-scale utilization of H_2. Rare earth hydrogen storage alloys have excellent dynamic performance and stability as well as high hydrogen storage capacity. This topic summarizes the research progresses of AB5 type rare earth hydrogen storage alloy in reducing cost, improving the magnification performance and widening the using temperature range in the past two years, summarizes the studies on the relationship of structure and performance on the rare earth-magnesium-nickel based AB3~3.8 rare earth hydrogen storage alloy and the cost reduction of the alloy by reducing Pr and Nd contents. The development of new components and new structures of rare earth hydrogen storage materials is an important way to expand the application of materials. The research and development of rare earth hydrogen storage materials in different application directions are introduced.

(9) Rare earth ceramics materials. The development on the synthesis and fabrication of rare earth ceramics is introduced, including several advanced sintering approaches obtainable and suitable for particular ceramics production, more for transparent ceramics. Three dimensional electron diffraction technology and synchrotron source based measurements are then taken as the samples for modern pioneering characterizations. The most important subject in modern material research is computational calculation and simulation and their applications. Nano-technology, grain engineering for ceramics are improved for the development of rare earth ceramic materials. At

present, the scintillator of rare-earth doped garnet has been rapidly used for the advanced medical equipment.

(10) Rare earth superconducting materials. Rare earth oxide high temperature superconductivity single domain bulk material has the ability to capture or freeze the external magnetic field. Magnetic support, high speed magnetic bearing and other technical fields have important application prospects. In order to develop the large-scale application based on the rare earth superconducting materials, the preparation cost of materials should be reduced. To fulfil the above objects, the technical maturity needs to be further improved, the scale of industrial capacity needs to be enhanced, and the overall performance of materials needs to be further improved.

(11) Rare earth cast iron materials. The recent development on application of rare earth in cast iron such as nodular graphite iron and compact graphite iron has been summarized in this topic, and the basic research progresses are introduced. The characteristics and differences of rare earth cast iron in China and abroad are compared. This topic points out the distance in the application of rare earth cast iron between China and the foreign countries. Prediction and prospect in the development of rare earth cast iron materials has been made.

(12) Rare earth in non-ferrous alloys. The application of rare earths in non-ferrous alloys involves in various fields of the metal processing and manufacturing. The design and development of biodegradable alloys, automotive die casting alloys and obdurability alloys have become an important research direction for rare earth magnesium alloys. Microalloying is an important route to develop a new generation of aluminum alloy. Improving the performance of aluminum alloy by means of adding trace amounts of rare earth elements is an effective way to improve the comprehensive performance of aluminum alloy. Rare earth doped in TiAl-based alloy can improve the microstructure of the alloy and the adhesion of the surface oxide film and improve the anti-stripping ability of the oxide film, which can play an important role in improving the heat resistance and the oxidation resistance.

(13) Rare earth compounds. The research and development status of electrothermal materials, rare earth additives, rare earth base target and polishing material are summarized. The types, preparation methods and applications of chromic lanthanum high-temperature electrothermal materials are mainly introduced. The research and application of rare earth additives in polymer materials are presented. The application area of rare earth base target is mainly introduced. The preparation methods of nano cerium oxide, the technical parameters, the industry standard and its application in the field of polishing are reviewed. The main existing problems and prospects are addressed.

14. Corrosion Electrochemical

This subject in recent years, the latest research progress in corrosion electrochemical is mainly concentrated in the spatial resolution of micro area corrosion electrochemistry, corrosion electrochemical noise analysis and electrochemical corrosion behavior relationship, thin liquid film corrosion, corrosion electrochemistry and new nanoscale, extreme environment corrosion electrochemical and so on five aspects. Developing rapidly in local corrosion mechanism research, published in the international academia more paper, affected gradually, but in the study of theoretical depth and research methods of innovative progress there is still space. Research the latest progress in the high temperature corrosion is mainly in research and the mechanism of high temperature corrosion protective coating in the design, development and the performance study, obtained has the characteristic, has the international influence and international peers was concerned, a series of achievements. In the study of natural environment corrosion, China has entered into the study of natural environmental corrosion research, and it has maintained a good momentum of development. In anti-corrosion technology research has made some new research achievements, such as a new type of corrosion resistant steel research and development, high wear-resisting corrosion resistance of modified ultra-high molecular weight polyethylene special pipe research and development, aluminum/magnesium alloy surface treatment technology, to develop high performance space technology with lubrication film, using laser cladding technology in copper alloy was prepared on the surface of high hardness, high wear resistance, thermal shock, metallurgical bonding of metal ceramic composite coating and matrix, was prepared by thermal decomposition method IrTa mixed metal oxide (MMO) coated titanium anode, designed to rich in oxygen, nitrogen, phosphorus, heterocyclic compounds and alcohol/condensation product of amine of new technology, such as rust inhibitor and dispersant and realize industrialization. In the aspect of corrosion test methods developed a variety of types of scanning probe microscopy techniques, such as scanning ion conductance microscopy (SICM STHM) and scanning acoustic microscope, scanning thermal microscope SAM PSTM, photon scanning tunneling microscopy (STM) and scanning near-field optical microscope SNOM technology. The national material environment corrosion platform realizes the corrosion state of continuous dynamic monitoring materials. Electrochemical noise technology, wire beam electrode technology, atomic fault analysis technique and laser electron speckle interference technology are also widely used in the study of corrosion. The emergence of these new technologies and the application of development in the field of corrosion science have greatly expanded the technological means of corrosion research and discovered a lot of new theories of corrosion.

After more than 30 years of construction and development, this subject in the academic organizational system, personnel training, basic research platform construction has been basically formed the personnel training of colleges and universities, and state key laboratory, a number of basic research and applied basic research, the high level of experiment platform and the platform of scientific research in higher educational institutions to carry out the basic research, enterprise R&D center for applied research framework model of multi-level crossing each other. Disciplines in recent years, the corrosion state key laboratory, the basic research platform and corrosion science and technology workers, national goals and focusing on national requirements, solve the national economic construction, social development and national security in many fundamental, key technical problems, in the social and economic development and national defense construction in our country has made the important contribution. In surface coating, surface treatment, corrosion resistant material, rust-proof oil, corrosion inhibitor and electrochemical protection and other fields have achieved fruitful research results, anti-corrosion materials industry already has a large scale. Undertake in corrosion science and research, and completed a large number of national science and technology research projects, to produce a batch of innovative research results, solved the economic construction, social development and national defense construction of many major key technical problems, has obtained the remarkable social and economic benefits, in the construction of the contingent of personnel training and play an important role. It provides strong support for basic research and basic research and applied basic research of China's corrosion science base on that advantages of talents, the advantages of the condition and the superiority of the funds.

In the past five years, China has replaced the United States as the largest publication of SCI papers, reaching 21.41%. The United States dropped from first to second, making up 15.42 percent of the total, indicating that with the rapid development of China's economy, the corrosion discipline of our country has accordingly entered the fast lane of development, leading the rest of the world. In the last five years, the number of articles published in the science of erosion has increased by the year, and by 2015, there have been 1500 scientific articles on the science of corrosion. The growth of domestic corrosion science related SCI papers is more obvious than that of international corrosion science, which indicates that China plays an increasingly important role in the recent development of international corrosion science.

At present, China still has a certain gap with the United States in the field of corrosion and protection, and we suggest that in the future development plan of the subject of corrosion and protection discipline, we should construct a collaborative research and development platform for corrosion prevention and protection discipline, a collaborative research and development platform for application technologies, a collaborative research and development platform for protection technology engineering, and strengthen the work of corrosion and protection, and promote the legislation of corrosion and protection.

Development trend and prospect of this discipline: In conclusion, while strengthening the above basic research, it is the basic principle to develop China's corrosion and protection discipline. On this basis, the development of the international first-class level of corrosion resistant materials anti-corrosion engineering, surface treatment and coating anti-corrosion anti-corrosion engineering, environmental engineering, electrochemical protection medium processing or process corrosion protection and corrosion resistant special equipment engineering of new materials, new technology and new equipment, improve the original innovation and application of corrosion and protection discipline ability, development is the fundamental purpose of corrosion and protection science in China.

15. Nuclear Technology

Nuclear Technology (NT) refers to science or technology that studies how to apply the laws of atomic nuclei reactions together with their inherent and adjoint physical phenomena revealed by nuclear science. NT is generally divided into nuclear weapons technology, nuclear power technology, non-power military nuclear technology and non-power civilian nuclear technology (also known as nuclear technology applications). The application of nuclear technology described in this report mainly refers to non-power civilian nuclear technology, which is further divided into accelerator technology, radioisotope technology, nuclear detection & electronics technology, nuclear instrumentation and applications of nuclear technology in the fields of medicine, agronomy, industry and social security.

As a comprehensive strategic industry, Application of Nuclear Technologies (ANT) has won great attention of different countries for its knowledge-intensity, cross-discipline, irreplaceability, widespread application and cost-effectiveness. Global nuclear technology and its application are showing higher level of industrialization, scale and high-tech factors. It is still at a stage of fast development, maintaining a grow rate higher than that of the world economy. The industries of nuclear technology application in developed countries have become quite large in scale and have played a significant role in promoting economic and social development.

ANT in China is also steadily accelerated its industrialization, and is now at a period of fast development. As of the end of December, 2016, more than 400 research institutions, colleges and enterprises had been engaging in R&Ds of ANT. According to statistics from China Isotope & Radiation Association, ANT in China had yielded an output of 300 billion RMB Yuan in 2015, 3 times the figure in 2010, and still maintains an annual grow rate of around 20%. China's ANT is currently keeping a momentum of accelerated growth, and is hopeful to become a new growth source of the Chinese economy under current economic background. ANT has created up to 100000 jobs in China, and has played an irreplaceable role in improving people's livelihood, and in boosting social and economic development. Moreover, China's ANT is breaking new ground in its application areas such as industry, agriculture, medicine and environmental protection, and has formed a fairly high level of industrialization scale in areas such as irradiation-induced material modification, radiation processing, radiation-based equipment, public health, public security and environment protection. Related products have witnessed rapid upgrading in their performance and technological indicators, and are gaining more and more recognition of the international peers. Some products such as electronic accelerator for irradiation, Cobalt-60 and large container/vehicle inspection system have been successfully exported to many countries in the world.

15.1 Accelerator

The number of medical linear accelerators in China has increased by 26 times from 71 in 1986 to 1931 in 2015, and the number has doubled since 2006. In addition, development and manufacture of medical accelerators have also seen rapid development and have formed comprehensive capabilities. Accelerators such as high-current proton cyclotron have possessed excellent technological indicators performance, and China has become one of the few countries in the world to own a new generation of accelerators for radionuclide production. China also makes breakthroughs in high-energy electron linear accelerator (>5 MeV) for industrial radiation, which has served as a more effective tool for radiation-induced material modification, new material development and environmental protection. Specialized, large-scale production bases of electron accelerator for radiation have taken shape, and have seen drastic growth in the number of the products. The number of electron accelerators for irradiation grew from 140 at the end of 2008 to more than 500 in 2016, accounting for nearly one-fourth of the global total of 2000. The total power of the accelerators have exceeded 35000 kW, accounting for 43.8% of the global total of 80000 kilowatts, which still maintains an annual growth rate of over 20%. Accelerators for non-destructive inspection have won international recognition of their performance. R&D capability is following, and to some extent, leading the international trend. The accelerators not only meet the domestic demand, but also are exported to foreign countries, holding a relative big share in the international market.

15.2 Isotope

China has greatly improved its technical capabilities and levels in isotope production, and has mastered a number of techniques for isotopic preparation and capability to produce some isotopes. It has also mastered the technology to prepare million-Curie Co-60 on CANDU reactor and established corresponding production capability. Relying on CMRR, a platform for preparing fission Mo-99 using HEU has been built, thus establishing a complete process of target preparation, irradiation, dissolution, extraction and separation. Large quantities of F-18 and C-11 have been produced using about 100 imported medical cyclotrons to meet domestic clinical needs. The radioactive source preparation has made remarkable progress. China has mastered the source design, target preparation, irradiation, transportation, source preparation and inspection technology related to production of Co-60 of 10000 Curies, and has established an annual capacity of 60 million Curies of Co-60 as radiation source, which can meet about 75% of the domestic market demand. The technical indicators and quality of the Co-60 products have reached, and even exceeded the levels of similar products in foreign countries. Radiopharmaceuticals preparation has also seen breakthroughs. With the support of the major national project "New Drug Creation", research platforms for H-3 and C-14 labeling have been restored to provide labeling reagent and related services for domestic scientific research.

15.3 Nuclear Detection and Nuclear Electronics

After years of development, China has made some achievements in nuclear detection and nuclear electronics. Early-developed detectors such as NaI scintillators, BGO and zinc sulfide have achieved mass-production. Out-core detectors and core detectors have realized localization, and have been used in domestic experimental reactors and defense-related reactors. A part of out-core detectors and core detectors used for commercial nuclear power plants have also realized localization.

15.4 Application of Nuclear Technologies

The level of research and production of nuclear medicine has been increasing year by year. The application in medical sciences has been expanding comprehensively. The diagnosis and treatment technology based on nuclear imaging and treatment have been developing rapidly. Institutions engaging in tumor treatment have risen sharply in the past 30 years, from 264 in 1986 to 1413 in 2015. Market value of radiotherapy increased from 5.83 billion Yuan in 2008 to 26.9 billion Yuan in 2015, and shows a sign of accelerated development. Radiation mutation breeding has made remarkable achievements. According to statistics from the International Atomic Energy Agency (IAEA), 27% of the 2316 new crop varieties bred by radiation-induced breeding worldwide are contributed by Chinese scientists. Among the national crop acreage, cultivated varieties have accounted for 20%. Maximum annual planting area of mutant varieties have exceeded 130 million mu, contributing 3.5~4 billion kilograms of the grain increase to the country. Oil crops induced by radiation mutation yield an annual output of more than 1 billion kilograms, with the annual output value reaching about 1.4~1.7 billion Yuan. Research and industrialization of irradiation processing in China are now at the leading position in the world. By the end of 2015, the output value had reached more than 14.4 billion U.S. dollars, which maintains an annual growth rate of about 15%. Industrialization has been achieved in irradiation-induced material modification, sterilization of medical appliances and radiation-based equipment, scale of the industries reaching 7.2 billion, 5.8 billion and 1.4 billion US dollars respectively. Radiation technology is expanding its applications in environmental protection. Ray-based non-destructive inspection technology is close to the international advanced level. Large-scale container/vehicle inspection system and other security equipment have provided safeguarding services for many major international events such as leader's summit, the Olympic Games, the World Expo and the annual meeting of Boao Forum for Asia (BFA). The systems have been equipped in customs and border crossings in more than 140 countries around the world.

Application of Nuclear Technologies (ANT) 2016—2017 was drafted by Chinese Nuclear Society (CNS). Carefully selected experts in various areas of ANT participated in the drafting of the report. ANT in China has achieved fruitful results in recent years, and is now at a stage of tremendous opportunities. The Report fully reflects the development features of this historical stage to provide related leaders, peers and readers with readable information and deep insights concerning the development of ANT. It is consisted of 1 general report and 8 topic reports in which we firstly provide an introduction of status, progress and trends of ANT in foreign countries, then review and summarize research results achieved in China's ANT in recent years, and scientifically assesse status and trends of ANT development in China. By identifying gaps in various ANT areas between China and advanced foreign countries, we push forward proposals for promoting development of the ANT related disciplines and the industry in China.

16. Oil and Gas Storage and Transportation

Oil and gas storage and transportation is an indispensable part of the petroleum industry. As in the midstream of the industrial chain, it serves as a link among petroleum exploitation, processing and sale, and mainly deals with production systems such as oil and gas gathering and transferring treatment, long-distance pipeline transporting, storage and loading, supply and distribution. It plays a major role in the oil and gas industry, as well as in the national economy development and the national security maintenance. Oil and gas storage and transportation engineering is a comprehensive application-oriented discipline, which requires a good knowledge of many other related disciplines.

Coming into the twenty-first century, China's oil and gas storage and transportation industry is not only faced with unprecedented prosperity, but also has entered the advanced international rank. The national oil and gas storage and transportation network has not only spanned the whole country, but also has been connected to many overseas pipelines. The mileage of oil and gas transmission pipelines has added up to 120000 km. Nine national oil reserve bases have been built. Also, China has become a leading importing country in both amount and total receiving capacity of LNG terminals.

During the "twelfth five year plan" period, the discipline of oil and gas storage and transportation engineering developed rapidly. Scientific achievements and breakthrough technologies continued to spring up and have been efficiently transformed into productivity, thus becoming a strong support for the great development of the oil and gas storage and transportation industry.

Oil and gas gathering and processing technologies have been upgraded in a comprehensive manner. Major breakthrough has been made in the treatment techniques for chemical flooding produced fluid. Novel progresses have been made in the low-temperature gathering technology of waxy crude oil and the multiphase transportation technology. Gathering and treating processes for low-permeability natural gas and unconventional natural gas have been innovatively upgraded. The surface engineering technology of high sour gas fields has been formed. All the above supporting technologies have contributed to the development of the complicated old and new oil and gas fields of various kinds.

Breakthroughs have been made in the construction technique and the equipment development of oil and gas pipelines. Innovative theories and approaches for pipeline design were developed. The manufacture technique of large-diameter and high-pressure pipes has been promoted to a new level. Asia's largest full-size steel pipe blasting experiment sites was built. The supporting techniques for pipeline construction under complicated geological environment have been further improved. Great progresses have been made in the indigenization of large equipments serving in oil and gas pipeline systems. The constructions and commissions of many great projects have thus been guaranteed.

The technology for oil and gas pipeline network operation has been continually upgraded. Reliability-based method for the safety assessment of waxy crude pipeline startup was established. Simulation software packages which apply to complicated transportation systems, such as batch transportation of cool and hot oil and intermittent transportation were designed. Centralized control technology for large-scale oil and gas pipeline network was developed. The 50000 km pipelines owned by CNPC are now monitored, dispatched and managed in a unified manner. As a result, the insurance capacity for oil and gas supply has been improved.

Significant progress has been made in the oil and gas underground storage technology. Technical issues on the construction of underground water-sealed oil storage caverns have been solved. China's first 300×10^4 m^3 strategic underground water-sealed oil reserve was constructed and commissioned. Technologies for two types of natural gas storages – gas storage reservoir and salt cavern reservoir-have been further improved, which guarantees the efficient and safe operation of 11 underground gas storages.

Natural gas liquefaction technology has leapt forward to a new level. The complete set of LNG technologies, including natural gas liquefaction, LNG regasification, large LNG carrier construction, small LNG carrier design and construction, have already been thoroughly mastered. The construction technology of large-scale LNG storage tanks with intellectual-property right has been formed. Significant improvements have been made in the technology and equipments of floating liquefied natural gas vessels (FLNG). Heretofore, an integrated industrial chain has come into being. Offshore oil and gas storage and transportation technology has been brought to a new stage. Achievements have been made in theoretical and technological researches on the offshore flow assurance issue. The ability of risk assessment, stability analysis, and risk forecast for submarine pipeline has been upgraded. The design technology of flexible pipe and riser with independent intellectual property rights and research and design capability of proposing overall plan for the underwater production system were obtained. The 1500m-deep Liwan 3-1 oilfield submarine project was constructed and commissioned.

Oil and gas pipeline safety technology and management has been continually improved. Pipeline integrity management, which was once implemented only during operation, has now been extended to the whole lifecycle of the project, including the investigation, design, construction and abandon of pipelines. Breakthroughs have been made in the pipeline inner inspection, risk warning, and monitoring technology. The pipeline risk assessment method which suits the actual conditions and situations in China was established. The pipeline maintenance technology and relative equipments were developed. More complete standard systems, laws and regulations for pipeline safety technologies were established. Sound guarantees have been provided for the safe operation of pipelines.

Overall, China's current technology of oil and gas storage and transportation is rather advanced at the global level. Some aspects, such as the technologies of the pipeline transportation of waxy crude oil, the application of X80 steel in large-diameter pipes, and the treatment of chemical flooding produced fluid, are playing an international leading role, while other aspects, such as the area of the scale of oil and gas pipeline network, the reliability-based design of pipeline system, the large-scale production and detection equipments, and the underwater oil and gas production system of fields at depths greater than 1500 meters, still cannot match those of the developed countries.

Today our country is faced with great challenges and opportunities in the development of oil and gas storage and transportation discipline. In May 2017, the National Development and Reform Commission and the National Energy Administration issued the "Mid- to Long-Term Plan for the Oil and Gas Pipeline Network", which provides a comprehensive strategy for the development of China's oil and gas pipeline network in the coming decade. According to the Plan, China's pipeline mileage should add up to 24×10^4 km by 2025, which doubles that of 2015. In order to meet the necessities of the major development of the oil and gas storage and transportation industry, the discipline should devote more efforts in brewing original theories and innovative technologies, carrying on interdisciplinary studies in combination with production and education, as well as in indigenization of important equipments and the application of novel technologies, materials and equipments.

17. Land Reclamation and Ecological Restoration in Coal Mining Areas

Land reclamation and ecological restoration is a new field in China. the research group has identified 10 special topics on this discipline study, which contains the discipline charcteristics and position, basic theory, monitoring and evaluation of ecological environment in mining areas, control technology of mining subsidence in mining area, planning and design of land reclamation and ecological restoration , soil and landscape reconstruction in mining areas, revegetation and ecological restoration technology, protection and construction technology of construction in mining area, restoration of water ecology in the subsidence area, policy and management of land reclamation and ecological restoration.

With the three industrial revolutions during the 18—19 century and the rapid development of the mining industry, a large number of land and the environment have suffered tremendous damage and need remedial measures. The discipline of land reclamation and ecological restoration in coal mines was emerged in the beginning of twenty century.

The United States and Germany are the earliest countries to study land reclamation and ecological restoration,and they began systematic research in 1970's. China formally put forward and began to pay attention to land reclamation and ecological restoration in 1980's. Through more than 30 years of systematic research and development, the content and main development fields of this discipline have been established. The object of the discipline is the land and environment damaged by mining activities. The tasks of the discipline are: characteristics of damaged lands and the environment and its generation mechanism, principle and method of repairing the damaged land and environment, and monitoring and management technology of land reclamation and ecological restoration.

After more than 30 years of research and practice, the land reclamation and ecological restoration of coal mines in China have made great progress, which has raised the land reclamation rate to more than 25%. The main functions of this discipline are as follows: (1) Land reclamation and ecological restoration in mining area is an effective measure to protect arable land. (2) Land reclamation and ecological restoration in mining area are effective measures to protect ecological environment and improve people's quality of life. (3) Land reclamation and ecological restoration in mining area are effective ways to alleviate the contradictions between workers and peasants, solve employment and maintain harmonious development of society. (4) Land reclamation and ecological restoration in the mining area is the powerful guarantee for the sustainable development of the mining industry.

The main research fields and contents of this discipline are as follows: (1) Monitoring and diagnosing the impact of coal mining on land and environment. (2) Engineering technology of land reclamation and ecological restoration in mining area (soil and landscape reconstruction). (3) Biological technology of land reclamation and ecological restoration in mining area (revegetation and ecological restoration). (4) The supervision and policy of land reclamation and ecological restoration in mining area. (5) Basic theory and principle of land and environment restoration in mining area.

Therefore, based on the comprehensive analysis of literatures and achievements over the past 30 years, this report expounds the development process and achievements of land reclamation and ecological restoration in China's coal mines. And compared with foreign research, we put forward the future direction of the development of this subject in China. This report is the first one on the discipline development of land reclamation and ecological restoration in coal mines, including a synthesis report and ten thematic reports which is discipline orientation, basic theory of discipline, monitoring and evaluation of ecological damage in mining area, source control of land and environment damage in mining area, planning and design, soil and landscape reconstruction, revegetation and ecological restoration, prevention and restoration of damaged buildings in mining area, water ecological restoration in the subsidence area and policy and management.

The development history of land reclamation and ecological restoration in China's coal mining area has formed four stages.

(1) Germination Stage (1980—1989): To introduce land reclamation and ecological restoration of foreign countries as experience. To begin organized governance of coal mining subsidence as a feature. The stipulation of land reclamation, which was enacted by the State Council in 1989, was regarded as a sign.

(2) Initial Stage (1990—2000): Taking the discussion on basic concepts and basic problems of the subject as the core. Taking a wide range of practice of land reclamation and ecological restoration as a feature. And taking the concept of land reclamation and ecological restoration, and the first

International Symposium On Land Reclamation in China in 2000 as a symbol.

(3) Development Stage (2001—2007): With the further improvement of the concept of the discipline and concept and method of soil reconstruction as the core. It is characterized by a comprehensive and in-depth study of technology which has achieved the technical breakthrough of land reclamation in the Eastern Central coal mine area. Land reclamation is formally incorporated into the mining license and the approval for land use of 2007 as a sign.

(4) High-speed Development Stage (2008—now): Taking the new ideas and new technology of land reclamation and ecological restoration proposed as the core. Characterize by groundbreaking achievements of land reclamation and ecological restoration technologies in Ecological fragile area in western China. And taking the promulgation of Land Reclamation Ordinance and related standards and the emergence of regulatory methods as a sign.

Through the quantitative analysis of research results of land reclamation and ecological restoration in coal mining area. It can be seen that the publication volume of periodical Papers is rising in waves, and the general trend are developing in a better direction. The number of patent applications, scientific and technological awards, and postgraduate training showed a jump-growth, indicating that the technical research and development of land reclamation and ecological restoration in coal mining area were increasing. At present, 7 National Science and technology awards have been obtained in this subject, and formed a relatively stable research team and platform. Through comparative analysis with foreign related research, It is found that the technology of land reclamation and ecological restoration for surface coal mines is basically equal to the foreign level. And the land reclamation and ecological restoration technology of open-pit mine in Loess Plateau has Chinese characteristics. The technology of ecological restoration in the ecological fragile area of the West, subsided land reclamation with higher water level, and building protection and restoration have Chinese characteristics. However, our country is weak in the legislation of reclamation, and the maintenance cost after reclamation and management is high. The professional equipments, products and industrialization for land reclamation are also needed to learn from other countries.

Looking forward to the future, the land reclamation and ecological restoration in coal mining areas will show vigorous development trends. It is mainly manifested in the following aspects. As development of coal resources moves westward. The influence of land ecological environment and its reclamation in the Western ecological fragile mining area will be the focus in the future. Reclamation of cultivated land and ecological restoration technology of water area is still one of the key focus on land reclamation and ecological restoration in eastern China. With the increasing number of closed mines, the land reclamation and ecological restoration of closed mines are becoming more and more important. The coordinated development technology of resources and environment in coal mine area is imperative. To this end, the proposals for the development of this subject are: (1) To make the subject scientific attribution and positioning, and improve connotation and system. (2) Increase science and technology investment. (3) To build national or even international scientific research platform. (4) Strengthen the establishment of scientific research team and personnel training. (5) To expand the construction of experimental and demonstration bases. (6) Expand international and regional cooperation and communication of science and technology.

18. Mineral Materials

18.1 Introduction

Mineral materials refer to the functional materials using the natural minerals or rocks as raw materials. The research on mineral materials includes structure, physicochemical characteristics, function and application performance of natural minerals from the perspectives of material science and application, which aims to make full use of raw mineral materials and enhance its application values to meet the needs of the technological progress and industrial development. The mineral materials discipline is an extension and development of the mining engineering discipline, and a multidisciplinary field combining with the disciplines of rock and mineralogy, materials science and engineering, chemistry and chemical engineering, biomedicine, environmental science and engineering, machinery, metallurgy, power and electronics, energy science and engineering, etc.

Mineral material is not only an indispensable functional material in the modern high-tech industries, but also new material for restructuring and upgrading of traditional industry such as metallurgy, chemical industry, light industry and building materials, and functional materials of ecological restoration and energy conservation-environmental protection. They are widely used in modern industries such as aerospace, electronic information, machinery, metallurgy, construction, building materials, biology, chemical industry, food, environmental protection, ecological restoration, rapid transit and telecommunications and so on. Nowadays, mineral material plays a more and more important role in the development of natural mineral resources, optimizing the structure of traditional industries, saving energy and resources, environmental protection, ecological restoration, and promoting the development of high-tech industry.

In this report, the main research progress and hotspots of mineral materials research in recent years are reviewed. The development status of mineral materials at home and abroad are discussed and compared. Moreover, the suggestions on promoting the development of mineral materials are also put forward.

18.2 Important achievements in recent years

Carbonaceous mineral materials: In recent years, the important research results, which is world-advanced level, have been obtained in the field of preparation and application of carbonaceous mineral materials in China, especially graphene. The main progress of natural graphite material is the processing and application of the battery cathode materials, graphite intercalation composites, flexible/expanded graphite, new energy materials and other functional materials. It has ranked the first or forefront in the world for the mumber of published literatures and patent applications in the field of graphene in China. The main progress of graphene in recent years is the large-scale preparation and its application in the fields of composite materials, microelectronics, optics, energy, biomedicine and so on. These studies have made breakthroughs in the laboratory stage. The pilot-scale preparation and application of graphene have also made some important progress.

Clay mineral materials: In recent years, the remarkable progress has been made in scientific research and technological development of montmorillonite based modified and intercalated composites, nano-TiO_2/montmorillonite composites, mycotoxin adsorption material, bentonite based environmental and ecological materials, two-dimensional nano-kaolinite functional filler, coal-based kaolinite ultrafine and calcined whitening technology, kaolinite-based molecular sieves and catalysts, the application of attapulgite nanorods crystal, sepiolite adsorption environmental friendly materials and halloysite bio-functional materials.

Porous mineral materials: In recent years, porous mineral materials have made remarkable progress in high-performance molecular sieve, nano-TiO_2/diatomite composite photocatalytic materials, zeolite antibacterial composite materials, diatomite based environmental protection materials and expanded perlite fire insulation materials.

Calcareous mineral material: In recent years, significant progress has been made in the aspects of gypsum decorative materials, surface modification and functional compounding of grinding calcite carbonate (GCC), control of crystal form of precipitated calcium carbonate (PCC), particle size control of nano-calcium carbonate, natural marble shaped profiles and artificial marble solidification technology.

Magnesia mineral materials: In recent years, the main progress of magnesia mineral materials are the efficient preparation and clean production of fused magnesia, the preparation and application of magnesia fire-retardant materials, the surface modification of talcum enhancer, the ultrafine active fire-retardant brucite filler, as well as the structural properties and application of hydro-magnesite resources.

Fiber mineral materials: In recent years, the main progress of fiber mineral materials are the preparation and surface modification of wollastonite mineral fibers and the key technology of large-scale and stable preparation of basalt fiber.

Siliceous mineral materials: In recent years, the main progress of siliceous mineral materials are high-purity quartz sand, spherical silica powder and high-purity metal silicon preparation technology and the optimization of polysilicon production technology and equipment, high-purity melted silica material and optical quartz material preparation and application technology.

Mica mineral materials: In recent years, the main progress of mica mineral materials are the insulation and functional fillers of the mica and the surface modification of mica based pigments.

18.3 Development status at home and abroad

Nowadays, compared with developing countries, the developed countries still possess most of the advanced technologies and obvious technical advantages in the field of mineral materials. The developed countries have started the researches on the preparation of mineral materials, material structure, performance and application from the 1970s, and gradually formed large-scale production and application in the 1980s. Since the 1990s, it has been widely used in fields of industry, agriculture, telecommunications, aerospace, marine, military, environmental protection, ecology and health, which not only promoted the structural adjustment and industrial upgrading of traditional industries, but also accelerated the development of high-tech and new material industries. By the turn of the century, some industries, such as bentonite mineral materials, have formed an industrial cluster with annual sales of over 10 billion dollars. Although China has started to develop mineral materials in the 1980s, our basic science research was still weak with lagging technologies and equipments. Since the 21st century, the gap between China and developed countries in mineral materials discipline, whether the basic research or technology development, is narrowing and some achievements are even international-advanced in the fields of graphene, ultra-fine calcined kaolinite, diatomite health and environmental protection materials, attapulgite mineral materials, artificial marble and quartz, magnesian mineral materials, functional mineral fillers, mineral-polymer composites and so on. Moreover, some fields in this area have even reached the world-leading level. In the past decades, domestic R&D institutions have ranked top 10 in the world on the number of published papers of graphite mineral materials, graphene, clay mineral materials, porous mineral materials, calcareous and magnesian mineral materials, fibrous mineral materials and siliceous materials. Especially, the domestic research institutions have ranked No.1 in the world in the field of graphene, graphite mineral materials, bentonite mineral materials, diatomaceous earth mineral materials, zeolite mineral materials, attapulgite mineral materials and quartz materials. However, the overall development level in science and technology of mineral materials in our country still lags behind the world-advanced level. Especially, there are still big gaps in the industrialization of scientific and technological achievements. And these gaps should be the main directions where our researchers need to make breakthroughs in the future.

18.4 Research hotspots in recent years

In recent years, the research hot spots of mineral materials are all related to new energy and energy-efficient materials, environmental management materials, ecological health materials, high-hardness and high-strength materials, inorganic flame-retardant materials, electrical and electronic materials, and functional composite materials.

For the carbon mineral materials, battery cathode materials, flexible graphite, ultra-high purity graphite, isostatic graphite and so on are all research hotspots in recent years. Graphene is one of the most important innovations in the field of materials science. The main researches focus on large-scale and low-cost preparation and application technologies and graphene composites. For clay mineral materials, most researches focus on bentonite, kaolinite and attapulgite based mineral materials. In recent years, the research of bentonite mainly has focused on product varieties and application properties, especially environmental and ecological restoration materials, waterproofing materials, mycotoxin adsorption materials, composite materials and so on. The research of attapulgite focuses on the structure, properties and the nanorod crystal dissociation, decentralization and high value utilization. On the other hand, the research of kaolinite focuses on functional filler, petrochemical catalytic materials and coal-based calcined kaolinite. For porous mineral materials, most researches focus on zeolite and diatomite mineral materials. The research of zeolite focuses on the synthesis of zeolite and its application in petrochemical catalysis and environmental protection. The research of diatomite focuses on the environmental functional materials, humidity-control materials, energy-efficient materials and high-performance filter materials. Gypsum mineral materials, the preparation and application of calcium carbonate are the research hotspots of calcareous mineral materials. The main research direction of gypsum mineral materials are varieties and applications. The research of calcium carbonate focuses on the preparation and application of nano-calcium carbonate, especially grain size and crystal regulation and optimization during the production process. For magnesia mineral materials, the research hotspots are mainly the preparation, modification and application of talc composites, the technology of energy-efficient and cleaning preparation of electric fused magnesia, and the preparation and application of the flame-retardant brucite materials. For fiber mineral materials, the research hotspots are the preparation, composition and surface modification of high-length diameters wollastonite fiber, and the preparation and application of basalt fiber. For siliceous mineral materials, high purity and ultra-high purity quartz sand, spherical silica micro-powder, monocrystalline silicon, polysilicon silicon and metallic silicon are the research hotspots in recent years. The main research direction of quartz is high purity processing technology. The main research directions of silicon materials including monocrystalline silicon, polysilicon and metal silicon, are the preparation process and equipments, material structure and properties and its application. For mica mineral materials, the high-performance mica insulation materials of modern electrician, electric appliance and electronic mica paper and products are the main research hotspots in recent years.

18.5 Suggestions and perspectives

Chinese mineral materials discipline was independently created in the geoscience and mining colleges around 2000, which has master's and doctoral degree at present. The college has trained a large number of professional talents in the past decades, but the mineral material enterprises in our country still lack undergraduates and the corresponding personnel seriously. It is recommended to establish "mineral materials engineering" or "mineral materials and applications" undergraduate majors under the first-class disciplines of "mining engineering" or "material science and engineering". In the geoscience and mining engineering, material science and engineering research institutes, the special mineral materials research centers or key laboratories need to be set up. Meanwhile, large enterprises in mineral materials industry should be encouraged to establish mineral materials and application or engineering technology research centers.

Despite of significant progress of mineral material discipline in our country, there are still a number of the challengs and problems for the discipline development in the coming years. Compared with developed countries, there is a large gap in basic mineral material science and application research, which still needs to continue to strengthen the scientific research. At present, Chinese mineral material industry still lies in the middle or lower levels development stage in the international mineral material industry. The lags in production and application technologies should be the main reason. In the long run, the industrialization of key technologies, especially the application technologies, should be still the main research directions in Chinese mineral material industry.

19. Architecture

The discipline of Architecture studies built facilities and their spaces and environments, both inside and outside. Normally, it refers to the integration of the art and technology related to design and tectonics, which distinguishes itself from the technology of construction. In different eras and with different social needs, Architecture continuously evolves and serves a fundamental function to improve our living environment. Traditional Architecture studies the design of built facilities, building interiors, furniture, landscaping, cities, towns, and villages. As the discipline evolves, Urban and Rural Planning and Landscaping separate from Architecture and become two parallel Level-1 disciplines.

The discipline of Architecture covers architectural design, interior design, architectural tectonics, architectural history, architectural theory, urban design, building physics, and building service systems. Although the realm of Architecture keeps evolving, its position as the fundamental discipline for the improvement of the living environment remains unchanged.

Based on the latest development of Architecture in China and abroad, this project is divided into six subjects. For each subject, the state-of-the-art is reviewed, followed by an analysis of the future developmental trend. These subjects are: urban design in the context of new-type urbanization, architectural culture in globalization, sustainable development of building environments with a view toward global climate change, rural construction based on the coordinate development of urban and rural areas, digital technology for architecture in the information age, building industrialization in the context of industry modernization.

Urban design studies urban space and its creation, including people, nature, society, culture, space, and form. In the 21st century, urban design breaks away from its tradition and limitation of primarily studying urban space and its visual effects and starts focusing on culture, society, and urban vitality.

Currently, the development of urban design in China is unprecedentedly active. Theoretical and methodological explorations are in parallel with the western world. Design and engineering practices and technological development show signs of surpassing. In terms of urban design theories and methodologies, China was first following the steps of developed countries such as the US and Japan and started establishing its own framework in the 1990s. Since 2000, the rapid urbanization and the change of needs for city and social development has provided plenty of practicing opportunities to urban design. Some new and systemic developments have emerged, including the paradigm shift to sustainability and low-carbon cities, the advancement of digital technology and its application, the influence of modern arts on urban design, etc. Compared with the developed countries, the urban design practice in China shows more varieties and tends to be larger in scale.

The discussion on the architectural culture has three levels of meaning: (1) Patterns and features shown by the internalization of social and cultural characteristics in architecture. (2) How architecture affects people's spatial behavior and relationship. (3) Systematic thinking, spreading, and theorizing of architectural tradition and modernity. Lately, the most discussed subjects include how to rethink critically the modernism and radical architectural culture and how to inherit and revive critically the architectural vernacularism and traditional architectural culture.

To cope with the climate change has become one of the central themes of Architecture. In the planning and design of urban spaces, measures include efficient development of land, energy efficiency, green transportation, ecological urban design, etc. In the urban environment, measures include emission reduction, flood management, sustainable landscape, etc. In the building environment, measures include comfortable indoor physical environments, energy efficient design, etc. While emphasizing on sustainability, the technology advancement should be parallel with and integrated into architectural and urban planning design. The keywords in this field include green, low-carbon, ecological, sustainable, etc. Although they are not exactly the same, they contain similar core ideas.

The problems of agriculture, farmers, and villages are one of the obstacles to China's economical and social development. The research on rural development is problem oriented and based on China's particular situations. The main research subjects include vernacular villages, new-type rural communities, rural management, rural tourism, relationship between urban and rural areas, rural revival, rural planning, etc. In the realm of buildings, the research is focused on vernacular residential buildings and expanded to cover groups of villages and green and energy efficient technologies. The theoretical and technological research in this field is heavily influenced by the central policy. Since 2006, China has been through a series of stages of rural development, which have made significant achievements and greatly promote the research and development.

Architectural design is closely linked with technological innovation. As the digital technology is introduced into the field of Architecture, its theory, methodology, and system are constantly evolving, bringing forth innovative and unique outcomes and explaining brand-new design ideas. Lately, the research is centered around these subjects, namely non-linear system theory and its integration into digital architecture, information and control theories and their influence on architectural design, digital fabrication, parametric design and algorithmic shape generation, Internet of Things, intelligent building environment, etc. With the full-scale and rapid urbanization in China, a large number of architectural projects have been designed and built with the assistance of digital technology. These projects have taken advantage of the latest development oi theories and technologies and practiced the ideas and beliefs of parametric design, digital tectonics, craftsmanship, vernacularism, environmental integration, etc. They have advanced the technology level of the entire building industry. Some of them have become landmark architecture.

The building industry is a pillar to the national economy. However, a series of problems still exist such as low production efficiency, high consumption of resources and energy, severe pollution, low durability, etc. Building industrialization is an effective means to solve those problems. Developed countries have taken different paths towards building industrialization, such as the highly mature wooden building system in the US and the pre-cast concrete building system in some European countries. Although the research and practice of industrialized buildings started in developed countries such as the US, Japan, Sweden, and Germany, China is rapidly catching up. Universities, research institutes, design companies, and construction companies have conducted research on design theory and methodology of building industrialization, industrialized structural systems, industrialized interior finishing systems, etc. The primary research tasks include the technical standards system of building industrialization, the industrialization of the main superstructure system, the industrialization of the building sub-systems, information and digital technology for building industrialization, etc.

20. Textile Manufacturing

China is the largest textile manufacturing, exporting and consuming country in the word. The textile industry is the backbone and livelihood industry in China, playing important roles in beautifying

people's life, driving the growth of relative industries, boosting consumer demands, enhancing self-confidence and promoting social harmony. The textile industry also possesses obvious superiority in international competitiveness. The number of production of fibers has begun to exceed 50 percent of the combined output of the word since 2009. The main business income of textile enterprises above designated size in China achieved more than RMB 7.33 trillion in 2016. Textile and garment exports reached 12.88 percent of domestic general export, and China's textile and garment exports account for 38 percent of those in the world, ranking the first in the world with the world's most complete industrial chain.

In recent years, the manufacture of differential and functional fibers was improved significantly; the industrialization and green production of biomass fibers made breakthroughs, and the development of high performance fibers was strengthened. Moreover, great improvement has been achieved in automation, serialization and high-speed production of textile equipments. Industrial textile manufacture was enhanced significantly; the clean production technologies, including printing and dyeing without water, shortened spinning process and application of ecological chemicals, were popularized; intelligent textile manufacturing is bringing great improvement for the textile industry based on internet of things technology. From 2013 to 2016, two projects were bestowed the second-class awards of national technology invention. Twelve projects achieved national science and technology awards, where one was first-class award and eleven were second-class awards. Until now, nine national research and development platforms closely relating to textile science have been established, in which there are two national key laboratories, six national engineering research centers and one national engineering laboratory.

In terms of talent cultivation, there are 40 colleges or universities with the bachelor degree of textile engineering, 60 colleges or universities with the bachelor degree of clothing design and engineering, 20 colleges or universities with the bachelor degree of light chemical engineering in dyeing and finishing field. In addition, 7 universities in China have the qualification for awarding doctoral degree of first level textile discipline, and 19 universities possess first level master's degree awarded qualification for textile discipline.

This report includes two parts: the first part summarized the situation and development of textile industry and discipline, comprised the progress of textile research at home and abroad, and analyzed the development trend of textile discipline; the second part detailed the achievement and development trend of textile industry and discipline in eight sections, that is, fiber material engineering, spinning engineering, weaving engineering, knitting engineering, nonwoven material and engineering, textile chemistry and dyeing & finishing engineering, fashion design and engineering, and industrial textiles.

In the area of the fiber material engineering, biomass fibers, chemical fibers, high-performance fibers and functional fibers keep developing; the Lyocell fibers have reached tens of thousands of tons productivity, and the first production line for high performance carbon fibers T800 with kiloton productivity has been used in China since 2016. For the spinning engineering, great progress has been achieved in high-efficiency combing, specific fiber spinning, extra high count spinning and intelligent and sequential spinning techniques. In weaving engineering field, the environmental friendly yarn sizing, high-quality cotton fabric weaving, functional structured fabrics have developed quickly; the projects in terms of foam sizing and cotton fabrics of super-elasticity achieved first-class science and technology awards issued by China national textile and apparel council, respectively. In the field of knitting engineering, significant advances are made in intelligent jacquard technology, seamless technology, high-speed knitting equipment and knitting MES technology; our self-developed MES technology passed the scientific and technical identification organized by China national textile and apparel council in April 2017. For nonwoven material and engineering field, the new melt-blown technology, medical and protective material, environmental protection and dedusting material, geotextile material, and filtration material are in the rapid development; the project named "the key technology and industrialization of medical protection materials" achieved the first-class award issued by China national textile and apparel council in 2016, and the first-class award was also granted to the project namely "key techniques for the preparation of filter materials for high temperature incineration flue gas" in 2016. In the field of textile chemistry and dyeing & finishing, great progress of textile dyestuff, pre-treatment, dyeing and printing, finishing and resource recovery techniques is made; the project "production and application of new reactive dyes strengthened by functional groups" was granted the second-class awards of national technology invention in 2016. In fashion design and engineering, high and new technologies are applied to 3D body measurement, virtual garment design, intelligent clothing, etc.; the functional protection clothing achieved a series of progresses in fireproof and heat insulation, water resistible and breathable as well as antibacterial applications; moreover, automation, intelligent and information technologies were also applied to fashion design and engineering field. In the field of industrial textiles, research on fiber-based composite materials, filtration textiles, medical and hygiene textiles, intelligent and health textiles and protection textiles has achieved rapid progress.

Reports on the special topics detail the comparison analysis of the textile progress, discussing the gaps between domestic and overseas technologies. The developmental level of domestic nanofibers, biological fibers and high performance fibers has improved greatly, approaching the international advanced level in many aspects. The stability of the spinning machine still needs to be enhanced, but the speed of the knitting and weaving machine achieved great improvements in recently years. The nonwoven technology developed quickly recently, but the corresponding production quality and manufacturing techniques are yet to be improved. Moreover, there is still room for the improvement regarding environmental friendly dyes and auxiliaries. The gaps of CAD system for clothing design are clear between domestic level and international advanced levels, such as the CAD system developed in Europe, America, Japan and Korea. Although there are complete types of industrial textiles in domestic, the property of the products are still behind the advanced level in the world. Based on the analysis and discussion of the gaps between domestic and overseas technologies, the keystone of further research and the development trends of textile industry and discipline are also reported.

21. Pulp and Paper Science and Technology

During the period of 12th Five-Year, Chinese paper industry has entered a new phase of development that is replacing the capacity expansion with S&T innovation as the driving force, seeking potentials with the sustainable development. The subject of Pulp and Paper Science and Technology has formed a sophisticated S&T innovation system with abundant S&T resources and reasonable layout. In the key technology areas of fiber resource utilization, energy saving and emission reduction, environmental protection and circular economy development, a large number of innovative achievements have been accomplished, with significant improvement in the overall level of industrial technology.

In pulping science and technology area, especially in chemical pulping, some technological progress has been made in recent years, including reducing cooking temperature, shortening cooking time, strengthening oxygen delignification and so on. The compact continuous cooking adopted as mainstream technology by large pulp mills and the DDS displacement cooking with better adaptability were both promoted. At the same time, non-wood pulping technology and non chlorine bleaching technology, the key deinking technology to improve the recycling efficiency of waste paper, recycling of pulping and papermaking process water and advanced treatment of wastewater to standard discharge technology have approached to or achieved at the advanced level of international standard.

In papermaking science and technology area is mainly reflected in the cost control and reduction, and the improvement of product quality. The era of industrial 4.0, characterized by digitalization, networking and intelligentization, has played a positive role in promoting the development of paper science and technology, enhancing the operation of international mainstream technologies such as medium consistency refining, advanced press section, new technology of modern drying system for high speed paper machine, up to date coating technology and so on; To meet the demands of energy saving and emission reduction, high efficiency and functional chemicals for pulping and papermaking process has been developed.

For the water pollution prevention and control technology of papermaking industry, it is adopted source control, mainly including the extraction of black liquor, extended delignification and oxygen delignification, elemental chlorine free bleaching (ECF), total-chlorine-free bleaching (TCF), unbleached pulp, low brightness bleaching etc. as the current domestic cleaner production technologies for pulping and papermaking wastewater. As for the end of treatment, the treatment technology of domestic papermaking wastewater is generally divided into the level of processing technology, i.e., first level, second level and third level processing technology; so the persistent organic pollutant abatement in the paper industry was mainly performed by source control, applying the end of treatment as supplementary approach, so far the governance has made breakthrough progress.

The technological innovation and product development in the field of paper-based functional materials has made breakthroughs, some high-tech products have filled the gaps in the domestic market, of which have achieved exportation with improvement in both production scale and quality; The development in nationalization of pulping and papermaking technology equipment has noticeably accelerated, the technology gap in continuous cooking equipment for chemical pulping, waste paper pulp papermaking equipment and high-speed paper machine between China and other countries is gradually narrowing, through the complete introduction or introduction of key components, to promote Chinese pulp and paper equipment manufacturing capacity, and S&T level; meanwhile, part of the large and medium-sized pulping equipment has been initially localized.

However, currently in pulping and papermaking area, the development of S&T independent innovation in China is quite backward with few original technology achievements, mainly referring to the pulping technologies including solvent pulping, bio pulping, waste paper deinking process, biorefinery process based on conventional pulping and papermaking process, and energy saving and cost reduction equipment; the original technology invention in specialty paper industry, such as papermaking technology innovation development of paper based functional material, lags behind compared to the developed countries; the development and implementation capacity of new technology engineering, such as new generation of coating machine, and industrialization of processing technology and engineering in process control, and functional additives in pulp and paper field are still at a low level.

The subject development of Pulp and Paper Science and Technology will be implemented with the goal of low resource consumption, low emissions during process, renewable products production and recycling of waste disposal, majoring in developing a recycling economy, innovation and development mode, building a resource-saving paper-making industry, focusing on development of highly efficient and recycling of resources, pollution control, energy saving and emission reduction technology, and research on equipment.

The future development trend of pulp and paper science is focusing on the shortage of fiber resources in Chinese papermaking industry, the lack of energy and water resources, and the increasingly strict environmental protection requirements and gradually more strict industrial policies for pulp and paper industry; therefore the pulp and paper S&T will be based on the current situation of raw material structural characteristics and technical level of enterprises in Chinese paper industry, to further increase the scientific research and technology by tracking and making efforts to study the development of international advanced technology, to initiate developing advanced technology and innovation independently, to speed up science and technology progress in the pulp and paper industry, eventually to promote the sustainable development of Chinese pulp and paper industry.

The report on the subject development of Pulp and Paper Science and Technology includes comprehensive report and research report of six key areas, including pulping, paper making, equipment, chemicals, pollution control, and paper-based functional materials, which basically reflects the overall S&T progress of this subject. In addition, the report focuses on the analysis and outlook of the future trends and key technology directions of related technologies as follows: (1) Efficient utilization and recycling technology of fiber resources. (2) The key technologies in environmentally friendly pulping and papermaking area. (3) The production technology of high performance paper based functional material. (4) R&D in the large, advanced and special technical equipment with high integration and improved performance. (5) Development and application of efficient, specific and functional paper chemicals, including the development and application of green chemicals, biomass chemicals and nanomaterials. (6) The key technology development of circular economy and low carbon economy in paper industry.

22. Optical Engineering

Optical engineering has now developed as a subject mainly of optical and optoelectronic technology, which also have close intersecting and interpenetration with information science, energy science, space science, material science, life science, precision machinery and manufacturing, computer science and microelectronic science etc.The new principles of modern optics, new technologies and new materials have accelerated the development of science, engineering and applications.

On the basis of science and technology, optical engineeringis expanding in the scale of space, time and spectrum and approaching or even breaking through the theoretical limits. Benefited from the developing of new devices, new materials and new technologies, it has formed and penetrated into thefield of basic interdisciplinary.In terms of engineering and application, the depth and breadth of application continue expanding. Optical engineering is affecting and changing the scientific, industrial and military forces of each country in the fields of communication, remote sensing, industrial manufacture and national security.

The global research hotspot and development trend of optical engineering: (1) The development of solid laser technology to multi polarization. (2) New optical systems and optical instruments have been invented. (3) Light wave precision regulation is going to get a breakthrough. (4) Quantum computing technology begins to take shape. (5) Laser additive manufacturing promotes industrial manufacturing to a new level. (6) Space laser communication technology is becoming an important way to realize the integration of space and ground. (7) Optical observation has been more and more important in space monitoring.

In recent years, under the guidance of national scientific and technological innovation strategy, native optical engineering has made a series of scientific and technological achievements with great international influence. Some of the technologies also achieve the great-leap-forward development internationally. However, there are still some gaps between the developed countries and the developed countries in terms of application base, mid-to high-end optoelectronic devices and optical instruments and equipment. Many high-precision instruments still need to be imported from abroad. The scientific research of optical engineering is mainly developed by universities, institutes of CAS and part of the Central Enterprise Research Institute. In recent years, China has strengthened basic research and has been collaborating closely with international research teams. Some subjects have achieved significant breakthroughs, and made international leading achievements.

Our research such as laser physics, femtosecond spectroscopy and application, optics and photonics, mesoscopic nano materials and applications, specific optical manipulation and precision spectroscopy of some concrete problems in the scientific research personnel have made important achievements. In the laser and matter relativistic and highly nonlinear interaction, laser particle acceleration, afcond pulse laser, laser fusion basic physics, new laser plasma light source and other aspects of our country researchers have been at the forefront of the international, many domestic research institutions have Some ultra-fast superlattice devices have been built, ranging from terabytes to hundreds of tera watts and even to wattage.

In 2016, for the first time in our country, we realized the multi-degree-of-freedom quantum teleportation and quantum communication secure transmission for the first time in an ordinary single-mode optical fiber using ultra-large-capacity ultra-dense wavelength division multiplexing to transmit 80 kilometers of information. Resolution single-molecule optical Raman imaging. For the first time, the solar atmospheric visible to near-infrared 7-band high-resolution tomographic images were obtained for the first time. On the silicon-based optical waveguide chip, an array of "swirling beam" emitting devices was first integrated and the world's only practical Deep ultraviolet all-solid-state laser.

In recent years, the state has given strong support to the development of optoelectronics industry and promoted the upgrading of manufacturing industry. With huge market demand and scientific research and production capacity, some domestic optoelectronic optoelectronic devices also have a higher technical level, and have a certain international market share, the domestic development of optoelectronic technology better regions have also formed optical devices The industrial chain, has a relatively perfect industrial cluster.

In the overall research level of lasers, our country has a certain gap compared with advanced countries, but in some aspects, domestic lasers also have their advantages, such as domestic UV lasers, which are not lost in import lasers in terms of stability and reliability. At the same time, lasers in our country have greater flexibility in customization tailored to the specific needs of our customers.

The overall development of China's modern optoelectronics industry has led to the development of the domestic optical lens industry. Since 2000, the domestic security and Internet industries have developed rapidly. Some world-class manufacturers have risen rapidly. Our country has become the country with the most demand for optical lenses in the world. The surge in market demand has led to the emergence of a number of outstanding lens manufacturers in China. Product development, design and production processes continue to invest in the introduction of higher resolution, more stable imaging quality of the optical lens, the optical lens industry has quickly become a new force, the domestic lens also led the development of the domestic optical cold-processing industry Rapid development, our country has become the world's largest optical cold-processing capacity to undertake and gather.

Although China has become the largest optical and optoelectronic market in the world, the development of China's high-end optoelectronics industry is still at a low level, and many high-performance devices also rely on imports. Many of the core optical products in the high-end optical instrument industry, the laser manufacturing industry and other optical radiation industries are controlled by foreign countries. At the same time, some optical engineering industries are seriously lacking the supporting key equipment and industrial support. For engineering applications and product industries in the field of large optical engineering There is also a lack of control and direction.

In the future, the development of China's optical engineering will show the following characteristics: (1) China will form a number of important achievements in basic scientific research. (2) China's laser additive manufacturing industry will enter a rapid development stage. (3) China's space laser communications will be a step into practical. (4) Optical fiber communication technology and application level will achieve a number of system-level innovations. (5) Optical technologies will further enhance China's military and counter terrorism capabilities.

23. Basic Agronomy

23.1 Significant research progress in basic agronomy disciplines

Thanks to continuous support of national scientific and technological plans/projects, along with unremitting efforts of scientific researchers, and joint efforts of domestic agricultural colleges and institutes to tackle major issues in the area of agricultural environment protection, agricultural products processing, farming and cultivation, significant research progress has been made in sub-disciplines of basic agronomy, including control and restoration of environment quality in agricultural production area, agriculture to cope with climate change, agrometeorology and agricultural disaster reduction, facility cultivation, food processing, agro-products quality and safety, food nutrition and function, fruit and vegetable fresh-keeping, crop cultivation and physiology, and crop ecological and cultivation and so on.

Major issues in basic agronomy discipline has been tackled. Targeting the forefront of international science and technology, breakthroughs have been witnessed in the basic theories and methods in mechanism of increasing green house gas sequestration sink and emission reduction in agriculture, regionalization and utilization of agroclimatological resources, the formation mechanism and control technology of fungal toxin during the agricultural products storage processing, safety control theory of agricultural products processing, high-yield and high-efficiency crop cultivation theory etc. The progress enables the discipline to seize the commanding heights of modern agricultural science and technology, and significantly enhanced the academic level and international influence.

Oriented for national demands, in view of the major bottleneck problems which restricted agricultural transformation and upgrading in China, a series of core technologies, common technologies and a complete set of matched technologies in agriculture research and technology development were settled, including of utilization of microbial fermentation bed for breeding waste, energy efficiency improvement and nutrition quality control of intelligent plant factory, nutritional and functional components analysis and detection, postharvest preservation of fruit and vegetable effectively, physiology and technology for high fertilizer and water use efficiency, etc. they had obtained a series of core independent intellectual property rights with significant industrialization prospect, and have a high degree of leading, wide coverage and high correlation for agricultural development and high-tech industries.

Serving the main battle of "issues of agriculture farmer and rural area", for solving the overall, orientational, and critical technical problems in agriculture and agro-economic development, a series of technology system with great scientific significance, highly effective progresses and obvious times feature were achieved, among in agrometeorological disaster monitoring and early warning, risk assessment and prevention-controlling, prevention and control of continuous cropping obstacle facilities vegetables, crop high-yield cultivation and so on. They are leading the leap-forward development in agricultural science and technology. The ability of innovation in basic agronomy research was improve dramatically, which is helpful in effective scientific support, strong technical reserve and technical guarantee for sustainable development in agriculture and country.

23.2 Comparative analysis of domestic and international development status in basic agronomy disciplines

Based on the analysis of international major research plan and research project in agronomy discipline, and study on the new development trend and character, the development status and overall level of basic agronomy disciplines are compared and analyzed from agricultural environment protection, agricultural products processing, farming and cultivation. The international standard and development state can be made clearly. It is helpful to provide the necessary foundation support on development strategy and determination of future direction.

With the development of international modern agriculture, biology, medicine, engineering, and informatics, the progress of basic agronomy discipline generally showed the trend of internationalization and globalization, such as agricultural environment protection, agricultural products processing, and farming and cultivation. Cross, integration and infiltration between each disciplines are increasingly strengthened. The development direction of discipline has been deep into the micro structure, extended to the macroscopic and comprehensive relationship, while the theory innovation continuous accelerated, and technique innovation continuous developed in mechanization, informationization, standardization, quantization, large scale and intensification. Industry upgrade was obviously accelerated by technology innovation.

In general, a series of national scientific plans on the projects of agricultural environment protection, agricultural products processing, farming and cultivation have been carried out in recent years. A number of research institutes and agro-experimental stations were established, a large team of researchers has been formed. Researches on carbon fix and emission reduction in agriculture, economical solar-greenhouse, food functional components detection analysis, crop disease control by inter-cropping are the world leading level. However, compared with the development of developed countries, legislation on agricultural environment protection were imperfect, while the provisions on legal liability were vague. The overall input of basic agronomy scientific research and development was insufficient. There were great gaps on research equipment and methods, accumulation of basic information and data, and basic theory study, etc. Some new frontier research had started late and are in catch-up stage. The supply capacity of core technology is still inadequate. The product development of many high-tech or sets of equipment are high dependent on developed countries, while lack of supporting technology and standard. The achievement transformation is low efficiency, and the enterprise independent research and development ability is obviously insufficient.

23.3 Development tendency, key areas, and future prospects in basic agronomy disciplines

Based on the analysis of future strategic needs and development trend in basic agronomy discipline, related with agricultural environment protection, agricultural products processing, and farming and cultivation, key areas and priorities direction of basic agronomy discipline in the following 5 years are definited, while it is helpful to explicit which are the major scientific problems and key technical problems for the new strategic needs.

On the aspect of agricultural environment protection, the agricultural environment risk management theory should be prioritize. The environment friendly fertilizer, biodegradable film, residual film recycling machines, and new type ecological fodder are researched. The recycling technology for pollutant resource recycling, the control and repair technology of soil pollution, new technologies and products of the harmless treatment and resource utilization for livestock, technology of agricultural carbon fixing and emission reduction, technology of agricultural meteorological disaster assessment and risk prevention, technology of facility cultivation are Integrated and innovated.

On the aspect of agricultural products processing, food processing and manufacturing theory should be strength, along with new processing technology exploring, intelligent and digital core equipment developing, food nutrition quality evaluation developing , control and monitor technology and equipment researching. the key technology difficulties of Chinese style main meal industrialization need to be breaking, and analyzing the mechanism of food nutrition components on health. We should vigorously strengthen nutrition functional food quality improvement and manufacturing technology research, actively promote agricultural fresh-keeping related scientific problems and the key technology and equipment research and development, and to carry out the integrated whole chain technologies application.

On the aspect of farming and cultivation, researches on crop ecology, physiology, and biochemistry should be strength. The focuses are the innovative technologies on high efficiency, high quality, green, mechanization, simple and disaster resistance crop cultivation techniques. Researches on fallow model research to build new crop rotation, inter-cropping mode, intelligent crop cultivation techniques should be carried out. Furthermore, more researches should be focused on soil cultivation system and the matched machine, and establishing a technical system for reducing carbon emission in farmland.

In conclusion, more supporting programs should be rolled out, for example to strengthen the legal guarantee, disciplines planning layout, to improve the relevant technical standards, implement different management in different region. More attentions should be paid to long term basic science and technology research, to continuously performed the efficiency of science and technology innovation, and so on. Furthermore, the integrated key techniques and its application should be popularized. Capacity building and international networking should be encouraged. Strategies and counter measures should be proposed to promote the development of agro-environment protection, agricultural products processing, crop cultivation and other basic agronomy disciplines.

24. Forestry

Forestry holds historical responsibilities for maintaining survival security, fresh water security, homeland security, species security, climate security and building a beautiful China. The 19th National Congress of the Communist Party of China creates Xi Jinping Thought on Socialism with Chinese Characteristics for a New Era, which emphasizes that ecological civilization construction is a long-term, tough systemic project for sustainable development of Chinese nation.

Forestry mainly takes forests and woody plants as research objects to reveal substantial rules of biological phenomena and focuses on forest resource cultivation, protection, management and utilization, etc. Cultivation area of plantation has reached 69.33 million hectares in China. Based on the theories of suitable trees and suitable sites, fine variety and good cultivation, "Five controls" cultivation system, which includes genetic control, site control, stocking density control, vegetation control and soil fertility control, have proposed to improve quality of plantation. The research on physiological and biochemical mechanism of stress-resistance afforestation has entered the functional genes analysis and regulation stage. The nutrient relationship and action chain theory among species in the mixed forest has been innovated and formed. The theory and technology of precise water and fertilizer regulation in the timber forests have been developed. Micro-site evaluation on afforestation in tough sites was on going. Natural secondary forests with low-price and low-productivity have a great potential of the yield and quality improvement, which could be excavated though the silvicultural measures. In addition, forest tending has innovated the theory and technical system of precise tending in ecological public welfare forests. A near-natural forest management technology, which was characterized by target trees management, was established. In the meanwhile, multi-functional cultivation of forests focused more on areas such as carbon sink forests, urban forests and contaminated soil bioremediation. The maintenance of biological diversity plays a key role in healthy and stable forest cultivation, and the study and utilization of biological diversity will get deeper.

Following the development of techniques and theories in modern biology, tree genome sequencing, construction of gene regulatory network for wood quality and resistance, and function of key genes and regulatory factors have become frontiers in forest genetics and tree breeding. Molecular breeding and biotechnology applications research in China are close to the developed countries in forestry, while ploidy breeding technology is in the leading position of the world, which lead to breakthroughs in efficiency and effect of improved varieties. Molecular maker-assisted breeding, somatic embryogenesis and clone system have become research hotspots in order to making full use of multi-generation improvement strategies of excellent germplasms in various generations. However, regular breeding techniques are still the most effective methods for breeding forest varieties.

Forest management has made a series of important advances in the theory and technology of forest resource investigation and monitoring, stand growth and yield prediction, forestry resources information management, forest management plan and structured management of natural forest by closely following the international forefront and applying new theories and new methods. The theory and technology system of forest sustainable management, such as the precision prediction of forest growth and the management response, the aero-space-ground integrated monitoring and evaluation technology of the forest quality, have been constructed. It will be an important task to explore the forest management theory with various functions of forest.

Forest ecology has made remarkable progress in the application of basic research and technology research and development. The application of new theories, new technologies and new methods, such as neutral theory, high-throughput sequencing, spatial analysis in forest ecosystem research, has greatly promoted the research development of forest biodiversity formation, coexistence, maintenance and conservation. The interdisciplinary and integration of forest ecology and biology, geosciences and social sciences have made the research on the fields of degraded forests adaptive ecological restoration and reconstruction, plantation ecosystem management, and forest ecosystem services scientific assessment developing more and more deeply. Microscopic study of Forest insects mainly focus on function identification and gene editing and validation of specific genefragments of pests. The scope of macro research hasextendedfrom individual to community and from ecosystem to landscape scale.It has become a trend to establish a long-term standardized and standardized monitoring platform for forest ecosystem. The hot topics of forest soil science are reasonable using bad site forest soil resources and improving soil productivity. In the field of forest meteorology,long-term and dynamic observation and research will be emphasized furtherly. More attention will be paid to quantitative research on influence mechanism meteorology on the key ecological, physiological and biochemical processes related to tree growth at the micro-scale, and to monitoring and evaluating the regional climate effect of forestry eco-engineering on the micro- and macro-scale.

Gratifying achievements have been obtained in the fields such as the relations on wood and its water, wood heat transfer and drying mechanism, wood elasticity mechanics, and wood viscoelasticity mechanics, forming an interdisciplinary research fields which include nano-scale characterization, functionalized wood theory, wood bionics. Especially, important breakthroughs have made in some key technologies of wood modification, artificial board, and wood products and adhesives, etc. Wood science and technology will further systematically study on wood cell wall structure and chemical composition and its impacts on the physical and mechanical properties to build a physical and mechanical model of wood cell wall and to clarify the variation law of wood structure under external environment excitation and its response mechanism to physical and mechanical properties. The main contents of wood and bamboo research will have synergistic mechanism of natural multi-scale microstructure and macroscopic function of wood and bamboo, the formation method and principle of bionic intelligent nano-interface of wood and bamboo. The fields of Forest chemical industry have extended to the science fields of bio-energy, bio-materials, bio-chemicals, extracts of biomass, and pulping and paper-making etc. The achievements in biomass utilization and technology innovation have promoted the sustainable development of biomass industry.

Key research domains in forestry of China: Forest ecology include: (1) key ecological process and effects of forest ecosystem; (2) formation, maintenance and conservation of forest biodiversity; (3) response, adaptation and restoration of forest ecosystem under changing environment; (4) forest health and ecological regulation; (5) forest ecosystem management and multi-objective major forestry ecological engineering; (6) forest and vegetation structure in arid regions; (7) pattern and process of main biological elements in forest soil; (8) ecological process and mechanism of forest soil drive by roots; (9) response and adaptation of forest soil to global environment change; (10) forest soil health and ecosystem services; (11) construction of digitization, information and modeled of dynamic management system of forest soil; (12) water carbon and nitrogen coupling relationship of forest ecosystems; (13) feedback effect of forest growth to climate change; (14) micro-meteorological mechanism ofoutput and quality of important economic forest; (15) physical characteristics of urban forest boundary layer and mechanism of mitigating the heat island effect, and (16) the regional climate effect of forestry ecological engineering and its influencing mechanism should be key priorities.

Silviculture and Forest Genetics and Tree Breeding include: (1) cultivation of forest with high carbon storage; (2) quality increase of forests; (3) afforestation in tough sites; (4) long-term maintenance of plantation productivity; (5) construction of green and high-efficient economic forests; (6) development of high-efficient bioenergyforests; (7) realizing the multi-functions and diversification of forests; (8) strengthening the layout of forest types; (9) accurate evaluation of growth, quality, resistance and adaptation in new germplasm; (10) breeding new varieties met multiple objectives; (11) mechanisms of polyploidy and heterosis for growth, quality traits and resistance in improved trees; (12) analysis of complex economic traits by GWAS and QTL;

(13) innovation in genetic engineering and cell engineering technology; (14) construction of breeding population and advanced seed orchards, and (15) developing key techniques of vegetative propagation, large-scale propagation and standardized propagation for improved varieties.

Forest management and protection include (1) site quality assessment and potential analysis techniques; (2) multi-objective forest collaborative optimization techniques at different scales; (3) precise forest forecasting and management response; (4) the aero-space-ground integrated monitoring and evaluation technology of forest quality; (5) construction of high-precision forest resources information system; (6) major forest pest monitoring and early warning techniques; (7) physiological, biochemical and molecular mechanisms of forest pest disaster; (8) bioecological mechanism and control techniques of forest pest disaster; (9) occurrence and damage characteristics and control techniques of forest biological hazards under climate change conditions, and (10) alien pest risk assessment, early warning and control techniques.

Wood science and forest chemical industry include (1) biological and chemical basis for wood formation and material improvement; (2) key technologies of production and management in material saving and consumption reducing, safe production, and pollution prosecution for wood industry; (3) key technologies for green production in surface decoration, environmentally friendly adhesives, wood protection and modification, and pulping of low value materials; (4) research on wood plastic composites and wood bamboo composites; (5) Health and safety performance testing and evaluation techniques of wood residential material; (6) chemical processing and energy conversion of woody resource; (7) key technology development and demonstration in high-efficiency utilization of non-woody resources; (8) technology integration of new bio-based materials and chemical products; (9) screeningactive ingredients and effective utilization of forestry resources, and (10) clean pulping technology using low-quality and combined materials.

25. Plant Protection Science

The report on the development of Plant Protection Science (2016—2017) mainly outlined the current status and research progresses achieved in plant pathology, entomology, biological control, invasive biology, allelopathy, pesticide science, weed science and rodent science in China since 2014, in order to promote the development and innovation of crop protection in the country, to provide pure and applied supports for safeguarding ecosystem, bio-safety, food security, qualified and safe agricultural products in the country, to promote efficient agriculture and farmers increasing income, to achieve our magnificent goal of building comprehensive well-off society.

Under the grim situation of continuous sever infestation of agricultural pests, firmly established and carried out the concept of innovation, coordination, green development, opening up and sharing, kept promoting agricultural supply-side structural reform promotion as the main line, held ensuring national food security, effective major agricultural product and increased farmer's income as the main task, took enhancing the quality and efficiency and competitiveness as the center, focused on cost efficiency, quality and safety, green development, and constantly improved our independent innovation ability, raised collaborative innovation to its new heights, speeded technology transformation and application, a number of important innovative achievements have been generated in plant protection science since the 12th Five-Year Plan. Remarkable progresses have been made in the basic and frontier research, crucial technologies for pest management, technology transfer and industrialization, technological innovation base and platform construction, innovative talent brought up, significantly advanced China's overall stands of plant protection science, enhanced supporting capabilities in managing agricultural pests. The main characteristics of its development are reflected as followings:

Firstly, significant breakthroughs in basic research have been made. Especially, substantial successes and fruitful results have been achieved in the areas of the toxicity variation and virulent mechanisms of agricultural pests, the origin and unique domestication process of weedy rice, molecular mechanisms underlying the competition and ecological adaptation of Echinochloa crus-galli, the tolerant mechanisms in host plants, the exploration and utilization of new resistant genes, the pesticide molecular synthesis, information technology, and other emerging technologies related to the basic and applied research for the prevention and management of agricultural pests, a good number of the research findings have been published in *Cell*, *Nature*, *Science*, *Nature Communication*, *Nature Biotechnology*, *Nature Genetics*, *Proceedings of the National Academy of Sciences* and *the ISME Journal*, *PLoS Pathogens*, *Current Biology* and *Trends in Plant Science*, further internationally expand the academic influence and discourse in some disciplines of Crop Protection Science. Considering the breakthrough achievements, Chinese plant pathologists, entomologists, and invasive biologists were invited to publish review articles in Annual Review of Entomology and Annual Review of Phytopathology, that certainly further enhanced our independent innovation capability, demonstrated our standing and discourse power in international crop protection, brought Chinese crop protection from the trace mainly into the tracing, parallel and pioneer coexisted new era.

Secondly, hi-tech research has been accelerated with extraordinary success. Researches in discovering novel genes resistant to plant disease and their mechanisms, genetic control of important insect pests, resistance management of target pests (e.g. Helicoverpa armigera) on Bt cotton, and volatile pheromone from female cotton bollworm, have explored and provided new approaches for managing vital agricultural pests. New approaches oriented on screening full-length functional proteins directly from microorganisms had been created. Targeting on inducing crop to raise its immunity, 8 plant immune inducting proteins, which were able to promote the immunity and growth of crops, were discovered from over 10000 proteins isolated from more than 1000 fungi. An integrated screening and evaluating system with multiple functions and quotas for plant immune inducting proteins had been established, that promoted the research and development of plant immune inducting protein products. Put forward the diapauses regulation technology for important natural enemies, significantly extended the storage period of natural enemy products; serialized and standardized some biological control products, laid the firm foundation for production expansion and application. The discipline of Invasive Biology provided a technological basis for the management of major invasive insect pests such as oriental Bactrocera dorsalis Hendel, Cydia pomonella (L.) and Bemisia tabaci (Gennadius) by using gene editing technology, and pointed out the direction for the continued innovation of environment-friendly management. Allelopathy, good successes in regulating behavior of pests and their natural enemies by using crop volatiles were achieved; with allelochemicals and signal molecules as leading compounds, a number of novel agrochemicals for pest management were developed by structure modification and activity evaluation. Rice varieties with chemical regulation functions were creatively bed, and approved, which showed important significance for weed management and herbicide reduction in paddy fields. Significant progresses in the establishment of high throughput screening and computation chemi-biology platform, in vitro and in vivo testing system, and the application of nanotechnology, Gig data and gene editing technology in pesticide discovery have been made. Such as the establishment of PEG mediated model for screening South Rice black-streaked dwarf based on virus (SRBSDV), which can quickly and effectively screen anti SRBSDV chemicals; building of computation chemi-biology platform for molecular designing of green pesticides; the realization in the macro regulation of pesticide droplet assembly adhesion by weak interaction of dual molecular surfactants, provided a new approach for pesticide reduction and efficiency. In the pesticide application techniques, new productive technology for producing dry seed coating agents and a series of new products were succeeded, using of high efficient catalyst, first prepared microcapsule suspension seed coating agent of fipronil and chlorpyrifos, breaking the difficulty of microencapsulation for solid active ingredients; developed capsule fumigant application technology, and comprehensively evaluated the efficacy and environmental behavior of fumigant capsule preparation; created a high efficiency and low risk technique system of pesticides, awarded the second-place prize of National Science and technology progress in 2016; developed China's oil-powered, single rotor unmanned helicopter, with large carrying capacity and long endurance, remote controlled low altitude and low volume pesticide application for pest management was achieved with the cooperation of different organizations. The research in weed science has clearly defined the herbicide resistant weeds in rice, wheat and other crop fields, as well as non cultivated lands, provided scientific basis for herbicide resistant weed management. In rodent management, through further standardize the collection of basic data, adjust the basic research, the basic realization of the rodent pest control technology combined with monitoring and early warning, firmly upheld the gradually realized and data based precise prediction and forecast for rodent management.

Thirdly, notable progresses have been made on key technologies. It has expounded that the infestation reason and the outbreak mechanism of rice planthopper, rice black-streaked dwarf viral disease (RBSDV) and Stripe leaf blight in the main cropping areas in China, integrated the sustainable prevention and management techniques; established high efficiency and low risk technology systems for pesticide application, discovered a new anti-virus fungicide with independent intellectual property rights, that enhanced the immunization prevention technology on managing rice virus disease and its vector insects in China; established a highly efficient system for agro-antibiotic strains, excavated and cultivated a batch of new bio-control resources and natural enemies, for the first time revealed rheochrysidin, a natural anthraquinone fungicide, from the plant metabolic products for crop disease management, pest management using biological control as the core has begun to take shape. In view of biological invasion, a series of early warning, monitoring and intercepting techniques have been developed, confirmed the invasive species and their risk levels, innovated quantitative risk assessment for invasive alien species, a breakthrough in molecular detection, species identification, and port quarantine for important plant pathogens has been made. Broken out the confined uses of metolachlor herbicide, having long been believed that can only be used as dryland herbicides at home and abroad. Overcome the key technologies for safety evaluation of preemergence herbicide mixtures, R&D in botanical safeners and mycoherbicide for paddy field weed management, achieved safe and high effective paddy weed management in South of China, creatively put forward the safe and effective system. By developing of bait trapping system suitable for different environments and requirements, improved the safe utilization, efficacy, validity and manual placing efficiency of chemical rodenticides. Research on infertile and ecological regulation technologies further improved the proportion and role of environment friendly technology in the integrated rodent management system.

Comparing with the developed countries, a certain gap in crop protection remains, such as the systematic, integrity and prospective. We propose that top-level design, development layout and strategic planning for crop protection science should be taken into consideration from the national level during the 13th Five-Year Plan - 14th Five-Year Plan, in order to systematically evaluate virulence variations of agricultural pests, crop resistance mechanisms and multi-interaction mechanisms among crop-pest-mediator, chemical information communication and information feedback, from multiple dimensions of molecule, cell, individual and population, to identify, clone, and analyze virulent genes and pest resistant genes, further improving our independent innovation ability and international academic standing. Having made full good use of modern information technology, such as internet and mega data, to build three-dimensional, diversified and comprehensive platform of monitoring and early warning, and gradually achieve the network management for remote diagnosis, real-time monitoring, early warning and emergency management, to consolidate the foundation of improving the timeliness and accuracy epidemic monitoring and warning.

Pest management should be focused on green development in the future, by further strengthening the usage reduction and high efficiency of pesticides, to develop high efficiency and low risk of green pesticides, to pay more attention to the protection and utilization of natural enemies, to implement diversified measures of RNA interference, physical control, biological control, ecological control and other non chemical green techniques, in order to gradually realize sustainable pest management for the crops, to maintain the diversity of eco-system, to promote agriculture production coordinated with ecology development, to prevent of biological invasion and spread, to provide sound technical supports for the green agriculture development.

26. Soil and Water Conservation & Desertification Combating

Soil and water resources are at the foundation of humanity's multiplication and are the conditions of irreplaceable material base on the process of social development and progress. The sustainable use of soil and water conservation practices and maintain the sustainable development of ecological environment are the basic requirement for a sustainable development of our society and economic. The severe soil and water loss not only causes resource destruction, ecological and environmental deterioration, aggravated all kinds of natural disasters and poverty and the crisis of national ecology, but also strongly restricted the sustainable development of social economy. As the irreplaceable basic resources and critical precondition for sustainable development, soil and water resources are the major constraint for urgent solution with the implementation of the strategy of sustainable development. Soil and water conservation is closely related to human survival and development. Soil and water conservation has the function of disasters prevention, natural resources protection, ecology improvement, economic promotion, etc., making it has a unique advantage and important status in promoting ecological, economic and social sustainable development.

Soil and water conservation practices are indispensable and effective means of protecting and rationalizing soil and water resource, maintaining and improving the ecological environment, which are an important guarantee for sustainable development. The discipline of soil and water conservation and desertification combating (hereinafter referred to as soil and water conservation discipline) as a multidisciplinary and interdisciplinary subject, the research areas include soil erosion mechanism and process regulation, space allocation of shelterbelt system, occurrence and control of desertification, ecology protection and engineering afforestation of construction projects, and so on. The soil and water conservation discipline in our country starts as earlier as 1958. The first master program of soil and water conservation was established at 1981, and the first Ph.D program in 1984 in China. The soil and water conservation discipline was affirmed as the first national key discipline in 1989, and further affirmed as national key construction discipline in 2001.

During the 12th Five-Year Plan, the Chinese government has adopted a series of measures in improvement of soil and water conservation legislation, preventing and control of soil erosion and protection of water resources. Watershed management at local scale has been nationwide conducted mainly by using the small watershed unit and integrated planning work of different types of ecosystems and landscapes. This ensures the accelerated development of ecological civilization construction and sustainable economic and social development. In October 2015, the State Council approved the "national soil and water conservation planning (2015—2030) ". This planning thoroughly analyzed the state of the art and the developing trend of the soil erosion prevention and control in China. It raises an overall layout and the main monitoring tasks at the national level in soil and water conservation zoning, key water and soil erosion control area and soil and water conservation goals. The monitoring works of soil and water conservation and the capacity of related institutions has been accordingly well promoted, Furthermore, the inclusion of indicators of soil erosion warning, and integrating the increased area of soil and water conservation into the national green development index are the important basis for the evaluation of ecological civilization construction at national level. The report of the 19th National Congress of the Communist Party of China proposed many measures for ecological civilization construction, such as implementing comprehensive reclamation of river basin and offshore area, determining the boundaries for ecological protection red line, permanent capital farmland, and urban development, and improving the rehabilitating system of cultivated land, grassland, forest, river and lake. These measures were all based on the existing soil conservation polices and ecological environment management. The comprehensively systematic governance of mountain, water, forest, cropland, lake and grass in the report especially highlights the critical position of soil and water conservation in ecological civilization construction.

Soil and water conservation is not only the cornerstone of green development, but also an important part of China's ecological civilization. Soil and water conservation is regarded as "work in the contemporary, beneficial future generations". In the next 10~20 years, China will build a basically comprehensive prevention and control system for water and soil loss compatible with China's economic and social development and basically achieve prevention and protection. Soil and water loss in key areas will be effectively protected, and the ecology will be further improved. We will improve the comprehensive prevention and control system for soil and water loss compatible with China's economic and social development so as to achieve overall prevention and protection. Soil and water loss in key areas for prevention and control will be fully governed and a virtuous cycle of ecology will be achieved. The key areas are eight areas, namely, the northeast black soil area, the north windy and sandy area, the north rocky mountain area, the northwest loess plateau area, the southern red earth area, the southwest purple soil area, the southwest karst area and the Qinghai Tibet plateau area. We should vigorously strengthen prevention and protection, push forward to comprehensive management, improve the level of monitoring and informatization, build the demonstration area meticulously, and strive to build institutional mechanisms compatible with the requirements of ecological civilization construction, and promote the systematic governance of "landscapes, forests, lakes and grasslands".

In accordance with the norms of the Subject Development Report of the Chinese Association for Science and Technology, we systematically reorganize the follow contents including history, existing conditions, existing problems, the development of this subject and the tightness between cultivating talents and the social services of this subject. Moreover, we review the theory, technical research and major applying results of soil and water conservation, in aspect of soil erosion, karst rocky desertification, mountain disaster prevention, forestry ecological engineering. This report strives to cover the overall progress and advanced results of soil and water conservation subject.

Due to the limited time, omissions and improper descriptions are unavoidable. We earnestly request criticism from peer institutions and readers public.

27. Tea

Tea science is a comprehensive discipline researching on tea germplasm and breeding, tea cultivation, tea plant protection and tea processing, etc. The goal of this discipline is to lay the foundation for the development of tea industry.

Tea science in China has been developing rapidly in recent years. Researchers have made joint efforts in carrying out frontier research focusing on various fields of tea. This report is a review report on progress and advance in tea science in the past seven years (2010—2016), and a forecast of the development trend in near future.

In the field of tea germplasm resources, the national germplasm resources nursery Hangzhou and Menghai branch have collected more than 3400 tea germplasms from home and abroad. Following on the ex-situ conservation of tea germplasm, new advances have been achieved on in-situ conservation, establishment of evaluation and appraisal technical guides, analysis of the genetic diversity on the morphological, chemical and DNA level, sampling strategy in the conversation biology of tea germplasm. "Zijuan" (a tea tree with purple buds) in Yunnan province, "Zhonghuang" (a tea tree with yellow buds) series in Zhejiang province and "Huangjincha" (a tea tree with high content of amino acids) series in Hunan province are the successful cases of special tea germplasm development and utilization. The mining of stable and main QTLs, functional genes, and beneficial alleles of important agronomic traits, the biosynthesis pathway of catechins and alkaloids provides the possibility of molecular breeding in tea plant.

In tea plant molecular biological studies, Chinese scientists were the first to have completed the whole genome sequencing of tea tree in the world. Studies on tea secondary metabolism revealed two-step galloylation of a catechin molecule and its regulatory genes. Functional characterization and expression patterns of multiple structural and regulatory genes in phenylpropanoid and flavonoid pathways have been conducted. Genes encoding N-methyltransferases in tea have been examined for their roles in caffeine biosynthesis. Both glutamine and ethylamine were found as indispensable precursors for tea theanine biosynthesis, which can be catalyzed by glutamine synthase.

Molecular mechanisms controlling tea terpenoid volatiles have been studied and implementation of linalool/nerolidol synthase gene in the two terpenoid volatile biosynthesis in tea plants has been demonstrated, with involvement of alternative transcription splicing regulation. Signaling mechanisms underlying tea plant responses to stress conditions with involvement of miRNA regulation have started to be revealed. Transcriptomic changes due to herbivoery and pathogen damages have been explored in depth. In addition, some tea genes related to abiotic stress resistance such as E3 ubiquitin ligase and dehydrin genes have also been studied. Moreover, molecular mechanisms of reproductive biology in tea plants have started to be elucidated and ubiquitination pathway and Ca+2 signaling pathway were found to be involved.

In tea plant breeding, the diversified demand of tea consumers led to the changes in research objectives towards diversification, precision and specialization. In order to obtain broad and excellent mutants for breeding selection, both traditional and new technologies were applied to innovate breeding materials. F1 generations with prominent heterosis were obtained by applying "artificial crossing with double-clone tea plants". Mutants with excellent characters including resistances to pests and diseases, as well as drought and cold hardness, low caffeine, high fragrance were obtained by applying new technologies such as ^{60}Co-gamma ray radiation technique, space mutation technique and transgenic technique. The application of metabonomics, proteomics and chromatographic techniques allows quality identification in tea breeding to be more accurate. The pest-resistance tea breeding was promoted by applying chemical ecology technology based on the interaction between tea plant and insects, resistance enzyme identification and related gene labeling technology. The identification of photosynthetic vitality and computer modelization technology improved the prediction accuracy of new cultivars' yield. During 2010 to 2016, 37 new clonal tea cultivars were authorized by national level identification, 56 new clonal tea cultivars were authorized by approval, identification or registration of provincial level, and 37 new clonal tea cultivars were authorized with new plant variety rights. The rate of tea fields cultivated with improved clonal tea cultivars reached 58.6% of national total tea fields at the end of 2016.

In tea cultivation, the in-depth study of tea soils showed some positive results. Molecular biological approaches are increasingly applied in studies of the community and evolvement of microorganism in tea soils. Meanwhile advancements in the evaluation of soil quality, the mechanism and ameliorative measures of soil acidification have been investigated. Recent progresses in the area of nutrition have greatly deepened our understandings of the impacts of nutrients such as nitrogen, phosphorus and potassium on the metabolism of quality components. The characteristics of nutrient absorption were elucidated at molecular and physiological levels, and putative genes of several nutrient transporters were cloned. Adoption of mechanical fertilization, fertigation and controlled releasing chemical fertilizers were recommended as measures of proper nutrient management to improve utilization efficiency while the environmental impacts of fertilization such as emission of greenhouse gases gained considerable attention. On the other hand, progresses were made in the field of safety of tea products as heavy metals and rare earth elements in the soils and their accumulation in tea plants are concerned.

In tea plant protection, the main diseases and insect species of tea plant in China were reacquainted by molecular methods and morphological identification. The efficiency of physical control on pests was dramatically improved. By digitizing the most effective color, the standardized and efficient color sticky traps were developed for trapping tea leafhoppers. LED insecticidal lamp was developed according to the different spectrum of phototaxis between insect pests and natural enemies in tea plantation. Important progress was made in rational application of chemical insecticides on tea plants. The concept was first internationally proposed that the level of pesticide residue in tea infusion should be regarded as the evaluation index of pesticide safety and the drafting principle of maximum residue limits for pesticides in tea. A high-risk warning was proposed in China's tea industry about the high water-soluble pesticide, imidacloprid and acetamiprid. Meanwhile, several highly effective and low water-soluble pesticides were screened out as substitutes, and have been used broadly in the major tea areas of China.

In tea processing, systematic research have been made on the theory and upgrading of tea processing technology, developing new equipments, exploring new products and so on, in order to saving labor and improving product quality. A batch of continuous automatic tea processing production lines of green tea, black tea and oolong tea has been developed. The microbial fermentation technology of dark tea and color selection technology of made tea have made significant breakthroughs, and meanwhile the research on basic theory of tea processing has made massive progress.

In tea deep-processing, a series of new technologies such as enzyme engineering, ultrasonic extraction, microwave extraction, ultrahigh pressure extraction, subcritical, supercritical CO_2 extraction technology, reverse osmosis membrane enrichment, rotating centrifugal concentration, scraper film forming technology, low negative pressure evaporation technology, have been successfully used in the extraction of tea functional components. In terms of the production technology of instant tea, a series of new extraction, concentration, drying and forming technology and equipment have been developed, which has laid a better technical foundation for improving the flavor quality of instant tea. In the process of tea beverage production, various enzyme preparations have been successfully used to improve the concentration of tea juice and the taste quality and prevent the formation of sediment in tea beverages. The instant tea and RTD tea with some health benefits became the new development trend. EGCG, theanine, tea plant flower, tea seed oil were listed in the new food resource catalogue. Tea food, functional tea leisure food, tea personal care products, tea wine and other new products of tea deep processing have been developed successfully, and have become trendy modern consumption.

In tea and human health, the research on the preventing function and mechanism of tea drinking has picked up its pace in recent years. Active areas of research involves cancer prevention, prevention of cardiovascular disease, lipid lowering and weight reducing effect, prevention of diabetes, antibacterial and antivirus action and prevention of neuro-degeneration disease of human. And the health-protecting effects of tea are attributed to the groups of catechins, flavonoids and its glycosides, theaflavins, thearubigins, anthocyanidin, purines, phenolic acid and polysaccharides.

In tea quality and safety, rapid progress has been made in the quality standard, detection technology and risk assessment of tea. China's tea standards have been revised for three times, and the maximum pesticide residue limits in tea have increased to 48 items, which is more stricter than the CAC standard. New progress has been made in efficient and accurate detection of pesticide residue. The improved QuChERS method was used to improve the purification effect of tea matrix. Some novel materials such as carbon nanotubes, Fe_3O_4 nanoparticles, grapheme, were used in tea matrix purification. The determination of a total of 653 pesticide residues in tea was achieved by using GC-MS/MS and HPLC-MS/MS. The method based on high-resolution mass spectrometry (TOF and orbitrap) for the multiresidue analysis of 272 pesticides in tea was also established. The chiral separation of enantiomers of epoxiconazole, indoxacarb, acephate and methamidophos was completed. ICP-OES and ICP-MS have become routine methods for elemental detection and are also used in element morphology and valence state studies. At the same time, study on the aroma compounds in tea became a hot topic. The application of electronic tongue, electronic nose, near infrared spectroscopy and other technologies in the tea quality analysis has deepened. Risk assessment of tea quality and safety has been launched in China. The dietary exposure of Chinese tea has been established. Risk assessments of pesticides such as emamectin benzoate and chlorpyrifos have been conducted with low safety risks. The fluorine and rare earth elements in tea were assessed. The rare earth limit has been removed from China's new national standard.

In the tea industry, the e-commerce has developed fast, and the relation between online and offline channels has changed from competition to mutual integration, which has enhanced tea consumption. With the expansion of the tea consumption market in China, the tea consumption structure has been constantly adjusting, in which young consumers have become a potential driving force. In the international market, the health and plant quarantine measures (SPS) have become main obstacles to Chinese tea exports; and industrial technology upgrading is the fundamental way to break through trade barriers. Against the background of increasing tea production, the effect of tea production cost has become the common challenge of the world's major tea producing countries. As a result, they hope to improve tea production efficiency, and reduce production costs through technical innovation and tea mechanization.

In terms of global development of tea sciences, tea genetic transformation technology, functional genome studies, the precision breeding technology, and the mechanized production technology (such as mechanical picking, tillage and fertilizing, pest-controlling etc.) in China are still lagged behind. These research fields need to see reinforced interdisciplinary research and high-tech applications. Moreover, with China entering the new era of socialism with Chinese characteristics, it would be of the high priority to strengthen the research of green production, tea branding, tea circulation channels change, aiming to serve the industry development.

28. Toxicology

Toxicology is the study of the adverse effects of chemical, physical and biological factors on living organisms and environmental ecosystems and their mechanisms, assessing and predicting the hazards of these factors to human health and environmental systems, eventually providing a scientific basis to establish relevant standards, management protocols and preventive measures.

In recent years, with the rapid development of industrial and agricultural production, emerging technologies and socio-economy, various harmful factors have been significantly increased, posing a threat to human health and the environmental systems. These have greatly promoted the development of toxicology. With the progress of emerging technologies such as cell biology, biochemistry and molecular biology, system biology and translational medicine, as well as the interdisciplinary integration of toxicology and chemistry, physics, mathematics, materials science and computational science, modern toxicology has been deeply influenced and promoted, resulting in several new toxicology disciplines such as toxicogenomics, toxicoproteomics, computational toxicology and translational toxicology. In 2007, the National Research Council released a report entitled "Toxicity Testing in the 21st Century: A Vision and A Strategy" (TT21C). The report addresses the challenges posed by current animal-based toxicology research and risk assessment, and points out that the focus of toxicology development in the 21st century should shift from animal tests to non-animal testing methods.

28.1 Outline of the development of Toxicology

In China, the modern toxicology started from 1950s, and it has been experiencing a rapid development stage from 1990s. Especially in recent years, along with the increase of people's demand for public health and the progress in life science theories and technologies such as "-Omics", the development and applications of basic researches and the related technologies of toxicology have drawn more and more attention from the government and the public. At present, toxicology in China has covered many fields such as human health, environment, industry, agriculture and military, and has developed various sub-disciplines, including industrial toxicology, food toxicology, drug toxicology, environmental and ecological toxicology, etc.

Toxicology plays a more prominent role in the economic and social development and national security. It has become an important support for public health and environmental safety, responses to public health emergencies and disaster rescues, sustainable development of economy, as well as maintaining national defense and social stability. Researches in toxicology show rapid growth, and the research level and quality are steadily increasing. We have made our contribution to the development of international toxicology.

28.2 The Main Progress and Achievements of Toxicology

Toxicology focuses on the detrimental interactions between environmental factors (especially chemicals) and living organisms, and its development is synchronized with that of life sciences. New theories and technologies of life sciences promote the development of toxicology, and toxicology has also become the tool and key to open the health mysteries for life sciences. The impact of environmental factors on human health has also promoted the development of toxicology.

28.2.1 Overview of Toxicology Discipline Construction

In China, toxicology education and training has been booming in recent years. There are more than 100 universities particularly medical schools to provide toxicology education for undergraduate students, and more than 60 universities or institutes provide toxicology postgraduate education and training program. Every year, hundreds of postgraduates start their toxicology careers in toxicology research, education, application and management in different institutions. CST and its 26 professional committees also conducted their annual high-grade on-the-job trainings, such as the training course named "Modern Toxicology: Essentials and Progresses" organized by CST. CST, as the largest academic and research organization of toxicology globally, consists of 26 professional committees and almost fourteen thousand members.

Chinese toxicology researchers have made notable achievements in scientific researches in recent years. Papers published in the field of toxicology account for 20% of global publications during 2016—2017. More importantly, they have made great contributions in many aspects like poison first-aid after accidental big explosions, emergency treatment of acute mass poisoning, emergency response of nuclear weapon crisis in North Korea, prevention and treatment of ambient air pollution, and so on.

28.2.2 Advances in Basic Toxicology Research

In recent years, basic toxicology research has rapidly developed and made great scientific progress. Basic toxicology research aims to study the mechanisms of regulation, oxidative stress, toxicity pathway, autophagy, receptor-ligand interactions, DNA damage and repair, and biomarkers, which plays an important role in elaborating the toxic effects of environmental chemicals. The findings of the key regulatory pathways and targets provide critical clues for revealing the target organ toxicity and disease prevention and control. Over the years, this field has gained enormous progress, mainly in the following aspects.

Redox regulation is a crucial mechanism in a variety of physiological processes, and the imbalance of redox homeostasis plays a pivotal role in a battery of pathological processes. Studies have shown that in addition to classic antioxidant enzymes, some novel antioxidant enzymes like thioredoxins, glutaredoxins and metallothionein can also participate in the regulation of redox balance. Chinese scientists have done a number of important studies and made great contributions in this field.

At present, chemical management is largely based on results from a battery of in vivo tests. In the future, management should be based on the results from alternative methods. Adverse outcome pathways (AOPs) delineate the documented, plausible, and testable processes by which a chemical induces molecular perturbations and the associated biological responses that describe how the molecular perturbations cause effects at the sub-cellular, cellular, tissue, organ, whole animal, and population levels of observation. It gives rise to the sense of complexity in toxicity. Particularly in the case for longer term health endpoints where effects are the result of multiple events, accumulates over time or are particular to a life stage of the organism.

In recent years, Chinese investigators have carried out a series of studies on environmental endocrine disrupting chemicals (EDCs) -induced health hazards and mechanisms by the application of molecular biology, genomics, and metabonomics. They have found a series of important results, such as the effects and mechanisms of EDCs-caused reproductive toxicity, biotransformation and transportation of EDCs, the application of computational toxicology to investigate EDCs-induced health hazards.

Researchers have made significant progress in screening, identifying and applying related biomarkers of chemicals and toxicants by using "-omics" and other new technologies in recent years. For instance, it is found that Incenp, Rpgripl, Hapln2 and Cxadr genes can be used as susceptible markers of liver injury induced by acetaminophen in rats. Beta 2-MG, alpha 1-MG and RBP can be used as early diagnostic indicators of mercury poisoning. In the application of biomarkers, the integrated biomarker responses (IBR) method was applied to carry out risk assessment of marine, electronic waste, groundwater and other environmental pollution.

In the past two years, great effort and significant progress has been made in analytic toxicology in China. Featured analytical methodologies have been developed for resolving unidentified toxic compositions, while isotope tracers have been applied to study the environmental fate and bioaccumulation of metal-containing engineered nanoparticles. Bioassay methods with independent intellectual property rights have also been established and promoted to accomplish rapid screening of environmental health-threatened pollutants. Computational toxicology has emerged recently but already plays a key role nowadays in realizing the promise of the toxicity testing in the 21st century. Structure-activity relationship study is still thriving, and great effort has been made in toxicity prediction of chemical mixtures. Various molecular simulation techniques have been used to successfully elucidate the interactions of toxicants with proteins, nucleic acids and even pulmonary surfactant monolayer. Exposure science and AOP is badly in need of methodology innovation.

28.2.3 Advances in Applied Toxicology

With the rapid development of economy since the 21st century, environmental pollution, food safety problems become more and more serious, which resulted in the rapid development of toxicology science in China. In terms of environmental safety, researchers began to lay emphasis on studies about the native species toxicology and environmental exposure parameters. With the purpose to develop local environmental benchmark doses which accords with the situation of our country. Besides, extensive toxicological mechanism studies provide sufficient scientific evidences to the formulation and improvement of a variety of laws and regulations established recently. With the establishment of the national food safety risk assessment center, the food safety science in China developed fast. In recent 5 years, the clinical medicine adverse reaction monitoring system in China have been significantly improved, drug toxicology studies also achieved gratifying progress. In addition, the detection methods constantly updated lead to the toxicogenomics research entering a big data era. The rapid development of modern biotechnology, and the proposals of "the Toxicity test of the 21st Century: Vision and Strategy" and "adverse outcome pathways" promoted toxicology research to a higher level.

28.2.4 The Progress of Regulatory Toxicology

Regulatory toxicologists have heavily involved in the recent revisions of some national laws and regulations. In the last few years, several national laws and regulations in public health and environment safety have been revised significantly which include the "Food safety law of the People's Republic of China" in October 2015, the "Regulations for the Implementation of the Drug Administration Law of the People's Republic of China" in February 2016, "Regulations on the Safety Management of Hazardous Chemicals" in December 2011, "Measures for Environmental Management of New Chemical Substances" in October 2010, "Regulation on Pesticide Administration" in June 2017 and "Measures for registration of pesticides" in August 2017. Many of these revisions have reflected the progress of regulatory toxicology and regulatory science.

A series of guidance, guidelines and standards have been issued as well for implementation and execution of each newly revised laws and regulations. For examples, Ministry of Agriculture has issued more than 10 guidance, guidelines and standards for pesticide testing, risk assessment and data quality.

Among the significant changes in the field is the shift of Good Laboratory Practice (GLP) made by different agencies. CFDA has made many significant revisions on its "Quality management specifications for Non-Clinical Drug Studies" in 2017 so that the GLP is more consistent with US and EU, while the GLP of Ministry of Agriculture has broadened the field to cover efficacy, residues, field studies in addition to health toxicity and environment fate. In a different direction, Ministry of Environment Protection has discontinued its GLP precertification program.

28.3 Prospects and Developments of Toxicology in China

Toxicology discipline in China now has already a solid academic foundation and integrated scientific system of branches. In particular, the concept of "broad toxicology" has been deeply rooted in many fields such as environment and health, medicine, military and national defense safety, etc. And it is more relevant to people's daily life and has been endowed much more social responsibilities. Standing on the new historical point of social development in China, the development of toxicology disciplines has also entered a new era. It is obligated to further establish the integrate toxicology disciplinary system covered of research, education and management, and settle the toxicology a right and suitable discipline position in China.

Toxicology research is a holistic chain of research activity paradigm composed of toxicological effects, toxicological mechanisms, risk assessment, intervention and prevention. Toxicological research should aim at the key practical and scientific problems that affect the environment ecology, health and safety. It should be focused not only on the environmental pollutants but also on the key populations facing health risk with the toxicology related problems. Action should be taken to draw up the national program of toxicology, and the technology roadmap of toxicology research. Several important issues should be taken into consideration in future toxicology research planning, which include: to create and optimize new modern techniques of toxicity assessment; promote the innovation and development of toxicity testing alternatives techniques; continuing support the researches of mechanistic toxicology and targeting effects; deepening the research of important environmental pollutants; strengthening the assimilation of biological technology and knowledge and further raising the science and technology levels of risk assessment strategies and management.

29. Public Health and Preventive Medicine

Public Health and Preventive Medicine is critical to successfully achieving national goals of Healthy China. Marvelous progresses have been observed in this discipline in the past years. We briefly summarize some of the advances in the following.

In 2016—2017, a number of large prospective cohorts including nature population cohorts and specific disease cohorts have been initiated by Chinese government, which are expected to provide critical evidence of causes or prognostic determinants of various diseases especially NCDs. For example, with the rapid transformation of China's rapid economic growth, population aging and life style, the incidence of stroke has been rising year by year. The population epidemiological survey and the national sentinel surveillance data show that the mortality of stroke in both urban and rural areas in China has declined in recent years. The Chinese government attaches great importance to the prevention and control of stroke. In December 2016, the National Health and Family Planning Commission, the State Administration of Traditional Chinese Medicine lauched a "work plan" for comprehensive prevention and treatment of stroke. Several important studies provide evidences for precise treatment of cerebrovascular disease. In the meantime, significant progress has been made in the development of diseases surveillance system, public health information system and high-throughput omics techniques, which generates a substantial amount of populations based big data. Updated analytical methods and interpretation of big data have become the hottest research topics in the epidemiology, which further translates into public health policies and practical guidelines.

Emerging infectious diseases (EIDs) have risen substantially over time and are dominated by zoonoses which most originate from wildlife. After the SARS outbreak in 2003, substantial improvements for EIDs response and research in China and other countries have been made in surveillance, sharing of information, outbreak investigation, clinical management, and laboratory virology and discovery. The response by the Chinese authorities to H7N9 infections in China has shown how much the national and international response has improved, and the rapid progress on the research of Ebola, Zika, and MERS, etc., have been witnessed. However, epidemiological, pathological, etiological, immunological studies are needed to understand the origins, transmission dynamics, prevent and control measures of EIDs to human beings. The global spread of infectious diseases is also facilitated by the ability of infected humans to travel thousands of miles in short time spans, rapidly transporting pathogens to distant locations, therefore the interdisciplinary studies, including medicine, geography, statistics, mathematics, computer science, etc., are also needed to quantify and mitigate the risk of global dispersal and locally transmission. Specifically, rapid and precise identification of the causative pathogen for an EID is not only beneficial for the clinical treatment for the patients, but also essential for the development of prevention measures and control strategies. Recently, along with the fast development of various biomedical techniques, numerous new methodologies for pathogen determination have emerged. Some techniques have finished the laboratory test and evaluation, and entered into practical and clinical application. Those techniques mainly include isothermal amplification technology, high-throughput multiple pathogenic detection techniques, next generation sequencing, etc. In particular, great progress have been observed in the fields of detection and analysis of the whole genome of pathogens.

In the last two years, China has achieved significant progresses in environmental health studies. There were a number of studies on toxicological mechanism of air pollution on the respiratory system, cardiovascular system, as well as on carcinogenicity, glycometabolism, etc. There were numerous epidemiological studies on air pollution, including panel studies, time-series studies, interventional studies and cohort studies. These studies have preliminarily established national exposure-response associations in China and further provided abundant evidence on the biological plausibility on the health effects of air pollution. For the health effects of weather variations and climate change, studies have been expanded to the biological mechanisms of temperature-related health effects, the health effects of extreme weather events and the impacts on insect-borne infectious diseases. For the health effects of chemical toxicants in drinking water, studies focused on some disinfection by-products, toxic microelements, antibiotics, antibiotic resistance genes in drinking water. Some epidemiological studies evaluated the reproductive and developmental toxicity of disinfection by-products and the hepatotoxicity of microcystin. Notably, these studies have provided scientific basis for the formulation of standards for drinking water quality.

Toxicology is a basic science in public health. Significant progress in toxicological research in China has been achieved in risk assessment, toxicological effects of $PM_{2.5}$, nanomaterials, selective solvents and insecticides, as well as establishment of research platforms for toxicologinomics. Animal models and interventional trials have been introduced into toxicological studies.

Women and children's health care (MCH) ensures and promotes women and children's well-being and quality of life. "Healthy China 2030" blueprint highlights the strategies of Women and children's health care. In China, current researches of women's health are focusing on birth defects prevention, prenatal testing and diagnosis, management of maternal health care, developmental origins of health and disease (DOHaD), prevention and control of cervical and breast cancer, and women's mental health. The recent efforts on children's health care have been emphasized on lowering neonatal mortality, prevention and control of children's nutritional diseases, premature health care, early childhood development and mental health.

Child and Adolescent Health or School Health aims at protecting, promoting and enhancing the physical and mental health of school age children, which is a key component of preventive health. The "Health China 2030" planning outline highlights the importance of the prevention and control of myopia and obesity among students, and indicated that students should spend at least 1 hour in in-school physical activity per day in 2030. The United Nations Sustainable Development Goals included multiple indicators related to child and adolescent health.

Mental health is a cross-disciplinary area that focuses on improving mental health of the whole population. There have been remarkable progress in research, practice and promotion of mental health in China including diagnostic classification of mental disorders, psychiatric epidemiology, utilization of new technologies, delivery and evaluation of mental health services.

Health education and health promotion has played important roles in protection of human health. In 2016, the 9th WHO Global Conference on Health Promotion was held in Shanghai, China. In this conference, health promotion was regarded as the center of the Sustainable Development Goals (SDG). Two outcome documents of the conference, the Shanghai Declaration on Promoting Health in the 2030 Agenda for Sustainable Development and the Shanghai Consensus on Healthy Cities offer pathways for advancing health promotion and addressing the determinants of health through good governance, healthy cities, health literacy and social mobilization. The third edition of the textbook of Heath Education as one of national planned textbooks was published in 2017. China central government and China Commission of Health and Family Planning issued several important documents to promote the development of health promotion and health education. Under such political supportive environment, the China Institute of Health Education and China Center for Disease Prevention and Control organized and implemented a series of programs and big action plans of health promotion and health education such as healthy cities and healthy townships, monitoring and promoting health literacy, and China Healthy Lifestyle for All. China also actively applied internet and social media into health promotion and health education.

Social Medicine is an interdisciplinary of studying medicine and health from a social perspective. In the evolving momentum for Universal Health Coverage (UHC) and Sustainable Development Goals (SDGs), the research on social medicine has achieved new awareness and progress in primary health care, social determinants of health, and healthy aging, which further emphasizes the dynamic health of whole life cycle and intergeneration, transformation from medical care to preventive care, the coordination and integration of health services delivery at all levels, and the participation and governance of society as whole. Meanwhile, the research on social medicine also explores the mechanism and path of social determinants, as well as new research methods such as social network analysis. Chinese researchers of social medicine focus on the key issues and priorities of healthcare reform in China, including the healthcare equity of vulnerable and priority population; the prevention of chronic diseases and social diseases and integrated health services delivery model of medical care and elderly care for dealing with challenges of urbanization and aging; studies on community health management, system of general practitioners, tiered healthcare delivery system, and people-centered healthcare system for achieving "strengthening capacity of primary health facilities, ensuring the access to basic health services, establishing the mechanism of primary health care"; exploring the application of new methods in health monitoring and measuring and internet technology in tiered health services system and health management. Simultaneously, the Chinese researchers of social medicine have also explored the study topics of social development and health and transferred the research outcomes into policy recommendations and interventions.

Global health is an emerging area of public health. Global health discipline devotes itself to dealing with health problems that need the joint efforts of the international community across borders, such as climate, aging, health unfairness, burden of disease and so on, therefore it is necessary for multi-disciplinary and multi-field participants to work together. The global health discipline has been developing rapidly in Western countries. Since 2010, some top universities in China have also set up a global health research and teaching institutions, conducted a diversity of teaching and research activities, and set up exchanges and cooperation, published journals and textbooks, as well as conducted field studies and trial practice in the global settings. In general, China's global health discipline is closely linked to the global development agenda and national foreign strategy, and has great potential for development. However, the discipline is still in its infancy in China and needs tremendous efforts in theory, methodology and practice.

Public health informatics (PHI) focuses on health data collection, storage, analysis and usage with various subject characteristics and technology, thereby contributing to raising the level of health and sanitation, preventing disease and injury, protecting the susceptible as well. In China, information construction for public health in disease control and prevention, emergency, supervision and the praxis of regional population health information platform has been implemented with remarkable achievements. A series of public health information standards have been published so as to direct information collection and exchange, data management, quality control and statistical analysis normatively. More measures are made to solve the problems of network information security. Since People's Republic of China Network Security Law formally implemented on June 1st, 2017, it calls for a full range of the classified protection of information security, such as classification, filing, rectification and evaluation. However, there is still a gap in information security management, the implementation of systems and sustained funding in particular.

30. The Science of Science and Technology Policies

30.1 Introduction

The science of science and technology policies is a multidisciplinary on the nature of science and technology policies, the way they are produced and their effects on science, technology and innovation.

Since the reform and opening up, China's science and technology policy studies have been greatly developed. For 40 years, the development of Chinese science and technology policy studies unceasingly, has formed the theoretical foundation, the discipline system and a professional team,which and played an important role in science and technology, the social and economic development of China.

30.2 Recent advances in science and technology policy studies

This study identifies recent progress in scientific and technological policy research based on the following methods: (1) scientific and technological policy research projects funded by scientific funds; (2) articles published in academic journals.

30.2.1 The frontiers of international science and technology policy research

According to the U.S.National Science Foundation SciSIP analysis from 2007 to 2016 ,the frontiers of international science and technology policy research can be summarized in the following aspects: (1) Science and technology policy issues, including the return on investment of science and technology, the organizational structure of scientific knowledge production and productivity, the role of the university and government in technology transfer and innovation, innovation incentive mechanism and achievement transformation. (2) Science and innovation issues including innovation, patent, open innovation and technology transfer, innovation policy and regulation. (3) Governance of science and technology innovation ,includes political power, professional skills and ethics. (4) Method tools and data collection and processing.

According to 16 international academic journals related to science and technology policy selected, high frequency keywords are as following: innovation, this research selected bibliometrics and citation analysis, patents, patent R&D, cooperation, open innovation, H index, scientific measurement, etc.

30.2.2 The hotspots of international science and technology policy research in China

During 2011—2016, China's science and technology policy research has made great progress. The main research frontier topics include indigenous innovation, collaborative innovation, technological innovation and innovation, network strategic emerging industries, and innovative ecosystem.

According to the academic journals related to science and technology policy research selected from 2012—2016, the 10 frequency keywords are as following: technology innovation, influencing factors, innovation performance, collaborative innovation, strategic emerging industries, science and technology innovation, innovation, the industrial cluster, intellectual property and indigenous innovation.

The studies in the such major areas have made important progress as sociology of science, scientometrics, science evaluation, scientific and technological personnel, technical innovation, regional innovation, innovation and entrepreneurship, scientific infrastructure. For example, the China Regional Innovation Ability Evaluation Report beginning in 1999 issued by China university of science and technology development strategy research group and University of Chinese academy has published nearly 20 years, 2015 series to be incorporated in national innovation survey system series report.

30.3 Comparison of China and foreign researches on science and technology policy

Compared with other advanced countries, the development of Chinese science and technology policy research has made outstanding achievements, although it started late .In the selected 16 international academic journals, the numbers of papers by Chinese scholars published had increased from 21 in 2007 more than 100 articles of the last four years. Despite the small amount, the increase was faster and more than eight times more than a decade ago. At the same time, the research contents and methods of the papers also show the trend of diversification, indicating that the research results of China's science and technology policy research are growing in the international academic circle. A group of Chinese science and technology policy researchers have also obtained high quality research results, which have a great impact in the world of science and technology policy.

On international research hotspot, the Chinese science and technology policy research hotspot are similar to the international research hotspot, but there are also a lot of differences, reflecting an effort by chinese researchers to pursue international frontier science and technology has made considerable achievements, but also shows that there exists a large gap.

30.4 Development trends and prospect of science and technology policy research

30.4.1 Development trends of science and technology policy research

The future development of science and technology policy studies will present several trends as following: (1) Attaching importance to the frontier issue of science and technology policy research. (2) Intersection of science and technology policy research and innovation research is increasingly close. (3) The focus on social, environmental and sustainable development will be more and more important. (4) The trend of multi-scientific crossover research will be enhanced. (5) Combination of qualitative and quantitative methods will be enhanced. (6) The study of Chinese practices will be further strengthened.

30.4.2　Prospects of science and technology policy research

The future focus of the development of science and technology policy are as following: (1) Science and technology policy development mode under the strategy of scientific and technological powers. (2) The mode, management and benefit of scientific and technological input. (3) Construction and governance of national innovation system. (4) Cultivation, use and policy of scientific and innovative talents. (5) Subject foundation and methodology of science and technology policy.

30.4.3　Measures for development science and technology policy research

In the future, China needs take measures to promote the development science and technology policy research as following: (1) To study and formulate the long-term plan and road map of China's science and technology policy research from the long-term development needs of the country. (2) Setting up special funding schemes to support scientific and technological policy research. The National Natural Science Foundation of China (NSFC) should has established a special funding scheme to support scientific and technological policy research in a long-term and stable manner. (3) Strengthen the basic construction of the disciplines and promote the cross-integration of disciplines. (4) Strengthen the research on Chinese practice and promote the emergence of Chinese school. (5) Strengthen the construction of science and technology policy research team and strengthen the technology policy education. (6) Strengthen communication and cooperation with technology policy makers. (7) Strengthen the platform construction of science and technology policy. (8) Strengthen international exchanges and cooperation.

附 件

附件 1：118 个学科优先发展领域

学部	优先发展领域数量	优先发展领域
数理科学部	15	数论与代数几何中的朗兰兹（Langlands）纲领；微分方程中的分析；几何与代数方法、随机分析方法及其应用；高维/非光滑系统的非线性动力学理论、方法和实验技术；超常条件下固体的变形与强度理论；高速流动及控制的机理和方法；银河系的集成历史及其与宇宙大尺度结构的演化联系；恒星的形成与演化以及太阳活动的来源；自旋、轨道、电荷、声子多体相互作用及其宏观量子特性；光场调控及其与物质的相互作用；冷原子新物态及其量子光学；量子信息技术的物理基础与新型量子器件；后 Higgs 时代的亚原子物理与探测；中微子特性、暗物质寻找和宇宙线探测；等离子体多尺度效应与高稳运行动力学控制
化学科学部	13	化学精准合成；高效催化过程及其动态表征；化学反应与功能的表界面基础研究；复杂体系的理论与计算化学；化学精准测量与分子成像；分子选态与动力学控制；先进功能材料的分子基础；可持续的绿色化工过程；环境污染与健康危害中的化学追踪与控制；生命体系功能的分子调控；新能源化学体系的构建；聚集体与纳米化学；多级团簇结构与仿生
生命科学部	15	生物大分子的修饰、相互作用与活性调控；细胞命运决定的分子机制；配子发生与胚胎发育的调控机理；免疫应答与效应的细胞分子机制；糖/脂代谢的稳态调控与功能机制；重要性状的遗传规律解析；神经环路的形成及功能调控；认知的心理过程和神经机制；物种演化的分子机制；生物多样性及其功能；农业生物遗传改良的分子基础；农业生物抗病虫机制；农林植物对非生物逆境的适应机制；农业动物健康养殖的基础；食品加工、保藏过程营养成分的变化和有害物质的产生及其机制
地球科学部	12	地球观测与信息提取的新理论、技术和方法；地球深部过程与动力学；地球环境演化与生命过程；矿产资源和化石能源形成机理；海洋过程及其资源、环境和气候效应；地表环境变化过程及其效应；土、水资源演变与可持续利用；地球关键带过程与功能；天气、气候与大气环境过程、变化及其机制；日地空间环境和空间天气；全球环境变化与地球圈层相互作用；人类活动对环境和灾害的影响
工程与材料科学部	18	亚稳金属材料的微结构和变形机理；高性能轻质金属材料的制备加工和性能调控；低维碳材料；新型无机功能材料；高分子材料加工的新原理和新方法；生物活性物质控释/递送系统载体材料；化石能源高效开发与灾害防控理论；高效提取冶金及高性能材料制备加工过程科学；机械表面界面行为与调控；增材制造技术基础；传热传质与先进热力系统；燃烧反应途径调控；新一代能源电力系统基础研究；高效能高品质电机系统基础科学问题；多种灾害作用下的结构全寿命整体可靠性设计理论；绿色建筑设计理论与方法；面向资源节约的绿色冶金过程工程科学；重大库坝和海洋平台全寿命周期性能演变
信息科学部	15	海洋目标信息获取、融合与应用；高性能探测成像与识别；异构融合无线网络理论与技术；新型高性能计算系统理论与技术；面向真实世界的智能感知与交互计算；网络空间安全的基础理论与关键技术；面向重大装备的智能化控制系统理论与技术；复杂环境下运动体的导航制导一体化控制技术；流程工业知识自动化系统理论与技术；微纳集成电路和新型混合集成技术；光电子器件与集成技术；高效信号辐射源和探测器件；超高分辨、高灵敏光学检测方法与技术；大数据的获取、计算理论与高效算法；大数据环境下人机物融合系统基础理论与应用

续表

学部	优先发展领域数量	优先发展领域
管理科学部	15	管理系统中的行为规律；复杂管理系统分析、实验与建模；复杂工程与复杂运营管理；移动互联环境下交通系统的分析优化；数据驱动的金融创新与风险规律；创业活动的规律及其生态系统；中国企业的变革及其创新规律；企业创新行为与国家创新系统管理；服务经济中的管理科学问题；中国社会经济绿色低碳发展的规律；中国经济结构转型及机制重构研究；国家安全的基础管理规律；国家与社会治理的基础规律；新型城镇化的管理规律与机制；移动互联医疗及健康管理
医学科学部	15	发育、炎症、代谢、微生态、微环境等共性病理新机制研究；基因多态、表观遗传与疾病的精准化研究；新突发传染病的研究；肿瘤复杂分子网络、干细胞调控及其预测干预；心脑血管和代谢性疾病等慢病的研究与防控；免疫相关疾病机制及免疫治疗新策略；生殖—发育—老化相关疾病的前沿研究；基于现代脑科学的神经精神疾病研究；重大环境疾病的交叉科学研究；急救、康复和再生医学前沿研究；个性化药物的新理论、新方法、新技术研究；中医理论的现代科学内涵及其对中药发掘的指导价值研究；个性化医疗关键技术与转化研究；多尺度多模态影像技术与疾病动物模型研究；智能化医学工程的创新诊疗技术研究
跨科学部优先发展领域	16	介观软凝聚态系统的统计物理和动力学；工业、医学成像与图像处理的基础理论与新方法、新技术；生物大分子动态修饰与化学干预；手性物质精准创造；细胞功能实现的系统整合研究；化学元素生物地球化学循环的微生物驱动机制；地学大数据与地球系统知识发现；重大灾害形成机理及其减灾对策；新型功能材料与器件；城市水系统生态安全保障关键基础科学问题；电磁波与复杂目标/环境的相互作用机理与应用；超快光学与超强激光技术；互联网与新兴信息技术环境下重大装备制造管理创新；城镇化进程中的城市管理与决策方法研究；从衰老机制到老年医学的转化医学研究；基于疾病数据获取与整合利用新模式的精准医学研究

附件2：48个重大专项

序号	专项名称	序号	专项名称
1	重点基础材料技术提升与产业化	9	干细胞及转化研究
2	材料基因工程关键技术与支撑平台	10	高性能计算
3	畜禽重大疫病防控与高效安全养殖综合技术研发	11	公共安全风险防控与应急技术装备
4	大科学装置	12	国家质量基础的共性技术研究与应用
5	大气污染成因与控制技术研究	13	海洋环境安全保障
6	蛋白质机器与生命过程调控	14	化学肥料和农药减施增效综合技术研发
7	地球观测与导航	15	精准医学研究
8	典型脆弱生态修复与保护研究	16	粮食丰产增效科技创新

续表

序号	专项名称	序号	专项名称
17	量子调控	33	先进轨道交通
18	林业资源培育及高效利用技术创新	34	现代服务业共性关键技术研发及应用示范
19	绿色建筑及建筑工业化	35	现代食品加工及粮食收储运技术与装备
20	煤炭清洁高效利用和新型节能技术	36	新能源汽车
21	纳米科技	37	增材制造与激光制造
22	农业面源和重金属污染农田综合防治与修复技术研发	38	战略性先进电子材料
23	七大农作物育种	39	智能电网技术与装备
24	全球变化	40	智能机器人
25	深海关键技术与装备	41	智能农机装备
26	生物安全关键技术研发	42	中医药现代化研究
27	生物医用材料研发与组织器官修复替代	43	重大科学仪器设备开发
28	生殖健康及重大出生缺陷防控研究	44	重大慢性非传染性疾病防控研究
29	食品安全关键技术研发	45	重大自然灾害监测预警与防范
30	数字诊疗装备研发	46	深地资源勘查开采
31	水资源高效开发利用	47	政府间国际科技创新合作
32	网络空间安全	48	云计算和大数据

附件3：2016－2017年香山科学会议学术讨论会一览表

序号	会议号	主题会议	召开日期
		2016年香山科学会议学术讨论会一览表	
1	553	自旋波电子学物理、材料与器件	2016-2-23
2	558	聚集诱导发光	2016-4-26
3	562	中国地基大口径光学/红外望远镜的科学与技术发展战略	2016-5-24
4	568	实验动物科技创新的关键科学问题与技术	2016-8-30
5	570	高功率高光束质量半导体激光技术发展战略	2016-9-22
6	571	沉积学发展战略国际研讨会	2016-9-26
7	572	高能环形正负电子对撞机–中国发起的大型国际科学实验	2016-10-18
8	573	空间碎片监测清除前沿技术与系统发展	2016-10-20
9	574	发展人工智能，引领科技创新	2016-10-25
10	575	幼龄反刍动物早期培育的关键科学问题及实践应用	2016-10-30
11	576	数字科技文献资源长期保存的前沿及重大问题研讨	2016-11-3
12	577	农业生态功能评估与开发	2016-11-10
13	578	心理学与社会治理	2016-11-15
14	579	中药安全性研究基础与方法	2016-11-17
15	580	热电转换：分子材料的新机遇与挑战	2016-11-22

续表

序号	会议号	主题会议	召开日期
\multicolumn{4}{c}{2016年香山科学会议学术讨论会一览表}			
16	581	创新驱动新食品资源健康产业发展	2016-11-24
17	582	中国微生物组研究计划	2016-12-1
18	583	生命分析化学	2016-12-6
\multicolumn{4}{c}{2017年香山科学会议学术讨论会一览表}			
1	586	准极限高能光源带来的新科学新技术	2017-1-17
2	587	生物多样性:从数据驱动到知识前沿	2017-2-28
3	588	非人灵长类脑与认知	2017-3-11
4	589	区域生态学学科建设与环境问题解决路径探索	2017-3-28
5	590	沸石分子筛:等级特性、选择催化与分子工程	2017-3-30
6	591	植物特化性状形成及定向发育调控	2017-4-11
7	592	类脑智能机器人的未来:神经科学与机器人深度融合的关键科学问题探讨	2017-4-18
8	593	异种移植走向临床研究的关键科学问题	2017-4-20
9	S34	高重复频率硬X射线自由电子激光的科学机遇与技术挑战	2017-4-22
10	594	中医临床原创思维的科学内涵及应用	2017-4-26
11	596	精准医学视野下现代中医基础理论:肝脏象理论创建探索	2017-5-24
12	598	北极海洋在全球变化中的作用及其对中国的影响	2017-6-14
13	599	加强科技评估,助力创新驱动发展	2017-6-22
14	600	生物气溶胶与人类健康、国家生物安全及大气污染	2017-6-29
15	601	新时期国民营养与粮食安全	2017-7-22
16	602	营养健康科学前沿与产业服务	2017-8-29
17	603	中国海水提铀未来发展研讨会	2017-9-7
18	604	计算光学成像科学基础研究:机遇和挑战	2017-9-19
19	606	纳米酶催化机制与应用研究	2017-10-12
20	607	组织再生材料:从基础研究创新到临床转化应用	2017-10-15
21	608	化合物半导体器件的异质集成与界面调控	2017-10-23
22	609	脊髓损伤再生修复的关键科学问题	2017-10-26
23	S36	人工智能技术、伦理与法律的关键科学问题探讨	2017-11-11

附件4：2016—2017年未来科学大奖获奖者

年度	奖项	获奖者	获奖理由
2016	生命科学奖	卢煜明	奖励他基于孕妇外周血中存在胎儿DNA的发现，在无创产前胎儿基因检查方面做出的开拓性贡献
2016	物质科学奖	薛其坤	奖励他在利用分子束外延技术发现量子反常霍尔效应和单层铁硒超导等新奇量子效应方面做出的开拓性工作
2017	生命科学奖	施一公	表彰他在解析真核信使RNA剪接体这一关键复合物的结构，揭示活性部分及分子层面机理的重大贡献
2017	物质科学奖	潘建伟	奖励他在量子光学技术方面的创造性贡献，使基于量子密钥分发的安全通信成为现实可能
2017	数学与计算机科学奖	许晨阳	表彰他在双有理代数几何上做出的极其深刻的贡献

附件5：2016年度"中国科学十大进展"

序号	进展名称
1	研制出将二氧化碳高效清洁转化为液体燃料的新型钴基电催化剂
2	开创煤制烯烃新捷径
3	揭示水稻产量性状杂种优势的分子遗传机制
4	提出基于胆固醇代谢调控的肿瘤免疫治疗新方法
5	揭示RNA剪接的关键分子机制
6	发现精子RNA可作为记忆载体将获得性性状跨代遗传
7	研制出首个稳定可控的单分子电子开关器件
8	构建出世界上首个非人灵长类自闭症模型
9	揭示胚胎发育过程中关键信号通路的表观遗传调控机理
10	揭示水的核量子效应

附件6：2016—2017年度"中国十大科技进展新闻"

序号	进展名称
	2016年度"中国十大科技进展新闻"
1	成功发射世界首颗量子科学实验卫星"墨子号"
2	全球最大单口径射电望远镜（FAST）在贵州落成启用

续表

序号	进展名称
3	长征五号首飞成功
4	神舟十一号飞船返回舱成功着陆，2名航天员安全回家
5	领衔绘制全新人类脑图谱
6	我国首获超算应用最高奖
7	率先破解光合作用超分子结构之谜
8	"海斗"号无人潜水器创造深潜纪录
9	利用超强超短激光成功获得"反物质"
10	首次揭示水的核量子效应
2017年度"中国十大科技进展新闻"	
1	我国科学家利用化学物质合成完整活性染色体
2	国产水下滑翔机下潜6329米刷新世界纪录
3	世界首台超越早期经典计算机的光量子计算机诞生
4	国产大型客机C919首飞
5	我国首次海域天然气水合物试开采
6	我国"人造太阳"装置创造世界新纪录
7	中国科学家首次发现突破传统分类新型费米子
8	量子通信"从理想王国走到现实王国"
9	中科院推出高产水稻新种质
10	"悟空"发现疑似暗物质踪迹

附件7：2016—2017年度国家科学技术进步奖获奖项目目录

2016年度国家科学技术进步奖获奖项目

特等奖1项（通用项目）

序号	编号	项目名称
1	J-236-0-01	第四代移动通信系统（TD-LTE）关键技术与应用

一等奖8项（通用项目）

序号	编号	项目名称
1	J-21701-1-01	北京正负电子对撞机重大改造工程
2	J-234-1-01	IgA肾病中西医结合证治规律与诊疗关键技术的创研及应用
3	J-223-1-01	前置前驱8挡自动变速器（8AT）研发及产业化
4	J-236-1-01	DTMB系统国际化和产业化的关键技术及应用
5	J-21702-1-01	互联电网动态过程安全防御关键技术及应用
6	J-25201-1-01	航天重大工程的遥感空间信息可信度理论与关键技术

续表

序号	编号	项目名称
7	J-230-1-01	新一代国家时间频率基准的关键技术与应用
8	J-222-1-01	生态节水型灌区建设关键技术及应用

创新团队 3 项

序号	编号	团队名称
1	J-207-1-01	第四军医大学消化系肿瘤研究创新团队
2	J-207-1-02	浙江大学能源清洁利用创新团队
3	J-207-1-03	中国农业科学院作物科学研究所小麦种质资源与遗传改良创新团队

二等奖 120 项（通用项目）

序号	编号	项目名称
1	J-201-2-01	多抗稳产棉花新品种中棉所 49 的选育技术及应用
2	J-201-2-02	辣椒骨干亲本创制与新品种选育
3	J-201-2-03	江西双季超级稻新品种选育与示范推广
4	J-202-2-01	农林生物质定向转化制备液体燃料多联产关键技术
5	J-202-2-02	三种特色木本花卉新品种培育与产业升级关键技术
6	J-202-2-03	林木良种细胞工程繁育技术及产业化应用
7	J-203-2-01	我国重大猪病防控技术创新与集成应用
8	J-203-2-02	针对新传入我国口蹄疫流行毒株的高效疫苗的研制及应用
9	J-203-2-03	功能性饲料关键技术研究与开发
10	J-203-2-04	中国荷斯坦牛基因组选择分子育种技术体系的建立与应用
11	J-203-2-05	节粮优质抗病黄羽肉鸡新品种培育与应用
12	J-204-2-01	躲不开的食品添加剂——院士、教授告诉你食品添加剂背后的那些事
13	J-204-2-02	了解青光眼 战胜青光眼
14	J-204-2-03	《全民健康十万个为什么》系列丛书
15	J-204-2-04	《变暖的地球》（影片）
16	J-205-2-01	机械化秸秆还田技术与装备
17	J-205-2-02	变压器潜伏性缺陷的油中气体检测技术及应用
18	J-205-2-03	热轧带钢柱塞式层流冷却系统研发及应用
19	J-206-2-01	中国机械工业集团科技创新工程
20	J-206-2-02	中船集团高端海洋装备科技创新工程
21	J-210-2-01	南海北部陆缘深水油气地质理论技术创新与勘探重大突破
22	J-210-2-02	延长油区千万吨大油田持续上产稳产勘探开发关键技术
23	J-210-2-03	古老碳酸盐岩勘探理论技术创新与安岳特大型气田重大发现
24	J-211-2-01	果蔬益生菌发酵关键技术与产业化应用
25	J-211-2-02	金枪鱼质量保真与精深加工关键技术及产业化
26	J-211-2-03	造纸与发酵典型废水资源化和超低排放关键技术及应用
27	J-211-2-04	中国葡萄酒产业链关键技术创新与应用
28	J-212-2-01	支持工业互联网的全自动电脑针织横机装备关键技术及产业化
29	J-212-2-02	苎麻生态高效纺织加工关键技术及产业化
30	J-212-2-03	干法纺聚酰亚胺纤维制备关键技术及产业化
31	J-213-2-01	大型乙烯装置成套工艺技术、关键装备与工业应用

续表

序号	编号	项目名称
32	J-213-2-02	大型高效水煤浆气化过程关键技术创新及应用
33	J-213-2-03	阿维菌素的微生物高效合成及其生物制造
34	J-214-2-01	超薄信息显示玻璃工业化制备关键技术及成套装备开发
35	J-214-2-02	水泥窑高效生态化协同处置固体废弃物成套技术与应用
36	J-214-2-03	高性能玻璃纤维低成本大规模生产技术与成套装备开发
37	J-215-2-01	高效低耗特大型高炉关键技术及应用
38	J-215-2-02	重型装备大型铸锻件制造技术开发及应用
39	J-215-2-03	铵盐体系白钨绿色冶炼关键技术和装备集成创新及产业化
40	J-215-2-04	电弧炉炼钢复合吹炼技术的研究应用
41	J-215-2-05	底吹熔炼–熔融还原–富氧挥发连续炼铅新技术及产业化应用
42	J-215-2-06	红土镍矿生产高品位镍铁关键技术与装备开发及应用
43	J-215-2-07	材料海洋环境腐蚀评价与防护技术体系创新及重大工程应用
44	J-216-2-01	大型重载机械装备动态设计与制造关键技术及其应用
45	J-216-2-02	钎料无害化与高效钎焊技术及应用
46	J-216-2-03	航天大型复杂结构件特种成套制造装备及工艺
47	J-216-2-04	大功率船用齿轮箱传动与推进系统关键技术研究及应用
48	J-216-2-05	复杂表面热功能结构形貌特征设计与可控制造关键技术
49	J-21701-2-01	面向铀矿与环境的核辐射探测关键技术、设备及其应用
50	J-21701-2-02	250MW级整体煤气化联合循环发电关键技术及工程应用
51	J-21701-2-03	风电机组关键控制技术自主创新与产业化
52	J-21702-2-01	电网大面积污闪事故防治关键技术及工程应用
53	J-21702-2-02	新能源发电调度运行关键技术及应用
54	J-21702-2-03	配电网高可靠性供电关键技术及工程应用
55	J-21702-2-04	大型汽轮发电机组次同步谐振/振荡的控制与保护技术、装备及应用
56	J-219-2-01	用于集成系统和功率管理的多层次系统芯片低功耗设计技术
57	J-219-2-02	高动态星敏感器技术与工程应用
58	J-220-2-01	新一代立体视觉关键技术及产业化
59	J-220-2-02	高性能光伏发电系统关键控制技术与产业化应用
60	J-220-2-03	高安全成套专用控制装置及系统
61	J-220-2-04	网络交易支付系统风险防控关键技术及其应用
62	J-220-2-05	大型风电水电机组低频故障诊断关键技术及应用
63	J-220-2-06	高性能系列化网络设备研制与应用
64	J-221-2-01	大跨空间钢结构关键技术研究与应用
65	J-221-2-02	大型复杂结构在线混合试验关键技术及应用
66	J-221-2-03	城市高密集区大规模地下空间建造关键技术及其集成示范
67	J-221-2-04	深部隧（巷）道破碎软弱围岩稳定性监测控制关键技术及应用
68	J-221-2-05	广州塔工程关键技术
69	J-221-2-06	高速铁路标准梁桥技术与应用
70	J-222-2-01	高混凝土坝结构安全关键技术研究与实践
71	J-222-2-02	长距离输水工程水力控制理论与关键技术
72	J-223-2-01	国家内河高等级航道通航运行系统关键技术及应用

续表

序号	编号	项目名称
73	J-223-2-02	滨海地区粉细砂路基修筑与长期性能保障技术
74	J-223-2-03	基于耦合动力学的高速铁路接触网/受电弓系统技术创新及应用
75	J-223-2-04	跨江越海大断面暗挖隧道修建关键技术与应用
76	J-231-2-01	城市循环经济发展共性技术开发与应用研究
77	J-231-2-02	难降解有机工业废水治理与毒性减排关键技术及装备
78	J-231-2-03	国家环境分区－排放总量－环境质量综合管控关键技术与应用
79	J-231-2-04	国家环境质量遥感监测体系研究与业务化应用
80	J-231-2-05	三江源区草地生态恢复及可持续管理技术创新和应用
81	J-233-2-01	高危非致残性脑血管病及其防控关键技术与应用
82	J-233-2-02	基于磁共振成像的多模态分子影像与功能影像的研究与应用
83	J-233-2-03	恶性血液肿瘤关键诊疗技术的创新和推广应用
84	J-233-2-04	慢性肾脏病进展的机制和临床防治
85	J-233-2-05	中国脑卒中精准预防策略的转化应用
86	J-233-2-06	结直肠癌个体化治疗策略创新与应用
87	J-234-2-01	国际化导向的中药整体质量标准体系创建与应用
88	J-234-2-02	中草药DNA条形码物种鉴定体系
89	J-234-2-03	益气活血法治疗糖尿病肾病显性蛋白尿的临床与基础研究
90	J-234-2-04	中医治疗非小细胞肺癌体系的创建与应用
91	J-235-2-01	化学药物晶型关键技术体系的建立与应用
92	J-235-2-02	基因工程小鼠等相关疾病模型研发与应用
93	J-235-2-03	瑞舒伐他汀钙及制剂产业化新制备体系的构建与临床合理应用
94	J-235-2-04	复杂结构天然产物抗肿瘤药物的研发及其产业化
95	J-25101-2-01	设施蔬菜连作障碍防控关键技术及其应用
96	J-25101-2-02	农药高效低风险技术体系创建与应用
97	J-25101-2-03	南方低产水稻土改良与地力提升关键技术
98	J-25101-2-04	东北地区旱地耕作制度关键技术研究与应用
99	J-25101-2-05	水稻条纹叶枯病和黑条矮缩病灾变规律与绿色防控技术
100	J-25103-2-01	黑茶提质增效关键技术创新与产业化应用
101	J-25103-2-02	油料功能脂质高效制备关键技术与产品创制
102	J-25103-2-03	棉花生产全程机械化关键技术及装备的研发应用
103	J-25201-2-01	国家电子政务协同式空间决策服务关键技术与应用
104	J-25201-2-02	航空地球物理勘查技术系统
105	J-25201-2-03	国产陆地卫星定量遥感关键技术及应用
106	J-25201-2-04	国家地理信息公共服务平台（天地图）研发与系统建设
107	J-25201-2-05	大别山东段深部探测与找矿突破
108	J-25202-2-01	急倾斜厚煤层走向长壁综放开采关键理论与技术
109	J-25202-2-02	有色金属共伴生硫铁矿资源综合利用关键技术及应用
110	J-25202-2-03	煤层瓦斯安全高效抽采关键技术体系及工程应用
111	J-25202-2-04	智能煤矿建设关键技术与示范工程
112	J-25301-2-01	胰岛素瘤诊治体系的建立及临床应用
113	J-25301-2-02	炎症损伤控制提高肝癌外科疗效的理论创新与技术突破

续表

序号	编号	项目名称
114	J-25301-2-03	主动脉扩张性疾病腔内微创治疗的研究和应用
115	J-25301-2-04	心脏病微创外科治疗新技术及临床应用
116	J-25301-2-05	基于肛门功能和性功能保护的直肠癌治疗关键技术创新与推广应用
117	J-25302-2-01	头面部严重烧伤关键修复技术的创新与应用
118	J-25302-2-02	牙体牙髓病防治技术体系的构建与应用
119	J-25302-2-03	中国严重创伤救治规范的建立与推广
120	J-25302-2-04	视网膜疾病基因致病机制研究及防治应用推广

2017年度国家科学技术进步奖获奖项目

特等奖2项（通用项目）

序号	编号	项目名称
1	J-21702-0-01	特高压 ±800kV 直流输电工程
2	J-23302-0-01	以防控人感染H7N9禽流感为代表的新发传染病防治体系重大创新和技术突破

一等奖12项（通用项目）

序号	编号	项目名称
1	J-210-1-01	涪陵大型海相页岩气田高效勘探开发
2	J-213-1-01	煤制油品/烯烃大型现代煤化工成套技术开发及应用
3	J-223-1-01	蛟龙号载人潜水器研发与应用
4	J-21701-1-01	600MW超临界循环流化床锅炉技术开发、研制与工程示范
5	J-210-1-02	南海高温高压钻完井关键技术及工业化应用
6	J-223-1-02	复杂环境下高速铁路无缝线路关键技术及应用
7	J-212-1-01	干喷湿纺千吨级高强/百吨级中模碳纤维产业化关键技术及应用
8	J-206-1-01	中国电子网络安全与信息化科技创新工程
9	J-213-1-02	高效甲醇制烯烃全流程技术

序号	编号	团队名称
10	J-207-1-01	中国科学院寒区旱区环境与工程研究所冻土与寒区工程研究创新团队
11	J-207-1-02	袁隆平杂交水稻创新团队
12	J-207-1-03	西安交通大学热质传递的数值预测控制及其工程应用创新团队

二等奖118项（通用项目）

序号	编号	项目名称
1	J-201-2-01	多抗广适高产稳产小麦新品种山农20及其选育技术
2	J-201-2-02	早熟优质多抗马铃薯新品种选育与应用
3	J-201-2-03	寒地早粳稻优质高产多抗龙粳新品种选育及应用
4	J-201-2-04	花生抗黄曲霉优质高产品种的培育与应用
5	J-201-2-05	食用菌种质资源鉴定评价技术与广适性品种选育
6	J-201-2-06	中国野生稻种质资源保护与创新利用
7	J-202-2-01	基于木材细胞修饰的材质改良与功能化关键技术
8	J-202-2-02	竹林生态系统碳汇监测与增汇减排关键技术及应用
9	J-202-2-03	中国松材线虫病流行规律与防控新技术
10	J-203-2-01	重要食源性人兽共患病原菌的传播生态规律及其防控技术
11	J-203-2-02	青藏高原特色牧草种质资源挖掘与育种应用
12	J-203-2-03	民猪优异种质特性遗传机制、新品种培育及产业化

续表

序号	编号	项目名称
13	J-204-2-01	《湿地北京》
14	J-204-2-02	《阿优》的科普动画创新与跨媒体传播
15	J-204-2-03	"科学家带你去探险"系列丛书
16	J-204-2-04	《肾脏病科普丛书》
17	J-204-2-05	《数学传奇——那些难以企及的人物》
18	J-205-2-01	电能表智能化计量检定技术与应用
19	J-205-2-02	航空发动机叶片滚轮精密磨削技术
20	J-210-2-01	三元复合驱大幅度提高原油采收率技术及工业化应用
21	J-211-2-01	干坚果贮藏与加工保质关键技术及产业化
22	J-211-2-02	高性能纤维纸基功能材料制备共性关键技术及应用
23	J-211-2-03	食品和饮水安全快速检测、评估和控制技术创新及应用
24	J-211-2-04	鱿鱼贮藏加工与质量安全控制关键技术及应用
25	J-211-2-05	两百种重要危害因子单克隆抗体制备及食品安全快速检测技术与应用
26	J-212-2-01	工业排放烟气用聚四氟乙烯基过滤材料关键技术及产业化
27	J-213-2-01	高汽油收率低碳排放系列催化裂化催化剂工业应用
28	J-213-2-02	提高轻油收率的深度延迟焦化技术
29	J-213-2-03	锂离子电池核心材料高纯晶体六氟磷酸锂关键技术开发及产业化
30	J-214-2-01	吸附分离聚合物材料结构调控与产业化应用关键技术
31	J-214-2-02	建筑玻璃服役风险检测和可靠性评价关键技术与设备及应用
32	J-214-2-03	冶金渣大规模替代水泥熟料制备高性能生态胶凝材料技术研发与推广
33	J-215-2-01	压水堆核电站核岛主设备材料技术研究与应用
34	J-215-2-02	热轧板带钢新一代控轧控冷技术及应用
35	J-215-2-03	高效节能环保烧结技术及装备的研发与应用
36	J-215-2-04	高强高导铜合金关键制备加工技术开发及应用
37	J-215-2-05	球形金属粉末雾化制备技术及产业化
38	J-216-2-01	高性能数控系统关键技术及产业化
39	J-216-2-02	药剂高效分装成套装备及产业化
40	J-216-2-03	高铁列车用高可靠齿轮传动系统
41	J-216-2-04	重型压力容器轻量化设计制造关键技术及工程应用
42	J-216-2-05	高效切削刀具设计、制备与应用
43	J-21701-2-01	气液固凝并吸收抑制低温腐蚀的烟气深度冷却技术及应用
44	J-21701-2-02	强电磁环境下复杂电信号的光电式测量装备及产业化
45	J-21701-2-03	新一代超低排放重型商用柴油机关键技术开发及产业化
46	J-21702-2-01	支撑大电网安全高效运行的负荷建模关键技术与应用
47	J-21702-2-02	大规模风电联网高效规划与脱网防御关键技术及应用
48	J-21702-2-03	特大型交直流电网技术创新及其在国家西电东送中的应用
49	J-219-2-01	光网络用光分路器芯片及阵列波导光栅芯片关键技术及产业化
50	J-219-2-02	工业智能超声检测理论与应用关键技术
51	J-220-2-01	面向互联网开放环境的重要信息系统安全保障关键技术研究及应用
52	J-220-2-02	税务大数据计算与服务关键技术及其应用
53	J-220-2-03	密码芯片系统的攻防关键技术研究及应用

续表

序号	编号	项目名称
54	J-220-2-04	复杂路网条件下高速铁路列控系统互操作和可靠运用关键技术及应用
55	J-22101-2-01	工业建筑抗震关键技术研究与应用
56	J-22101-2-02	超高层建筑钢骨高强混凝土结构体系抗震关键技术及其应用
57	J-22101-2-03	城市大型地下结构抗震设计理论与方法及工程应用
58	J-22102-2-01	超深等厚度水泥土搅拌墙成套施工装备与技术研发及应用
59	J-22102-2-02	山区大跨度悬索桥设计与施工技术创新及应用
60	J-22102-2-03	高速铁路狮子洋水下隧道工程成套技术
61	J-22102-2-04	新一代运载火箭力学试验与发射测试厂房建造关键技术
62	J-222-2-01	锦屏二级超深埋特大引水隧洞发电工程关键技术
63	J-222-2-02	泥沙、核素、温排水耦合输移关键技术及在沿海核电工程中应用
64	J-222-2-03	中国节水型社会建设理论、技术与实践
65	J-223-2-01	低能耗插电式混合动力乘用车关键技术及其产业化
66	J-223-2-02	深水板桩码头新结构关键技术研究与应用
67	J-223-2-03	新一代交流传动快速客运电力机车研究与应用
68	J-223-2-04	季冻区高速公路抗冻耐久及生态保护关键技术
69	J-230-2-01	超痕量物质精密测量关键技术及应用
70	J-230-2-02	气动元件关键共性检测技术及标准体系
71	J-230-2-03	干旱环境下土遗址保护关键技术研发与应用
72	J-230-2-04	我国检疫性有害生物国境防御技术体系与标准
73	J-231-2-01	填埋场地下水污染系统防控与强化修复关键技术及应用
74	J-231-2-02	流域水环境重金属污染风险防控理论技术与应用
75	J-231-2-03	膜集成城镇污水深度净化技术与工程应用
76	J-231-2-04	高铝粉煤灰提取氧化铝多联产技术开发与产业示范
77	J-231-2-05	危险废物回转式多段热解焚烧及污染物协同控制关键技术
78	J-231-2-06	嵌套网格空气质量预报模式（NAQPMS）自主研制与应用
79	J-23301-2-01	单倍型相合造血干细胞移植的关键技术建立及推广应用
80	J-23301-2-02	肺癌分子靶向精准治疗模式的建立与推广应用
81	J-23301-2-03	肺癌精准放射治疗关键技术研究与临床应用
82	J-23301-2-04	内分泌肿瘤发病机制新发现与临床诊治技术的建立和应用
83	J-23301-2-05	缺血性脑卒中防治的新策略与新技术及推广应用
84	J-23302-2-01	红斑狼疮诊治策略及其关键技术的创新与应用
85	J-23302-2-02	疟疾、血吸虫病等重大寄生虫病防治关键技术的建立及其应用
86	J-234-2-01	中药大品种三七综合开发的关键技术创建与产业化应用
87	J-234-2-02	寰枢椎脱位中西医结合治疗技术体系的创建与临床应用
88	J-234-2-03	中药和天然药物的三萜及其皂苷成分研究与应用
89	J-234-2-04	神经根型颈椎病中医综合方案与手法评价系统
90	J-235-2-01	坎地沙坦酯原料与制剂关键技术体系构建及产业化
91	J-235-2-02	艾滋病诊断、治疗和预防产品的评价关键技术建立与推广应用
92	J-235-2-03	大血管覆膜支架系列产品关键技术开发及大规模产业化
93	J-236-2-01	新型光纤制备技术及产业化
94	J-236-2-02	大规模接入汇聚体系技术及成套装备
95	J-236-2-03	高精度高可靠定位导航技术与应用

续表

序号	编号	项目名称
96	J-251-2-01	花生机械化播种与收获关键技术及装备
97	J-251-2-02	大型灌溉排水泵站更新改造关键技术及应用
98	J-251-2-03	高光效低能耗 LED 智能植物工厂关键技术及系统集成
99	J-251-2-04	全国农田氮磷面源污染监测技术体系创建与应用
100	J-251-2-05	番茄加工产业化关键技术创新与应用
101	J-251-2-06	作物多样性控制病虫害关键技术及应用
102	J-251-2-07	大型智能化饲料加工装备的创制及产业化
103	J-25201-2-01	国家海岛礁测绘重大关键技术与应用
104	J-25201-2-02	航空航天遥感影像摄影测量网格处理关键技术与应用
105	J-25201-2-03	全球 30 米地表覆盖遥感制图关键技术与产品研发
106	J-25201-2-04	全国危机矿山接替资源勘查理论创新与找矿重大突破
107	J-25201-2-05	空间高动态卫星精密定位及其综合测试理论与关键技术及重大应用
108	J-25202-2-01	矿山超大功率提升机全系列变频智能控制技术与装备
109	J-25202-2-02	煤矿深部开采突水动力灾害预测与防治关键技术
110	J-25202-2-03	煤层气储层开发地质动态评价关键技术与探测装备
111	J-25202-2-04	超大规模微细粒复杂难选红磁混合铁矿选矿技术开发及工业化应用
112	J-253-2-01	肝移植新技术——脾窝异位辅助性肝移植的建立与应用
113	J-253-2-02	脑胶质瘤诊疗关键技术创新与推广应用
114	J-253-2-03	胃癌综合防治体系关键技术的创建及其应用
115	J-253-2-04	免疫性高致盲眼病发生的创新理论、防治及应用
116	J-253-2-05	配子胚胎发育研究与生育力改善新方法的应用
117	J-253-2-06	外科术式改变脑血流的基础与临床创新
118	J-253-2-07	骨质疏松性椎体骨折微创治疗体系的建立及应用

附件 8：2016—2017 年度国家自然科学奖项目名单

2016 年度国家自然科学奖项目

一等奖 1 项

序号	编号	项目名称
1	Z-102-1-01	大亚湾反应堆中微子实验发现的中微子振荡新模式

二等奖 41 项

序号	编号	项目名称
1	Z-101-2-01	自适应与高精度数值方法及其理论分析
2	Z-101-2-02	共振情形哈密顿系统的稳定性
3	Z-101-2-03	奇点量子化理论研究
4	Z-101-2-04	Ricci 流理论及其几何应用
5	Z-102-2-01	重离子碰撞中的反物质探测与夸克物质的强子谱学与集体性质研究

续表

序号	编号	项目名称
6	Z-102-2-02	磁电演生新材料及高压调控的量子序
7	Z-103-2-01	生物分子界面作用过程的机制、调控及生物分析应用研究
8	Z-103-2-02	碳–碳键重组构建新方法与天然产物合成
9	Z-103-2-03	有机场效应晶体管基本物理化学问题的研究
10	Z-103-2-04	高效不对称碳–碳键构筑若干新方法的研究
11	Z-103-2-05	氧基簇合物的设计合成与组装策略
12	Z-103-2-06	化学修饰石墨烯可控组装与复合的基础研究
13	Z-103-2-07	具有重要生物活性的复杂天然产物的全合成
14	Z-104-2-01	亚洲季风变迁与全球气候的联系
15	Z-104-2-02	显生宙最大生物灭绝及其后生物复苏的过程与环境致因
16	Z-104-2-03	变化环境下生物膜对海洋底栖生态系统的影响
17	Z-104-2-04	中国东部板内燕山期大规模成矿动力学模型
18	Z-104-2-05	地球动物树成型
19	Z-104-2-06	高风险污染物环境健康危害的组学识别及防控应用基础研究
20	Z-105-2-01	植物小RNA的功能及作用机理
21	Z-105-2-02	水稻产量性状的遗传与分子生物学基础
22	Z-105-2-03	猪日粮功能性氨基酸代谢与生理功能调控机制研究
23	Z-106-2-01	大肠癌发生分子机制、早期预警、防治研究
24	Z-106-2-02	乳腺癌发生发展的表观遗传机制
25	Z-107-2-01	图像结构建模与视觉表观重构理论方法研究
26	Z-107-2-02	视觉场景理解的模式表征与计算理论及方法
27	Z-107-2-03	复杂动态网络的同步、控制与识别理论与方法
28	Z-107-2-04	氧化物阻变存储器机理与性能调控
29	Z-107-2-05	碳基纳米电子器件及集成
30	Z-107-2-06	微波毫米波新型基片集成类导波结构及器件
31	Z-108-2-01	非金属基超常电磁介质的原理与构筑
32	Z-108-2-02	长寿命耐高温氧化/烧蚀涂层防护机理与应用基础
33	Z-108-2-03	基于晶体缺陷调控的铁性智能材料新物理效应
34	Z-108-2-04	纳米结构单元的宏量制备与宏观尺度组装体的功能化研究
35	Z-108-2-05	荧光传感金属–有机框架材料结构设计及功能构筑
36	Z-108-2-06	储能用高性能复合电极材料的构筑及协同机理
37	Z-109-2-01	新型核能系统的中子输运理论与高效利用方法
38	Z-109-2-02	工程结构抗灾可靠性设计的概率密度演化理论
39	Z-109-2-03	高增益电力变换调控机理与拓扑构造理论
40	Z-109-2-04	超快激光微纳制造机理、方法及新材料制备的基础研究

序号	编号	项目名称
41	Z-110-2-01	求解力学中强非线性问题的同伦分析方法及其应用

2017 年度国家自然科学奖项目

一等奖 2 项

序号	编号	项目名称
1	Z-105-1-01	水稻高产优质性状形成的分子机理及品种设计
2	Z-103-1-01	聚集诱导发光

二等奖 33 项目

序号	编号	项目名称
1	Z-101-2-01	微分几何中的几个分析问题研究
2	Z-102-2-01	新型半导体深能级掺杂机制研究
3	Z-103-2-01	低维碳材料的拉曼光谱学研究
4	Z-103-2-02	若干有机化合物结构性质关系及反应规律性
5	Z-103-2-03	新型分子基铁电体的基础研究
6	Z-103-2-04	芳香化合物立体及对映选择性直接转化新策略
7	Z-104-2-01	华北克拉通破坏
8	Z-104-2-02	青藏高原及东北缘晚新生代构造变形与形成过程
9	Z-104-2-03	华夏地块中生代花岗岩成因与地壳演化研究
10	Z-104-2-04	卤代持久性有机污染物环境污染特征与物化控制原理
11	Z-104-2-05	流域径流形成与转化的非线性机理
12	Z-104-2-06	饮用水中天然源风险物质的识别、转化与调控机制
13	Z-105-2-01	飞蝗两型转变的分子调控机制研究
14	Z-105-2-02	植物油菜素内酯等受体激酶的结构及功能研究
15	Z-105-2-03	促进稻麦同化物向籽粒转运和籽粒灌浆的调控途径与生理机制
16	Z-106-2-01	细胞钙信号及分子调控
17	Z-106-2-02	胶质细胞 - 神经元功能耦合与缺血脑保护
18	Z-106-2-03	艾滋病病毒与宿主天然防御因子相互作用新机制的研究
19	Z-107-2-01	仿生机器鱼高效与高机动控制的理论与方法
20	Z-107-2-02	编码混叠成像与计算重建理论方法研究
21	Z-107-2-03	网络化动态系统的分析与控制
22	Z-107-2-04	若干低维半导体表界面调控及器件基础研究
23	Z-107-2-05	预测控制的原理研究与系统设计
24	Z-108-2-01	高质量石墨烯材料的制备与应用基础研究
25	Z-108-2-02	面向太阳能利用的高性能光电材料和器件的结构设计与性能调控
26	Z-108-2-03	金属材料强韧化的内在与外在微纳尺度效应
27	Z-108-2-04	新型磁弹性材料的功能调控、晶体生长和大磁致应变特性研究
28	Z-109-2-01	太阳能光催化制氢的多相流能质传输集储与转化理论及方法
29	Z-109-2-02	聚合物/层状无机物纳米复合材料的火灾安全设计与阻燃机理
30	Z-109-2-03	高速运动刚柔相互作用系统非线性建模与振动分析
31	Z-109-2-04	功能纳米材料和微生物修复难降解有机物和重金属污染湿地新方法
32	Z-110-2-01	先进梯度功能材料的断裂力学研究
33	Z-110-2-02	范德华层状介质的滑移行为和力学模型

注　释

研究与试验发展（R&D）：指在科学技术领域，为增加知识总量、以及运用这些知识去创造新的应用而进行的系统的、创造性的活动，包括基础研究、应用研究和试验发展三类活动。

基础研究：指为了获得关于现象和可观察事实的基本原理的新知识（揭示客观事物的本质、运动规律，获得新发展、新学说）而进行的实验性或理论性研究，它不以任何专门或特定的应用或使用为目的。

应用研究：指为获得新知识而进行的创造性研究，主要针对某一特定的目的或目标。应用研究是为了确定基础研究成果可能的用途，或是为达到预定的目标探索应采取的新方法（原理性）或新途径。

试验发展：指利用从基础研究、应用研究和实际经验中所获得的现有知识，为产生新的产品、材料和装置，建立新的工艺、系统和服务，以及对已产生和建立的上述各项作实质性的改进而进行的系统性工作。

研究与试验发展（R&D）经费：统计年度内全社会实际用于基础研究、应用研究和试验发展的经费支出。包括实际用于研究与试验发展活动的人员劳务费、原材料费、固定资产购建费、管理费及其他费用支出。

研究与试验发展（R&D）经费投入强度：全社会研究与试验发展（R&D）经费支出与国内生产总值（GDP）之比。

研究人员：指 R&D 人员中具备中级以上职称或博士学历（学位）的人员。

R&D 人员全时当量：是国际上通用的、用于比较科技人力投入的指标。指 R&D 全时人员（全年从事 R&D 活动累积工作时间占全部工作时间的 90% 及以上人员）工作量与非全时人员按实际工作时间折算的工作量之和。例如：有 2 个 R&D 全时人员（工作时间分别为 0.9 年和 1 年）和 3 个 R&D 非全时人员（工作时间分别为 0.2 年、0.3 年和 0.7 年），则 R&D 人员全时当量 =1+1+0.2+0.3+0.7=3.2（人年）。

规模以上工业企业：规模以上工业企业的统计范围是年主营业务收入 2000 万元及以上的工业法人单位。

发文量：指被 WOS 核心合集中的三大期刊引文数据库收录的且文献类型为论文（article）和综述（review）的论文数量。

被引频次：指论文被来自 WOS 核心合集论文引用的次数。

专利家族：具有共同优先权的在不同国家或国际专利组织多次申请、多次公布或批准的，内容相同或基本相同的一组专利文献称作专利家族。

影响因子：影响因子（Impact Factor，IF）是汤森路透（Thomson Reuters）出品的期刊引证报告（Journal Citation Reports，JCR）中的一项数据。即某期刊前两年发表的论文在该报告年份（JCR year）中被引用总次数除以该期刊在这两年内发表的论文总数。这是一个国际上通行的期刊评价指标。

国际合作论文：指由两个或两个以上国家和 / 或地区作者合作发表的被 WOS 收录的论文。本报告中，中国的国际合作论文特指中国大陆学者与国外学者合作发表的论文。合作论文的计数方式为,每一篇合作论文在每个参与国家和/或地区中均计作一篇发文。

高被引论文占比:指基于合作论文总量的高被引论文占比。若国际合作论文中高被引论文总数为 A,国际合作论文总量为 B,则高被引论文占比为 H index=A/B。

学科国际合作论文占比:指某学科的国际合作论文发文量在全部学科国际合作论文总量中的占比。若全部学科国际论文发文总量为 N,其中某学科的国际合作论文总量为 G,则学科国际合作论文占比为:G/N。

学科内国际合作论文占比:指某学科的国际合作发文量在该学科论文总量中的占比。若某学科论文发文总量为 M,而该学科的国际合作论总量为 G,则学科内国际合作论文占比为 G/M。

国际科研合作中心度:是用来测度某国在国际科研合作网络中地位和重要性的一个指标。计算方法如下:如果两个国家 A 和 B 合作的论文数为 P,B 国的国际合作论文总数是 N,P/N 代表 A 国在 B 国的所有合作国家中的活跃度。PN 比值越高,表明 A 国作为 B 国的合作伙伴的地位和重要性越高。A 国与所有国家合作的活跃度 P/N 值相加,即为 A 国的国际科研合作中心度。

学科国际合作相对活跃度(本报告中简称合作相对活跃度):通过计算某学科在某国国际科研合作中的相对规模,测度该学科在该国国际科研合作中的相对活跃度。计算方法如下:

$$PAI_j = \frac{P_j/P_{wj}}{P/P_w}$$

P_j——一国在某学科发表的国际合作论文数;
P_{wj}——全球在某学科发表的国际合作论文数;
P——一国发表的国际合作论文数;
P_w——全球发表的国际合作论文数。

合作相对活跃度指标消除了学科间国际合作论文发文总量差异带来的影响,使得同一国家不同学科之间具有了可比性,若 $PAI > 1$,说明该学科在该国中的国际合作程度高于该国所有论文的国际合作程度。

ESI 22 学科:ESI 设置的 22 个学科为生物学与生物化学、化学、计算机科学、经济与商业、工程学、地球科学、材料科学、数学、综合交叉学科、物理学、社会科学总论、空间科学、农业科学、临床医学、分子生物学与遗传学、神经系统学与行为学、免疫学、精神病学与心理学、微生物学、环境科学与生态学、植物学与动物学、药理学和毒理学。

Nature Index(自然指数):依托于全球 68 本顶级期刊,统计各高校、科研院所(国家)在国际上最具影响力的研究型学术期刊上发表论文数量的数据库。自然指数最近十二个月的数据都在指数网站上(https://www.natureindex.com/)滚动发布,以方便用户分析自己的科研产出情况。通过该网站,科研机构可根据大的学科分类浏览自己最近 12 个月的论文产出情况,各机构的国际和国内科研合作情况也有显示。

AC/article count(论文计数):Nature Index 里面的一个指标,不论一篇文章有一个还是多个作者,每位作者所在的国家或机构都获得 1 个 AC 分值。

FC/fractional count(分数式计量):Nature Index 里面的一个指标,FC 考虑的是每位论文作者的相对贡献。一篇文章的 FC 总分值为 1,在假定每人的贡献是相同的情况下,该分值由所有作者平等共享。例如,一篇论文有 10 个作者,则每位作者的 FC 得分为 0.1。如果作者有多个工作单位,则个人 FC 分值将在这些工作单位中再进行平均分配。

WFC/weighted fractional count（加权分数式计量）：Nature Index 里面的一个指标，即为 FC 增加权重，以调整占比过多的天文学和天体物理学论文。这两个学科有 4 种期刊入选自然指数，其发表的论文量约占该领域国际期刊论文发表量的 50%，大致相当于其他学科的 5 倍。因此，尽管其数据编制方法与其他学科相同，但这 4 种期刊上论文的权重为其他论文的 1/5。